Marine mineral resources

During the past century, scientists, world statesmen, and international entrepreneurs have become increasingly aware of the potential of the oceans as a source for minerals. This book provides an authoritative picture of the current state of marine mineral extraction. A major work of reference, it will be essential reading both for those engaged in maritime studies and for professional organizations involved in the extraction of underwater minerals.

Professor Earney gives an overview of our marine mineral endowment, and he details how this is being exploited. He examines present and future prospects of ocean mining, especially for the hard minerals, and considers programmes directed at expanding our ability to exploit the ocean's mineral wealth. He also identifies the economic, political, and technological problems which now hinder or may even prevent ocean mining, and examines in detail contemporary political problems concerning Law of the Sea negotiations and the resulting United Nations Convention. He reviews our present knowledge of deep seabed minerals and their exploitability, looking in particular at the continental shelves and including analyses of petroleum resources and the important but often overlooked placers, construction aggregates, industrial materials, and sea-water minerals.

The book is comprehensive in its coverage of most marine minerals now produced or those that may be produced in future. Much of the information is based on primary sources, such as letters, personal interviews, and field observations, and is unavailable elsewhere. *Marine Mineral Resources* will do much to increase awareness of the ocean's significance and to demonstrate the complexities we face in using its resources. It will encourage those in industry and government agencies to improve ocean resource management and enhance our sense of responsibility for the ocean's present and potential resources.

The Author
Fillmore C. F. Earney is Professor of Geography at Northern Michigan University.

OCEAN MANAGEMENT AND POLICY SERIES

Edited by H D Smith

Development and Social Change in the Pacific Islands
Edited by A D Couper

forthcoming

World Ocean Management
H D Smith and C S Lalwani

MARINE MINERAL RESOURCES

Fillmore C. F. Earney

Routledge
London and New York

TO
Lea, Andrew, Rachell, Sheena,
Alexander, Ashley, Brandon and Jo

First published 1990
by Routledge
11 New Fetter Lane, London EC4P 4EE
29 West 35th Street, New York, NY 10001

© 1990 Fillmore C. F. Earney
Typeset by J&L Composition Ltd, Filey, North Yorkshire

Printed and bound in Great Britain by
Biddles Ltd, Guildford and King's Lynn

British Library Cataloguing in Publication Data

Earney, Fillmore C.F., *1931–*
    Marine Mineral resources—(Oceans/Management and
    policy)
    1. Oceans. Mineral resources
    I. Title   II. Series
    333.8'5

ISBN 0–415–02255–X

Library of Congress Cataloging in Publication Data is available

ISBN 0–415–02255–X

Contents

Contents

Figures

Tables

Preface

On the bottom of the sea there
is the glitter of gold
Rubies and diamonds and treasures untold

Folksong

Before the dawn of history, ocean space and ocean waters were resources to humankind. Fishing in prehistoric times occupied most coastal peoples, providing them with fishbones, shells, sharks' teeth, corals, and pearls for tools, for weapons, and for adornment. Many of these gifts of the sea continue in use today.

Merchants and admirals of the distant past knew well the waters of the Mediterranean, the North Sea, and the East and South China Seas. Phoenician, Norse, or Chinese sailors traversing these seas today might soon recognise their twentieth-century counterparts. But they would not so quickly understand the herculean structures (oil production platforms) that jut from these waters. Neither would they recognise strangely shaped vessels (oil exploration rigs) that churn ahead with no familiar cargo or instruments of war. When beaching their crafts, these sailors might stand in utter puzzlement to view serpent-like structures (oil pipelines) emerging from the sea on to the shore. If our ancient mariners were to sail from their busy coasts to venture into the solitude of deep-sea waters of the Atlantic or Pacific Oceans, they could observe craft moving slowly along, seeming to seine for fish. To their surprise, they might see the ships' towing apparatus disgorge not wriggling, shiny creatures of the water but lifeless lumps of 'black gold' (ferromanganese nodules) from the deep seabed.

Although the ocean still functions as a transport mode and a food source, and although miners long have extracted minerals from beneath the sea, government leaders, industrialists, and scientists only recently have recognised the ocean's mineral-wealth potential. Pessimistic observers, however, might suggest that these mineral treasures are better described as 'fools' gold'. Which view is reality? Or, does the truth lie somewhere between?

With the depletion of many high-grade mineral reserves during the Second World War, with the post-war period of worldwide rapid population growth, and with increased industrialisation (in both technologically advanced and less developed countries) concern has arisen for the world's future mineral base. This concern is especially apparent in states dependent on imports of strategic minerals produced in politically unstable areas of Africa, Asia, and South America. Consequently consumers in recent decades have become interested in the oceans as a potential source of petroleum and of metallic and non-metallic hard minerals.

Although the oceans cover approximately 70 per cent of the earth's surface, their seabed mineral potential remains relatively unexplored. On the other hand, those in academia, govenment, and industry have done much basic and applied research to identify and assess the mineral-resource potential of both the continental margins and deep seabeds. But much more must be done if the oceans are to provide the mineral wealth the world's people either now or may eventually need.

I began this book in 1985, the 'Year of the Ocean'. The book's objectives parallel those identified by John Byrne, a former Director of the United States National Oceanic and Atmospheric Administration, when the Reagan Administration proclaimed 1985 as a year for each of us to focus on the ocean's importance. Thus, in the broadest sense, I seek to

1 increase awareness of the ocean's significance;
2 demonstrate the complexities we face in using its resources;
3 encourage better relationships among the general public, those in industry and governmental agencies that may help to improve ocean resource management;
4 enhance our sense of stewardship of the ocean's present and potential resources.

More specifically, I seek to

1 examine present and future prospects for ocean mining, especially for the hard minerals;
2 consider programmes directed at expanding our ability to exploit the ocean's mineral wealth;
3 identify economic, political, and technical problems now hindering or that may prevent ocean mining.

The book is divided into two parts. Part I examines contemporary political problems concerning Law of the Sea negotiations and the resulting United Nations Convention on the Law of the Sea. It also reviews our present knowledge of deep seabed minerals and their exploitability. Part II looks at the continental shelves, including analyses of petroleum resources and the important but often-overlooked placers, construction aggregates, industrial materials, and sea-water minerals.

Acknowledgements

Preparation of this book has provided me with an opportunity to expand my scientific vistas, to re-cement old bonds of friendship, and to initiate new professional associations. Without the help of these many professionals, my task would have been more difficult. I thank those who reviewed the manuscript to rid it of errors of fact and interpretation; errors remaining are mine. Those reviewing entire chapters or parts include: Robert J. Bailey, Outer Continental Shelf Co-ordinator, Department of Land Conservation and Development, State of Oregon (Chapter 10); Frederick R. Best, Associate Professor, Department of Nuclear Engineering, Texas A&M University (9); Gerald Blake, Professor, Department of Geography, University of Durham (12, 13); Elisabeth Mann Borgese, Professor, Department of Political Science, Dalhousie University and Chairman of the International Ocean Institute (2, 3); James M. Broadus, Director, Marine Policy Center, Woods Hole Oceanographic Institution (2, 3); Robert G. Burke, Associate Publisher/ Editor, *Offshore* (11); Byron B. Clow, Executive Director, International Magnesium Association (9); Tim Cornitius, Head, Ocean Oil Information Center, *Offshore* (11); Clive Cowley, Public Relations Manager, CDM (Proprietary) Limited (6); Peter H. Crorkan, Manager of Operations, Empresa Nacional del Carbon, Chile (8); Michael J. Cruickshank, Consultant in Marine Mining (7); Jack A. Draper II, ARCO Oil and Gas Company (13); Michael J. Driscoll, Professor, Department of Nuclear Engineering, Massachusetts Institute of Technology (9); David B. Duane, Director, Marine Minerals and Coastal Processes, National Sea Grant College Program, National Oceanic and Atmospheric Administration (7, 8); John E. Flipse, Associate Deputy Chancellor, College of Engineering, Texas A&M University (10); Duane K. Fowler, Professor and Head, Department of Physics, Northern Michigan University (9); John E. Frey, Professor, Department of Chemistry, Northern Michigan University (9); Martin Glassner, Professor and Chairman, Department of Geography, Southern Connecticut State University (2, 3); James W. Good, Associate Professor, Marine Resource Management Program, College of Oceanography, Oregon State University (6); Porter Hoagland III, Research Associate, Marine Policy Center, Woods Hole Oceanographic Institution

(2, 3); Alex Hogg, formerly of Aberdeen College of Education (12); Arild Holt-Jensen, Geografisk Institutt, Universitetet i Bergen (12, 13); Arnfinn Jorgensen-Dahl, Director, Polar Project, Fridtjof Nansen Institutt (12, 13); Deborah A. Kramer, Physical Scientist, Division of Nonferrous Metals, US Bureau of Mines (9); Leonard A. LeBlanc, Engineering Editor, *Offshore* (11); Bert Lundgren, President, PetroScan AB (11); Gregory McMurray, Marine Minerals Co-ordinator, Department of Geology and Mineral Industries, State of Oregon (10); Frank T. Manheim, Office of Marine Geology, Atlantic-Gulf Branch, US Geological Survey (4); Harry M. Mikami, Consultant in Magnesium Resources (9); J. Robert Moore, Professor, Department of Marine Studies, The University of Texas at Austin, and Editor, *Marine Mining* (6); Brent F. Nelsen, Fulbright Scholar, Norsk Utenrikspolitisk Institutt (12, 13); Willy Østreng, Director, Fridtjof Nansen Institutt (12, 13); Robert D. Palmore, Manager, Geology and Survey Department, Dravo Basic Materials Company, Inc. (8); Curt Peterson, College of Oceanography, Oregon State University (6); Ulrich von Rad, Bundesanstalt für Geowissenschaften und Rohstoffe, Federal Republic of Germany (8); Peter A. Rona, Marine Geologist, Atlantic Oceanographic and Meteorological Laboratory, US National Oceanic and Atmospheric Administration (4); J. Michael Uren, Chairman and Managing Director, Civil and Marine Ltd (7); Conrad Welling, Senior Vice President, Ocean Minerals Company, Inc. (3, 5).

Northern Michigan University librarians saved me countless hours of labour. Those helping to document or to obtain literature items include John Berens, Carolyn Cooper, Kathleen Godec, Jacquelyn Greising, Darlene Gruler, Roberta Henderson, Stephen Peters, and Tian-chu Shih. Janet Locatelli, at Michigan Technological University, and Tamara Brunnschweiler, at Michigan State University, also gave assistance in providing references, as did Mary Fisk, Law of the Sea Officer, in the Law of the Sea Office at the United Nations. Special thanks are due to Kumi Inaba for assistance in Japanese literature translations and those that completed typing or other chores – Lori Bishop, Nancy Burgander, Brenda Cilc, Linda Cleary, Phyllis DeWitt, Cyndie Koski, Debra Laliberte, and Roberta Niemann. I am especially indebted to Carol Etten and Linda Stille for their skill in deciphering my script and in typing numerous drafts of the entire manuscript, without a frown or a whimper. My colleague, Professor Patrick Farrell, provided cartographic counsel and supervised the excellent work done by David Johnston and Robert Regis, who prepared most of the figures. I appreciate the help of Donald Pavloski in duplicating photographic prints and also the many photographs provided by industrial firms and governmental agencies and the permissions granted by publishers to use tabular and illustrative materials. My thanks go to Perrin Fenske for his help in obtaining funds for preparation of the figures and to Cheryl Liubakka for monitoring my purse.

During the summer and autumn of 1986, the University provided a sabbatical leave and travel funds for field work in the United States,

South America, and Europe, where I observed seabed mining activities and visited those working in the industry. Their generous assistance, in granting interviews and in answering my letters, provided information unavailable elsewhere. Finally, I extend my gratitude to John Frey who, for more than two decades, has been a source of wit and encouragement.

Fillmore C.F. Earney, Ph.D.
Professor of Geography
Marquette, Michigan

Abbreviations and acronyms

ACE	Army Corps of Engineers
AFERNOD	Association Française d'Etude et de Recherche des Nodules
AIT	Agency of Industrial Technology
AMR	Arbeitsgemeinschaft Meertechnisch Gewinnbare Rohstoffe
ARCO	Atlantic Richfield Company
BGC	British Gas Corporation
BNOC	British National Oil Corporation
BRG	Federal Institute of Geosciences and Natural Resources
CBDC	Cape Breton Development Corporation
CCZ	Clarion-Clipperton Zone
CDM	Consolidated Diamond Mines (Proprietary) Limited
CIDS	Concrete Island Drilling System
CNOOC	China National Offshore Oil Corporation
CSO	Coastal States Organization
CT	Corporation Tax
DCFROR	Discounted Cash Flow Rate of Return
DEC	Department of Environmental Conservation
DEIS	Draft Environmental Impact Statement
DME	Department of Mines and Energy
DOC	Department of Commerce
DOI	Department of the Interior
DOMA	Deep Ocean Mining Associates
DOMCO	Deep Ocean Mining Company Limited
DOMES	Deep Ocean Mining Environmental Study
DORD	Deep Ocean Research and Development Company
DOS	Department of State
DSHMRA	Deep Seabed Hard Mineral Resources Act
EEZ	Exclusive Economic Zone
EIS	Environmental Impact Statement
EPA	Environmental Protection Agency
EPC	Economic Planning Commission
FRG	Federal Republic of Germany

G-77	Group of 77
GCCS	Geneva Convention on the Continental Shelf
GEMONOD	Groupment pour la Mise au Point Moyens Nécessaires a l'Exploitation de Technologies Préliminaires
GNYMA	Greater New York Metropolitan Area
GR	Gorda Ridge
GRTTF	Gorda Ridge Technical Task Force
GZA	Grey-Zone Agreement
HA-JA	Hawaiian Archipelago-Johnston Atoll
ICJ	International Court of Justice
IEA	International Energy Agency
ILC	International Law Commission
IRR	Internal Rate of Return
ISA	International Seabed Authority
ITA	International Tin Agreement
ITC	International Tin Council
JDZ	Joint Development Zone
KCON	Kennecott Consortium
KORDI	Korea Ocean Research and Development Institute
LDC	Less Developed Countries
LGD	Land-locked and Geographically Disadvantaged
LME	London Metal Exchange
LOS	Law of the Sea
LTC	Legal and Technical Commission
MF	Malvinas Falklands
MFCA	Magnuson Fishery and Conservation Act
MMS	Minerals Management Service
MPE	Ministry of Petroleum and Energy
NACOA	National Advisory Committee on Oceans and Atmosphere
NOAA	National Oceanic and Atmospheric Administration
NRDS	Natural Resources Discovery System
NZ	New Zealand
NZOI	New Zealand Oceanographic Institute
OCS	Outer Continental Shelf
OCSLA	Outer Continental Shelf Lands Act
OGEA	Oil and Gas Enterprise Act
OGS	Office of General Services
OMA	Ocean Mining Associates
OMCO	Ocean Minerals Company
OME	Office of Minerals and Energy
OMI	Ocean Management Incorporated
OMM	Office of Marine Minerals
OPEC	Organisation of Petroleum Exporting Countries
OSGMW	Offshore Sand and Gravel Mining Workshop
OSIM	Office of Strategic and International Minerals

PCP	Progressive Conservative Party
PEI	Prince Edward Island
PIP	Pioneer Investor Protection
POSMM	Panel on Operational Safety in Marine Mining
PPA	Petroleum Production Act
PRC	People's Republic of China
PREPCOM	Preparatory Commission
PROCAP	Technological Capability Program for Deep Water Oil Exploitation
PRT	Petroleum Revenue Tax
PSPA	Petroleum and Submarine Pipeline Act
ROV	Remotely Operated Vehicle
SLC	State Lands Commission
SPT	Special Petroleum Tax
SR	Sementsverksmiðja Rikisins
ST	Svalbvard Treaty
STATOIL	Den Norske Stats Oljeselskap
TA&M	Texas A&M University
UK	United Kingdom
UKOOA	United Kingdom Offshore Operators Association
UN	United Nations
UNCLOS	United Nations Conference on the Law of the Sea
US	United States of America
USBM	United States Bureau of Mines
USGPO	United States Government Printing Office
USGS	United States Geological Survey
USSR	Union of Soviet Socialist Republics
WHOI	Woods Hole Oceanographic Institution
bbl/d	barrels per day
cm	centimetres
ft	feet
gal	gallons
kg	kilograms
kl	kilolitres
m	metres
mi	miles
n.d.	no date
n.l.	no location
n.p.	no paging
nmi	nautical miles
sec	seconds
yd	yards
yr	year

Chapter one

Introduction

Throughout humankind's history of ocean use, economic and military functions have been paramount. The seas have functioned to separate and to join, depending on the time and circumstance of peoples. Whether for sustenance, for transportation, for communication, or for protection, use of the oceans has usually involved political interrelationships – good and bad – among peoples. This situation has been especially evident in recent decades, and it continues today. Indeed, one can neither study nor fully understand contemporary marine affairs without taking into account the politics of the oceans locally, nationally, and internationally.

As the world's population has grown and as our resource-use systems have pushed more and more into the oceans, societies have been forced into a maritime proximity with others. The consequence has been, oftentimes, a reaction to protect what one already holds or to take today what one fears may ‘not be there tomorrow. Such perceptions and relationships have led to waste, to frustration, to conflict, and to accommodation. A central thread throughout the fabric of this book is an examination of efforts by local, national, and international economic and political entities to obtain what they feel are, on the one hand, their oceanic needs and rights and, on the other, what the collective community is willing to give. Recent efforts at international accommodation reflect an evolutionary period of several decades and in 1982 emerged in a focused form as the United Nations Convention on the Law of the Sea, a complicated body of mandates designed to administer and allocate humankind's last resource frontier – the oceans. How this Convention impinges on the future of ocean mining is another important thread within this volume.

Why seabed mining

During the past century, scientists have gradually become more knowledgeable about the potential of the oceans as a source for minerals. From the late 1960s until the early 1980s world statesmen, international entrepreneurs, and mining industry personnel waxed enthusiastic about prospects for pushing seaward the frontiers of continental-shelf petroleum production and for mining the deep seabeds for metals. By the mid-1980s the mood had turned pessimistic, if not sour, regarding the deep seabed. What caused this dramatic turnabout in opinion which holds that deep-seabed mining will not go forward for at least the next fifteen years, and more likely not until well into the twenty-first century?

Advantages of ocean mining

Several advantages of ocean mining stand out in contrast to onshore production:

1 many seabed ores are richer than onshore deposits;
2 the water provides for relatively cheap transportation needs, both logistical and distributional;
3 facilities (ports) for loading mining supplies and unloading mineral products are already in place;
4 onshore processing operations may be built in politically stable areas and in specifically desired labour- and energy-supply regions;
5 fewer constraints exist in environmental regulations and zoning ordinances;
6 world states can gain greater independence in meeting their strategic mineral needs.

Disadvantages of ocean mining

Major disadvantages of ocean mining are:

1 distances from mine sites to markets may be several thousand km;
2 building the mining and processing equipment and mastering the engineeering technology require much time and large capital investments;
3 costs of weather-related work stoppages could be significant;
4 unknown environmental problems of the deep seabed and the water column must be solved;
5 present problems of glutted world mineral markets may be exacerbated, causing difficulties for onshore producers of minerals also mined in the oceans;
6 political and economic problems may occur in association with the establishment of an international body that will administer deep-seabed mining and that may mine in its own right.

Of these several disadvantages, present-day mineral-market conditions and political issues demand special scrutiny. Events in these sectors will be crucial to the future of marine mining.

Glutted mineral markets

In the mid–1980s most mining industries are not producing at capacity, some not even at 50 per cent. Only a few produce at more than 80 per cent of capacity. In 1983 among eighteen selected minerals now produced or potentially available in the oceans, annual production averaged not quite 75 per cent of capacity (Table 1.1). Therefore an evaluation of currently glutted mineral markets and the likelihood of this situation's continuing will be an important part of mining firms' decision-making process about whether to mine the deep seabed or to extend petroleum development programmes into deeper continental shelf areas. The mining of any given mineral from the seabed is dependent on

Table1.1 Total world mine production and capacity for selected minerals now produced (A) or potentially mineable (B) in the oceans, 1983

A	Unit of measurement (thousands of metric tonnes unless otherwise identified)	World capacity	World production	World production as a % of capacity
Bromine		499	224[a]	45.1
Diamond-industrial (Stone)	Thousands of carats	28,600	21,300	74.5
Gold	Kilograms	1,517,852	1,385,133	91.3
Magnesium	Thousands of tonnes of contained Mg	6,532	5,001	76.6
Platinum	Kilograms	—[b]	81,025	—[b]
Salt		205,027	163,033[a]	79.5
Sand and gravel	Millions of metric tonnes	9,072	7,167[c]	79.0
Silica sand	Millions of metric tonnes	236	181	76.9
Sulphur		65,800[a]	50,472	76.7
Tin		296	212	71.6
B				
Barite		7,725[a]	5,552	71.5
Cobalt		35[a]	23[a]	67.4
Copper		10,290	8,100	78.7
Feldspar		3,992	3,518	88.1
Lead		4,150	3,370[a]	81.2
Manganese	Thousands of tonnes of contained Mn	11,794	7,983[a]	67.7
Nickel		1,022	690	67.5
Phosphate rock		175,800[a]	135,000	76.8
Zinc		8,055	6,268	77.8

Source: Compiled from US Bureau of Mines, *Mineral Facts and Problems 1985*, USBM Bulletin 675 (USGPO, Washington, DC, 1985), S. G. Ampian, 'Barite', p. 66; P. A. Lyday, 'Bromine', p. 103; W. S. Kirk, 'Cobalt', p. 172; J. L. W. Jolly, 'Copper', p. 201; J. F. Smoak, 'Diamond–Industrial', p. 234; M. J. Potter, 'Feldspar', p. 256; J. M. Lucas, 'Gold', p. 324; W. D. Woodbury, 'Lead', p. 434; T. S. Jones, 'Manganese', p. 484; D. A. Kramer, 'Magnesium', p. 471; S. F. Sibley, 'Nickel', p. 536; W. F. Stowasser, 'Phosphate Rock', p. 580; J. R. Loebenstein, 'Platinum-group metals', p. 601; D. E. Morse, 'Salt', p. 681; L. L. Davis, 'Sand and gravel', pp. 690, 697; D. E. Morse, 'Sulfur', p. 783; J. F. Carlin, Jr, 'Tin', p. 848; J. H. Holly, 'Zinc', p. 924.
Notes: a. Estimated; b. No data available for platinum; c. Data for 1982.

1 how economically and technologically competitive it is with onshore resources;
2 the availability of substitutes that meet specific physical and chemical qualities needed by mineral consumers;
3 whether large amounts of already mined and recyclable materials are available;
4 changing strategic concerns of national governments.[1]

The political situation

Political issues associated with current efforts to establish an ocean-mining regime are a major deterrent to proceeding with deep-seabed mining. The less developed countries (LDCs) are determined to obtain a greater share of the wealth that might be provided by deep-seabed mineral resources. Numerous industrial states are equally determined that the welfare of their populace and national economy will not be jeopardised. Most interested parties agree, however, that these divergent positions must be reconciled so that ocean mining is managed efficiently and more equitably under policies which meet the economic, political, social, and environmental needs of all world citizens and nationals within states. Accordingly a primary focus of this book is to examine controversies and problems faced by world states and by industry in establishing equitable policies and in instituting wise management techniques for exploiting marine mineral resources.

The ocean basins

The oceans have two basic geological areas – a continental region and a deep-seabed region. Because we cannot see them directly, it is easy to envision the ocean basins as smoothly uniform and gradually deepening bowl-like structures. Nothing could be further from the truth, because seabed topography is often a mosaic of mountains, canyons, escarpments, and plains.

Earth's outermost and thinnest layer (about 0.3 per cent of its radius) is composed of two types of gigantic, moving plates – oceanic crust (mainly basalts) and continental crust (mainly granites). Most continental crust is geologically older, weighs less, and is thicker than oceanic crust. The continents average about 40km in thickness, but where high mountain ranges occur, they may measure 70km. A continental crust's seaward extension consists of a continental shelf, continental slope, and continental rise – collectively known as the continental margin. The continental shelf (average width 65km) is usually defined as those seabed areas with a water depth of less than 200m. At about 130m the shelf usually begins to angle downward rapidly, with an average slope of 4.3°. This zone is the continental slope. The continental rise begins when the gradient becomes less steep, usually at a depth of 1,400m to 3,200m. This region slopes downward with a relatively gentle gradient and extends to greatly varying distances until at about 4,000m the seabed becomes part of the so-called abyssal plain, the deep seabed (Figure 1.1).

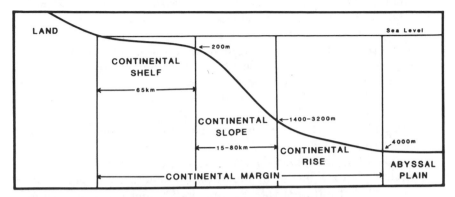

Figure 1.1 Schematic profile of the continental margin.

Sources: After R. W. Rowland, M. R. Goud, and B. A. McGregor, *The U.S. Exclusive Economic Zone – A Summary of its Geology, Exploration, and Resource Potential*, USGS Circular 912 (USGS, Alexandria, VA, 1983), p. 4; V. E. McKelvey, J. I. Tracey, Jr, G. E. Stoertz, and J. G. Vedder, *Subsea Mineral Resources and Problems Related to their Development*, USGS Circular 619 (USGS, Washington, DC, 1969), p. 2.

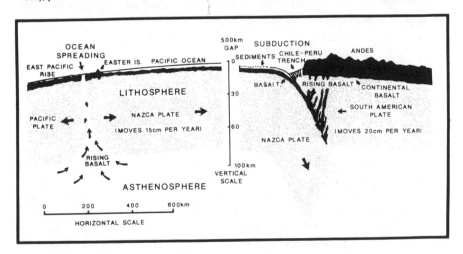

Figure 1.2 The Nazca Plate, off the west coast of South America

Source: After National Science Foundation, *Deep Sea Searches: The Story of the Seabed Assessment Program* (NSF, Washington, DC, 1975), p. 15. With permission.
Note: The Nazca Plate exemplifies crust formation processes along mid-oceanic spreading zones. Vertical and horizontal measurements are converted from mi to km and from in to cm.

Mid-ocean areas contain some of earth's youngest rocks, because it is here that geological processes form new oceanic crust in spreading (divergence) zones, composed of ridges and fractures (Figure 1.2). Spreading zones – forming originally as small, narrow fractures within continents – ultimately create (through plate tectonics) the ocean basins.

Geologists hypothesise that spreading zones are areas where portions of convection-like cells of internal heat and a migrating upper layer of the mantle (asthenosphere) emerge, forming hydrothermal polymetallic sulphide minerals when hot magma reacts with sea-water both below and on the seabed surface. As the magma pushes upward and cools, it is shunted to one side or the other of the spreading zone, becoming part of a migrating oceanic plate that, after many millions of years, is destined to collide (converge) with and subduct beneath a continent lying in its path, oftentimes helping to create mineralised volcanic coastal mountain ranges. During the plate's migration, mineral accretions may form on its surface as continuous crusts and as individual nodules.

Oceanic plates, in part because of their subduction beneath the continents, are overridden by the continental plates which are, in turn, embedded within rigid basaltic plates. The continents and their underpinning basaltic plates 'probably driven by plastic undercurrents' may also be moving slowly (at different rates) across an adjacent subducting oceanic plate.[2] Simultaneously with their lateral tectonic movements, the continents' rock materials are weathered and transported to adjacent oceans. Rivers, oceanic coastal surf, and longshore currents create placers (sorted mineral-bearing alluvial sands and gravels). Where the proper sediments, physiographical and petrological conditions occur, hydrocarbons (oil and gas) form and become trapped.

Types of ocean minerals

Oceanic minerals can be classified into several broad groupings – construction aggregates, industrial materials, placers, hydrocarbons, hydrothermal sulphides, polymetallic nodules and crusts, and sea-water (Table 1.2). The oceans' most ubiquitous mineral resource is sea-water. In future, sea-water could provide greater quantities of those minerals now extracted from it and others as well, such as uranium.

On the continental shelves, the main mineral types are construction aggregates, industrial materials, placers, and hydrocarbons. Miners have long obtained construction materials and hydrocarbons in shallow continental shelf areas, and are now pushing into deeper waters to extract petroleum. Marine placers provide tin, gold, and platinum and will become more important in future. Although not now produced, continental shelf phosphorites (so useful in fertiliser production) also will be exploited.

During the last two decades, mining consortia have turned their attention to the less accessible but potentially important minerals of the deep seabeds. They have spent millions of dollars, pounds, lira, marks, francs, and yen to locate and evaluate polymetallic (ferromanganese) nodules that cover the seabed in several of the oceans, especially at depths between 4,000m and 6,000m. Developing mining and processing technologies needed to recover the desired minerals from the nodules –

Table 1.2 Main types of oceanic minerals

Type of deposit	Materials or elements	Main geological setting
Construction aggregates[a]	Pebbles, quartz	Coast and continental shelf
Industrial materials[a]	Phosphorites, sulphur, aragonite, shells and silica sand[a]	Coast (locally a continuation of land resources) nearshore submarine plateaux
Placers	Iron, gold, platinum, tin, diamond, rare earth elements, zirconium, titanium, chromium, scheelite, and others	Coast and nearshore
Hydrocarbons	Petroleum (oil and gas), coal[a]	Mainly passive continental margins
Polymetallic hydrothermal sulphides[a]	Iron, copper, manganese, lead, zinc, silver, and others	Fracture zones and spreading centres
Polymetallic (ferromanganese) nodules and crusts[a]	Manganese, iron, cobalt, nickel, titanium, molybdenum and others	Deep sea (4,000 metres or more)
Sea-water[a]	Fresh water, bromine, magnesium, salt, uranium[a] potassium, gypsum	Ubiquitous, but salinities vary

Source: After P. Rothe, 'Marine geology: mineral resources of the sea', *Impact of Science on Society*, vol. 33, nos 3/4 (1983), p. 360. With permission of UNESCO.
Note: a. Author's additions or alterations.

nickel, copper, cobalt, and manganese – have required even larger investments. One enterprise is now in an advanced stage of preparatory work for extracting hydrothermal metalliferous muds from deep trenches of the Red Sea. Recently entrepreneurs have become interested in hydro-thermal polymetallic massive sulphides associated with other deep-ocean spreading centres, as in the eastern Pacific Ocean. Within only the last few years, polymetallic (ferromanganese) crusts that coat seamounts and the sides of volcanic islands have aroused the interest of both scientists and deep-ocean mining consortia.

Conclusions

Despite the rather dismal mineral market conditions and the unsettled political situation of the mid-1980s, in future we will become more dependent on the oceans as a mineral resource reservoir. In coming decades, as earth's population continues to increase and as peoples of the LDCs seek improved living conditions and utilise advanced technologies, demands for mineral resources will grow. Despite efforts to find new on-shore mineral deposits and to improve low-grade-ore processing methods, and despite energy diversification and conservation programmes, we may fail to meet the needs of a burgeoning world population that could total 6,200 million in the year 2000 and 8,200 million in 2020. To supply these consumers' energy requirements will be difficult enough, but to produce usable raw materials from increasingly lower grades of metal ores will require additional energy inputs, a form of diminishing returns. This process can proceed only so far until an ore becomes uneconomic. Thus, depleting high-grade onshore mineral sources and rising energy costs eventually will encourage mineral producers to look to the oceans as a new resource frontier. Both government and industry will face formidable oceanic management challenges requiring reasoned decisions based upon a detailed understanding of first, the distribution and physical properties of marine minerals and second, the environmental, political, and economic relationships of marine mining to the world community.

Notes

1 For a useful discussion of several of these points, see J. M. Broadus, 'Seabed materials', *Science*, vol. 235, no. 4,791 (1987), pp. 853–60.
2 P. Rothe, 'Marine geology: mineral resources of the sea', in J. G. Richardson (ed.), *Managing the Ocean: Resources, Research, Law* (Lomond Publications, Mt Airy, MD, 1985), p. 21.

Deep seabed politics and minerals

Introduction

The world's deep oceans contain a variety of potentially useful mineral deposits – ferromanganese nodules, ferromanganese crusts, and polymetallic sulphides. For more than three decades, marine geologists, marine engineers, mineralogists, oceanographers, and other specialists have been locating and delineating seabed mineral deposits and developing deep ocean mining and processing technologies. Much has been learned, but our understanding of these minerals has only begun. While the engineers and physical scientists have been at work, economists and entrepreneurs have been analysing the commercial potential of these deep sea minerals. Who should administer and use these minerals has been the focus of vigorous debate among world statesmen, specialists in international law, and industry leaders for nearly two decades. Their decisions will play a vital role in the future of marine mining and its management. These topics form the core of Part I.

Chapter two

Law of the sea

Throughout history, the oceans have been important to the political and commercial well-being of civilisations, but perceptions have varied about how the oceans should be used and by whom. Some empires have claimed the oceans as the domain of all peoples and states, open to every commercial and communication need. Others have sought control over large portions of the seas, attempting to exclude all but their own merchantmen and fishermen. It was with these differing views that international maritime law evolved from ancient times to the present. This evolution was sometimes contentious and, for the most part, piecemeal.

The nascence of ocean law

The cradle of western maritime commerce lay in the eastern Mediterranean Sea. Sea-kings of Crete, from 2500 BC to about 1000 BC, provided a nexus for commerce. During 1100 BC to 800 BC the Phoenicians dominated commerce and welded the coastal Mediterranean Basin into an interdependent trading area. By 600 BC Babylonia and India had significant oceanic links,[1] and by the fourth century BC, Mauryan Kings in India had developed regulations for oceanic trade, transport, and fishing.[2] By the first century AD, the Roman Empire and India pursued a lively maritime commerce in silks, spices, and jewels.[3] Throughout the ancient world, rules evolved for using the oceans, as in Rhodian Sea Law of pre-Christian origin which was codified in the eighth century.[4] These early seafaring civilisations and the Chinese and Arabs from the 1300s into the 1500s laid the groundwork for ocean management through customary maritime law, carrying out their activities in accepted ways, based on empirical experience. The seas were viewed as 'free' and open to commerce by all.[5]

During the late Middle Ages maritime law and ocean use experienced the hand of divine arbitration. In 1493 Pope Alexander VI, at the request of Spain and Portugal, issued several bulls that allocated to these colonialist states yet unknown lands touched by waters lying east and west of a meridian 100 leagues (about 550km) west of the Cape Verde Islands. Portugal received the right to explore and colonise all areas east of the demarcation line and Spain all areas west. After direct negotiations, the two adversaries in 1494 agreed (in the Treaty of Tordesillas) to move the demarcation line(s) farther west. For lands already known, the treaty established a line 200 leagues (1,100km) west of the Cape Verde Islands and for lands not yet discovered, the line was put at 370 leagues (2,050km) west.[6]

In 1580, nearly a century after Spain and Portugal's pre-emption of known and unknown oceanic areas lying west of Europe and Africa, Queen Elizabeth of England, despite Spain's protests, challenged the concept that the seas could be claimed as the domain of any one state. She held that the oceans, like the air, are common to all.[7] Although Queen Elizabeth died in 1603, support for the freedom-of-the-seas principle was in 1609 again forcefully stated by the Dutch jurist Hugo Grotius. At the request of the Dutch East India Company, which was then confronting Portugal's East Indies trade monopoly, Grotius penned an essay titled *De Jure Belli Ac Pacis*, a revised chapter of a larger then-unpublished manuscript.[8] In his treatise, Grotius said that, because the sea could not be contained, it could not be claimed; the seas were a *mare liberum* open to all users, whereby 'the benefits from the enjoyment of common things should be given to the entire human race rather than to one nation alone'.[9]

Grotius' work was challenged in relation to near-shore waters when in 1635 John Selden, a Britisher, argued for a closed-sea concept, a *mare clausum*, if a state's self-interest were at risk. Selden sought to justify excluding foreign fishermen from the offshore of the British Isles. Despite occasional efforts by states to declare portions of the oceans as their territorial domain, the principles of *mare liberum* reigned for the next three centuries,[10] with Great Britain a leading proponent.

During the early eighteenth century (1702), nearly a century after the appearance of Grotius' work, another Dutchman, Cornelius van Bynkershoek, published a volume titled *De Dominio Maris*. His thesis held that coastal dominance was the most appropriate outer limit for national territorial-sea sovereignty. This criterion represents the view that a state may project its power a distance equal to its technical capabilities. Theoretically, with technology the criterion for control, a state's capability for claiming increasingly greater distances from its shores becomes a distinct possibility. This concept, however, has never been generally accepted, although at least one contemporary international convention (Geneva Convention on the Continental Shelf) implies its utility.

Law of the sea conferences

Although no major conceptual changes occurred in international and national maritime-law policies from the seventeenth century to the early twentieth century, maritime legal affairs had become so complex by then that many statesmen felt some codification was needed. Thus the League of Nations called upon world states to send delegates to a meeting, scheduled for 1930, at The Hague in the Netherlands, for the purpose of codifying existing law-of-the-sea principles, especially those concerning the territorial sea. Not much came of this meeting, perhaps because only forty-eight states participated, not an adequate cross-section of world states.[11]

By the mid-1950s it seemed that a 'creeping jurisdiction' by many coastal states might close numerous straits that previously had been open to navigation. This fear led to the first truly broad-based efforts to codify law-of-the-sea principles. In 1958, after a seven-year study (1949–56) done by the United Nations' (UN) International Law Commission (ILC), eighty-six states met in Geneva, Switzerland, for the first Law of the Sea Conference – to discuss four draft conventions developed by the ILC.[12] The participants discussed and adopted these four important conventions that helped codify existing law-of-the-sea (LOS) principles. The Geneva Conventions (which entered into force at varying dates) dealt with

1 the territorial sea and the contiguous zone;
2 the high seas;
3 fishing and the conservation of living resources of the high seas;
4 the continental shelf.

These conventions helped clarify some issues, but they also contained ambiguities, and in at least one instance, failed in purpose.[13] For example, no adequate fisheries regime developed to protect coastal states from foreign fishing fleets along their shores. Especially difficult problems stemmed from ambiguities in provisions for maximum yields and apportionments to various parties exploiting coastal fisheries, that is no exclusive property rights were assigned.[14] Consequently a second UN conference on the LOS convened in 1960 to look again at fishing rights and the breadth of territorial sea problems, but the delegates failed to achieve substantive agreements.[15]

One of the most important of the 1958 Geneva Conventions dealt with the continental shelf. The convention defines the shelf as

> the seabed and subsoil of the submarine areas adjacent to the coast, but outside the area of the territorial sea, to a depth of 200 metres or, beyond that limit, to where the depth of the superjacent waters admits of the exploitation of the natural resources of the said areas.

Because the convention condoned coastal-state jurisdiction over technologically exploitable resources on the shelf, some observers saw significant implications for the future availability of marine resources (especially minerals) for many states.[16] Indeed the Convention on the Continental Shelf seemed to legitimise a long-debated action taken by US President Harry Truman, who in 1945 saw opportunities for petroleum resource development in the US offshore beyond its territorial waters. He assured these opportunities by issuing a proclamation which stated that the seabed resources of the contiguous continental shelf were under US jurisdiction.[17] Truman's precedent encouraged nine Latin American states to take similar actions in the late 1940s and early 1950s. Unlike Truman's claim, which was limited to the seabed, these states declared sovereignty over the ocean surface and fishing rights for a distance of 200nmi (370km – hereafter referred to as 200nmi, because of conventional usage). Other states also began to extend their territorial seas beyond the traditional but not universal 3nmi (5.5km). The trend toward creeping jurisdiction of offshore waters disturbed many governments, including the US and the USSR, as well as the geographically disadvantaged states, that is those with short coast lines, and the shelf-locked. The People's Republic of the Congo with a very short coast line and the Netherlands with no immediate access to the open sea exemplify the geographically disadvantaged. In addition, thirty land-locked states, as Bolivia, were also concerned.

In 1967 with the underdeveloped, the land-locked, and the geographically disadvantaged states in mind, Arvid Pardo (Malta's ambassador to the UN) called upon the UN

1 to reserve all nonterritorial seabed resources as a 'common heritage of mankind' (that is belonging to everyone or to no one)
2 to establish some form of international control in the use of the oceans' seabeds.

Pardo's initiative encouraged the UN General Assembly to organise a thirty-five member *ad hoc* committee to study the implications of his proposal for the deep seabed and its resources.[18] In 1968 the committee grew to forty-one members and took the name Committee to Study the Peaceful Use of the Sea-Bed and the Ocean Floor Beyond the Limits of National Jurisdiction (Sea-Bed Committee); its task was to consider the status of existing international seabed agreements and to decide the best procedure for establishing co-operative efforts to explore, to use, and to conserve the seabed and its subsoil.

As a result of the Sea-Bed Committee's work, the General Assembly in 1969 adopted several resolutions concerning the seabed. Three resolutions had special significance. They called for

1 the UN Secretary-General to determine the desires of UN members for assembling interested states for a third LOS conference;
2 the Secretary-General to submit to the Sea-Bed Committee (in 1970) proposals for the structure and operational procedures for a seabed regime;
3 the world community not to exploit the deep seabed, until an international regime could be put into force.[19]

By mid-1970 the Secretary-General had completed his assignments. That same year the General Assembly, to give more force to its mining moratorium resolution, proclaimed the seabed as a common heritage of mankind. Its action came via the adoption of a 'Declaration of Principles Governing the Sea-Bed and Ocean Floor, and the Subsoil Thereof, beyond the Limits of National Jurisdiction'. The General Assembly, also, called for the convening of a third LOS conference (UNCLOS III). An enlarged Sea-Bed Committee (ninety-one members) functioned from 1971 to 1973 as a Preparatory Committee to organise for this conference, and in 1973 presented its report. Delegates to UNCLOS III's first session, held in December 1973, elected conference officers – with Hamilton Shirley Amerasinghe of Sri Lanka taking office as president. Discussion began, too, on rules of procedure. A second session met in Caracas, Venezuela, in mid-1974, to adopt rules of procedure and to begin initial substantive (specific issue) discussions on alternate texts drafted by the Preparatory Committee.[20] The 148 states represented constructed negotiating texts that focused on a variety of topics, including (among others) ocean mining, living resources protection and exploitation, access to the high seas by shelf-locked and land-locked states and national jurisdiction of ocean space.

Many delegates at the conclusion of the Caracas meeting, although aware of the complexities of the task before them, were buoyant with optimism that both consensus and equity might be achieved. This buoyancy turned to frustration during the next eight years, as negotiators attempted to reach agreement on a final document that could be adopted as a package. Participants developed and debated various texts at sessions that alternated between New York City in the US and Geneva, Switzerland. Finally, during an eleventh session, held in New York in the spring of 1982, the UNCLOS III delegates felt they had put together a package that would receive approval by most developing and developed states. On 30 April 1982 the Draft Convention was put to a vote, with the US calling for a recorded vote, the first since the beginning of UNCLOS III negotiations. The count was 130 states for adoption, and 4 states (US, Israel, Turkey, and Venezuela) against adoption, with 17 states abstaining. Although the US voted against adoption, later that year it did sign the Final Act. On 10 December 1982 in Jamaica, 117 states signed the Convention; many others signed by 10 December 1984, the closure date (Table 2.1). The US, the United Kingdom (UK), and the Federal Republic of Germany (FRG) did not sign the Convention. But why, after having devoted a decade of effort to establish an LOS regime, should the US and others have refused to become Parties?

UNCLOS III Convention – dissent and compromise

To understand the problems of reaching a consensus among the more than 150 states that participated in the UNCLOS III negotiations, we should return to a point made earlier, that is Arvid Pardo's call for a recognition of the seabed (beyond the limits of national jurisdiction) as a common heritage of mankind. His plea exemplified a growing feeling within many

Table 2.1 States and other entities signing the United Nations Convention on the Law of the Sea, as of 10 December 1984

State signatory	Date of signature[a]	State signatory	Date of signature[a]
Afghanistan	18–03–83	Gabon	X
Algeria	X	Gambia	X
Angola	X	German Democratic	
Antigua and		Republic	X
Barbuda	07–02–83	Ghana	X
Argentina	05–10–84	Greece	X
Australia	X	Grenada	X
Austria	X	Guatemala	08–07–83
Bahamas	X	Guinea	04–10–84
Bahrain	X	Guinea-Bissau	X
Bangladesh	X	Guyana	X
Barbados	X	Haiti	X
Belgium	05–12–84	Honduras	X
Belize	X	Hungary	X
Benin	30–08–83	Iceland	X
Bhutan	X	India	X
Bolivia	27–11–84	Indonesia	X
Botswana	05–12–84	Iraq	X
Brazil	X	Ireland	X
Brunei Darussalam	05–12–84	Islamic Republic	
Bulgaria	X	of Iran	X
Burkina Faso	X	Italy	07–12–84
Burma	X	Jamaica	X
Burundi	X	Japan	07–02–83
Byelorussian SSR	X	Kenya	X
Cameroon	X	Kuwait	X
Canada	X	Lao People's	
Cape Verde	X	Democratic	
Central African		Republic	X
Republic	04–12–84	Lebanon	07–12–84
Chad	X	Lesotho	X
Chile	X	Liberia	X
Colombia	X	Libyan Arab	
Comoros	06–12–84	Jamahiriya	03–12–84
Costa Rica	X	Liechtenstein	30–11–84
Côte d'Ivoire	X	Luxembourg	05–12–84
Cuba	X	Madagascar	25–02–83
Cyprus	X	Malawi	07–12–84
Czechoslovakia	X	Malaysia	X
Democratic		Maldives	X
Kampuchea	01–07–83	Mali	19–10–83
Democratic People's		Malta	X
Republic of Korea	X	Mauritania	X
Democratic Yemen	X	Mauritius	X
Denmark	X	Mexico	X
Djibouti	X	Monaco	X
Dominica	28–03–83	Mongolia	X
Dominican Republic	X	Morocco	X
Egypt	X	Mozambique	X
El Salvador	05–12–84	Nauru	X
Equatorial Guinea	30–01–84	Nepal	X
Ethiopia	X	Netherlands	X
Fiji	X	New Zealand	X
Finland	X	Nicaragua	09–12–84
France	X	Niger	X

(Table 2.1 contd.)

State signatory	Date of signature[a]	State signatory	Date of signature[a]
Nigeria	X	Sri Lanka	X
Norway	X	Sudan	X
Oman	01–07–83	Surinam	X
Pakistan	X	Swaziland	18–01–84
Panama	X	Sweden	X
Papua New Guinea	X	Switzerland	17–10–84
Paraguay	X	Thailand	X
People's Republic of China	X	Togo	X
		Trinidad and Tobago	X
People's Republic of the Congo	X	Tunisia	X
		Tuvalu	X
Philippines	X	Uganda	X
Poland	X	Ukrainian SSR	X
Portugal	X	Union of Soviet Socialist Republics	X
Qatar	27–11–84		
Republic of Korea	14–03–83	United Arab Emirates	X
Romania	X		
Rwanda	X	United Republic of Tanzania	X
St Christopher and Nevis	07–12–84	Uruguay	X
		Vanuatu	X
Saint Lucia	X	Viet Nam	X
St Vincent and the Grenadines	X	Yemen	X
		Yugoslavia	X
Samoa	28–09–84	Zaïre	22–08–83
São Tomé and Principe	13–07–83	Zambia	X
Saudi Arabia	07–12–84	Zimbabwe	X
Senegal	X	Other Signatories	
Seychelles	X	Cook Islands	X
Sierra Leone	X	European Economic Community	07–12–84
Singapore	X		
Solomon Islands	X	Namibia (UN Council for)	X
Somalia	X		
South Africa	05–12–84	Niue	05–12–84
Spain	04–12–84		

Source: Compiled from 'Status of the United Nations Convention on the Law of the Sea', *Law of the Sea Bulletin*, no. 8 (Nov. 1986), pp. 1–6.
Note: a. Those states signing the Convention on 10 December 1982 are indicated by an X; those signing later are indicated by the date.

states that they were neither sharing in earth's resource wealth nor improving their economic condition. This feeling became especially intense in the late 1950s and 1960s with the emergence of many new states, as colonialism collapsed. A common bond of rising expectations and of indignation at their second-class position in world economic and political relations helped propel these states toward seeking a new international economic order. UNCLOS III became the locus of this drive for equity, as the less-developed countries (LDCs) envisioned it. How different their and the developed states' visions of equity were and how it should be achieved became clear at the onset of UNCLOS III deliberations, especially concerning seabed minerals.

The Convention (containing 320 articles and 9 technical annexes) encompasses oceanic-related problems of living resources; protection and preservation of the marine environment; access rights to and from the sea by land-locked states; international navigation; and marine scientific research, among others. These topics are examined here, however, only as they bear on marine minerals and the effort of the world community to establish an International Seabed Authority (ISA) to administer them.

Most LDCs and developed states came to the UNCLOS III negotiations with a deep distrust of the other. The LDCs' distrust of the developed states is not surprising, considering that only a decade or two before many of them still maintained a colonial empire. One difficulty encountered from the start in implementing an international seabed regime stemmed from a difference in preferences for its administrative organisation and functions. A coalition of LDCs that had worked together as early as 1964 in the UN Trade and Development Conference, continued co-operating within UNCLOS III and came to be called the Group of 77 (G–77).[21] The G–77 states (now numbering about 125) wanted a strong ISA with comprehensive powers. The developed (industrialised) states desired an ISA with well-defined but limited regulatory powers. Another early stumbling-block focused on voting within the ISA. Most of the G-77 states insisted that decision-making be done by numerical majorities: one country, one vote. Many developed states at first sought to have decisions made by using weighted voting, to parallel technological capabilities and investments in seabed mining. LDCs viewed this approach with grave scepticism, seeing it as a form of veto power over the will of the majority. Given the divergence of views held by the so-called North–South camps, it is surprising that anything of value came from these negotiations.

What, in fact, did evolve from UNCLOS III's marathon sessions from 1974 to 1982? Without a doubt, what came from this herculean effort was a document whose conceptual goals and delicate balancing of national interests represent the most important international agreement since the establishment of the UN itself. Although not a perfect document, it shows a remarkable capacity for national governments to compromise in an effort to meet the needs of the international community. It has also set aside a large portion of the world's seabeds as a common heritage of mankind and taken a step, if only a small one, toward establishing a new international economic order.

Although the UNCLOS III Convention has been signed by the majority of the world's states, it is not yet in force. Currently the national governments that signed the Convention are submitting the document to their constitutional processes of ratification. Even though a state did not sign the Convention, it may become a Party *by acceding to it*. The ratification process has been in progress since 1982. Not until one year after sixty states have ratified or acceded to the Convention will it become effective. As of late 1987, thirty-five states had ratified or acceded (Table 2.2). Assuming a similar ratification rate in coming years, it may be 1992 before the necessary sixty Parties is reached and 1993 before the Convention is officially operative. Considering the diversity of national interests of the

Table 2.2 Ratifiers of the United Nations Convention on the Law of the Sea

State	Date of ratification[a]
Antigua and Barbuda	02–02–89[e]
Bahamas	29–07–83
Bahrain	30–05–85
Belize	13–08–83
Brazil	12–12–88[e]
Cameroon	19–11–85
Cape Verde	10–08–87[b]
Côte d'Ivoire	26–03–84
Cuba	15–08–84
Cyprus	12–12–88[e]
Democratic Yemen	21–07–87[c]
Egypt	26–08–83
Fiji	10–12–82
Gambia	22–05–84
Ghana	07–06–83
Guinea	06–09–85
Guinea-Bissau	25–08–86
Iceland	21–06–85
Indonesia	03–02–86
Iraq	30–07–85
Jamaica	21–03–83
Kenya	02–03–89[e]
Kuwait	02–05–86
Mali	16–07–85
Mexico	18–03–83
Nigeria	14–08–86
Paraguay	26–09–86
Philippines	08–05–84
Saint Lucia	27–03–85
Sao Tome and Principe	03–11–87[d]
Senegal	25–10–84
Sudan	23–01–85
Togo	16–04–85
Trinidad and Tobago	25–04–86
Tunisia	24–04–85
United Republic of Tanzania	30–09–85
Yugoslavia	05–05–86
Zaire	17–02–89[e]
Zambia	07–03–83
Namibia (United Nations Council for)	18–04–83

Sources: Compiled from a. 'Status of the United Nations Convention on the Law of the Sea', *Law of the Sea Bulletin*, no. 8 (Nov. 1986), pp. 1–6.; b. 'International', *Oceans Policy News* (Sept. 1987), p. 1; c. 'Law of the Sea (LOS) Convention', *Oceans Policy News* (Aug. 1987), p. 1; d. Letter: M. B. Fisk, Law of the Sea Officer, UN, NY, 8 Dec. 1987; e. 'Law of the Sea Ratifications', *Special Report: the Preparatory Commission, February 27 – March 23 1989*, Council on Ocean Law, Washington DC, 1989, p. 1.

UNCLOS III participants and the variety of topics included in the Convention – not to mention the constant demands for delegates to have made compromises at the UNCLOS negotiations – it would be naive to expect all states to have an immediate national consensus for ratification. For the most part, their debates over ratification focus on problems associated with the ISA's administration of seabed mineral resources in what is termed 'the Area'.

The Area and the Exclusive Economic Zone

Under the Convention, the Area includes all oceanic regions (with some specifically defined exceptions) beyond 200nmi as measured from coastal states' territorial sea baselines. Within the Area, only the minerals on and in the seabed are controlled by the ISA, not other resources. When a coastal state's continental shelf extends beyond 200nmi, it may claim seabed biotic and mineral resources out to 350nmi (684km) from its territorial sea baseline or to a point 100nmi (185km) beyond the 2,500m isobath. Where submarine elevations occur as 'natural components of the continental margin such as plateaux, rises, caps, banks and spurs', the limit could be pushed beyond 350nmi. Waters and seabed lying within the 200-nmi limit are designated as an Exclusive Economic Zone (EEZ), with all biotic resources of the water column and seabed and all mineral resources on and in the seabed being controlled by the coastal state. All states, however, have the freedom of access (by water and overflight) to the EEZ and may lay submarine cables and pipelines on the seabed (Figure 2.1).[22]

Advocates of the freedom-of-the-seas principle view the acceptance of the EEZ concept (putting oceanic resources under the control of coastal states) as a step backward, a return to the era of John Selden. The world community is witnessing the enclosure of the oceanic commons much as the rural villagers of England in the eighteenth and nineteenth centuries witnessed the enclosure of their terrestrial commons. What has been oceanic space open to all capable of using it is now becoming the private domain of the coastal states – a revolutionary development, some would say. Or is it?

In reality, the EEZ concept was not spawned solely from the efforts of UNCLOS III; it has identifiable antecedents. During the past century, many states have developed 'contiguous zones'[23], areas that extend varying distances beyond their outer territorial sea limit, whose purpose is to protect the national integrity of the state from coastal piracy, drug running, smuggling, and illegal immigration. Within the contiguous zone, most states claim the right to place ships' passengers under quarantine and to detain, search, and seize illegal goods. Under the UNCLOS Convention, the contiguous zone extends 12nmi beyond a 12-nmi territorial sea. In time, states may increasingly exercise those same jurisdictional prerogatives within all of their 200-nmi EEZ,[24] until little distinction remains between it and their territorial sea and contiguous zone.

Another important antecedent is Truman's 1945 proclamation (noted earlier) which claimed the seabeds surrounding the US. His action initiated a trend toward increasingly more distant enclosures of the oceans. Mexico, one month after the Truman Proclamation, claimed ownership of all its adjoining continental shelf and its associated resources, focusing especially on fisheries. Argentina and Panama (1946),

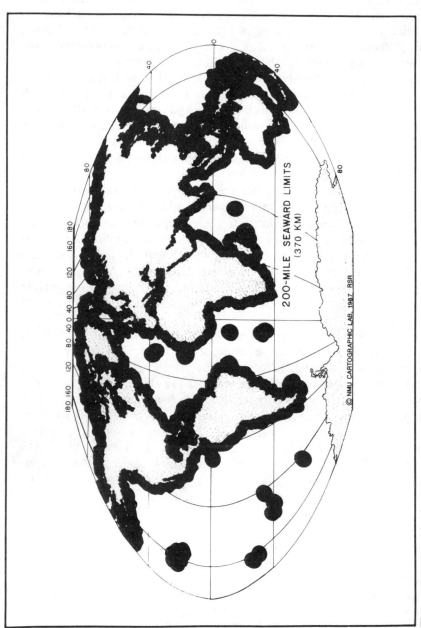

Figure 2.1 Worldwide 200-nmi Exclusive Economic Zones

Source: Author.

Chile and Peru (1947), Costa Rica (1948), Honduras and El Salvador (1950), and Ecuador (1952) made similar proclamations.

In 1952 the South American Pacific-rim states of Chile, Peru, and Ecuador met in Santiago, Chile, reaffirming their earlier actions. They adopted what came to be known as the Santiago Declaration, which proclaimed their exclusive sovereignty over waters and living resources within 200nmi of their coasts. In effect, they had established a new territorial sea. Branford Taitt suggests that, because of its adamant focus on economic issues, the Santiago Declaration was the 'direct antecedent' of the EEZ concept of today.[25]

Nearly two decades after the Santiago Declaration, a series of regional conferences began in response to LOS issues and the concern of many coastal states for their offshore resources and sovereignty claims. Concerned Latin American states met in Montevideo, Uruguay, and Lima, Peru (1970); African and Asian states convened in Colombo, Ceylon [Sri Lanka] (1971); and Caribbean states met in Santo Domingo, Dominican Republic (1972). Numerous ideas and points of view concerning patrimonial fishing zones and the concept of the EEZ emerged from these meetings. Many of their concerns were incorporated into the agenda of the first full working session of UNCLOS III in 1974, in Caracas, Venezuela.[26]

Of the 110 states attending the UNCLOS III meeting in Caracas, only 8 delegations opposed the EEZ concept. By 1976 seventeen states had made, implemented, and enforced unilateral 200-nmi EEZ declarations,

Table 2.3 States that have unilaterally declared and put into force an Exclusive Economic Zone, as of 1987[a]

Antigua and Barbuda	Gabon[c]	Oman
Bangladesh	Grenada	Pakistan
Barbados	Guatemala	Philippines
Burma	Guinea	Portugal
Cape Verde	Guinea-Bissau	Samoa
Colombia	Haiti	São Tomé and Principe
Comoros	Honduras	Senegal[b]
Cook Islands	Iceland	Seychelles
Costa Rica	India	Solomon Islands
Côte d'Ivoire	Indonesia	Spain
Cuba	Kenya	Sri Lanka
Democratic Kampuchea	Madagascar[b]	Surinam
Democratic People's	Mauritania	Togo
Republic of Korea	Mauritius	Tonga
Democratic Yemen	Mexico	Trindad and Tobago[c]
Djibouti[b]	Morocco	USSR[b]
Dominica	Mozambique	United States of America
Dominican Republic	New Zealand	Vanuatu
Equatorial Guinea[b]	Nigeria	Venezuela
Fiji	Niue	Viet Nam
France	Norway	

Sources: a. Office of the Special Representative of the Secretary-General for the Law of the Sea, UN, *Law of the Sea Bulletin*, no. 2 (March 1985), pp. ii–iv; b. Letter: A. Demarffy, Senior Officer, UN, NY, 14 Jan. 1986; c. Office for Ocean Affairs and the Law of the Sea, UN, *Law of the Sea Bulletin*, no. 9 (April 1987), pp. 3–17.

and by 1987 there was a total of sixty-one (Table 2.3). Numerous other states have proclaimed EEZs but failed to enact or to enforce implementing legislation. Another twelve states claim 200-nmi territorial seas and twenty-two others claim 200-nmi fishery zones. In addition, two states have claimed the seabed out to 350nmi (648km) – Argentina along its entire coast and Chile in waters surrounding its Easter Island and Isla Sala y Gomez. And Ecuador claims – as continental shelf – the seabed (centred on the Carnegie Ridge) that connects its mainland area with its Galápagos Islands, an offshore distance of 540nmi (1,000km). The Ridge lies in waters of less than 2,500m which puts it within the limits prescribed by Article 76 of the Convention. The US contends that neither Chile nor Ecuador has the necessary data to prove its claim.[27] Many similar claims will likely arise in future. Indeed, interpretations of Article 76 (Appendix A) will keep both lawyers and geologists busy for many years to come. Collectively claimants of various types of 200-nmi offshore zones totalled nearly 100 by 1986.[28] If all coastal and island states were to establish 200-nmi EEZs, 30 to 40 per cent of the world's oceans would be placed under national control.[29] Although a basic objective of the LOS negotiations was to provide a more equitable distribution of the world's resource wealth, what has been accomplished is to enhance the position of coastal states, especially those with exceptionally long coast lines and numerous archipelagos (Table 2.4).

Table 2.4 States with a large Exclusive Economic Zone

State	Approximate Area (thousands of km^2)[a]
United States	7,621
Australia	7,007
Indonesia	5,409
New Zealand	4,833
Canada	4,699
USSR	4,490
Japan	3,862
Brazil	3,169
Mexico	2,850
Chile	2,288
Norway	2,024
India	2,013
Philippines	1,890
Madagascar	1,293
Total	53,448

Source: L. M. Alexander and R. D. Hodgson, 'The impact of the 200-Mile Économic zone on the law of the sea', *San Diego Law Review*, vol. 12, no. 3 (1975), pp. 574–5.
Note: a. Converted from nmi^2; $1nmi^2$ equals $3.43m^2$.

Structure of the International Seabed Authority (ISA)

Some points of debate between the developed countries and LDCs over the structure and functioning of the ISA (the Area's overall managing agency) were noted earlier, such as its breadth of responsibility and voting procedures. In addition to these questions, delegates were concerned with (1) how and by whom potential seabed resources should be managed and exploited; (2) how much of certain seabed minerals should be produced; and (3) how mining profits could be equitably shared with LDCs, with land-locked and with geographically disadvantaged states. Through compromise, a tightly organised ISA (whose seat will be Jamaica) took form, consisting of an Assembly, a Council, a Secretariat, an Enterprise, and two special commissions. In addition, an autonomous International Tribunal for the law of the Sea was created (Figure 2.2).

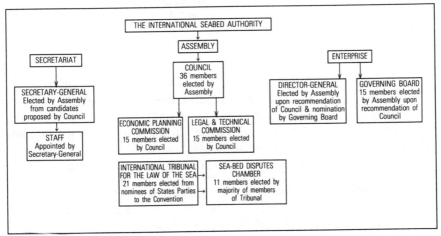

Figure 2.2 Schematic representation of the organisation of the International Seabed Authority and the International Tribunal for the Law of the Sea.
Source: After F. C. F. Earney, 'Law of the Sea, resource use, and international understanding', *Journal of Geography*, vol. 84, no. 3 (1985), p. 106. With permission.

The Assembly

The final arbiter of decisions within the ISA is the Assembly, of which every Party to the Convention is a member. It will meet annually at the designated seat of the Authority. A majority of members present and voting decides procedural questions. Matters of substance require a 'two-thirds majority of the members present and voting, provided that such majority includes a majority of the members participating in the session'. Developed states had argued for a three-quarters majority, a requirement that would have allowed them to defeat more easily unwanted legislation. Functions of the Assembly include the election of Council members, the Secretary-General of the Secretariat, members of the Governing Board of the Enterprise, and the Director-General of the Enterprise. The Assembly also

1 establishes subsidiary organs within the Authority;
2 assesses contributions of members to the Authority's administrative budget;

3 determines the distribution of financial and other economic benefits accruing from activities in the Area;
4 considers and approves the budget proposed by the Council;
5 examines all reports from the various organs of the Authority;
6 initiates studies and makes recommendations for promoting co-operation of states in all matters relating to the Area;
7 investigates problems that arise concerning the Area, especially as they relate to developing states, and with land-locked and geographically disadvantaged (shelf-locked) states;
8 suspends the rights and privileges of Authority membership;
9 discusses any matter within the competence of the Authority, and decides which of the Authority's organs should be responsible for action in a given problem.[30]

The Council

The Council (fulcrum of power in the ISA) shall consist of thirty-six members, elected by the Assembly. The US wanted a guaranteed membership, but this provision is not in the Convention. Its fear of exclusion, however, is probably not justified because of other membership provisions (see my emphases) as outlined in items 1 and 2 below. As quoted from the Convention, election to the Council occurs in the following order:

1 Four members from among those States Parties which, during the last five years for which statistics are available, have either consumed more than 2 per cent of total world consumption or have had net imports of more than 2 per cent of total world imports of the commodities produced from the categories of minerals to be derived from the Area, and in any case one State from the Eastern European (Socialist) region, as well as the *largest consumer*.
2 Four members from among the eight States Parties which have the *largest investments* in preparation for and in the conduct of activities in the Area, either directly or through their nationals, including at least one State from the Eastern European (Socialist) region.
3 Four members from among States Parties which on the basis of production in areas under their jurisdiction are major net exporters of the categories of minerals to be derived from the Area, including at least two developing States whose exports of such minerals have a substantial bearing upon their economies.
4 Six members from among developing States Parties, representing special interests. The special interests to be represented shall include those States with large populations, States which are land-locked or geographically disadvantaged, States which are importers of the categories of minerals to be derived from the Area, States which are potential producers of such minerals, and least developed states.
5 Eighteen members elected according to the principle of ensuring an equitable geographical distribution of seats in the Council as a whole,

provided that each geographical region shall have at least one member elected under this paragraph. For this purpose, the geographical regions shall be Africa, Asia, Eastern European (Socialist), Latin America, and Western European and Others.[31]

Council membership stipulations are designed to proportionally reflect interest groups (blocs) within the Authority, giving special consideration to onshore producers concerned about competition from seabed minerals. For example, Zaïre has a significant dependence on sales of copper and cobalt, which are among the main minerals that may be extracted (should economic conditions allow it) from ferromanganese nodules. Membership priorities also give an assured inclusion of some states with corporate investors in seabed mining, as the US, the FRG, and Japan. Equally important, the land-locked and the so-called disadvantaged states are accommodated. One important characteristic of the Council's composition, however, has recently been questioned by the US Government, that is three seats are guaranteed to the Eastern European (socialist) states. This provision, however, was designed to provide a balanced regional membership, because the preponderance of potential members are western states. The complaint by the US comes *ex post facto*, because initially the US was satisfied with this provision (Article 161) when it was first drafted.

Council members serve four years, may be re-elected, and meet at the seat of the ISA. Passage of most decisions requires a two-thirds majority of those members present and voting. Several specifically identified questions are decided by a three-quarters majority of those present and voting, 'provided that such majority includes a majority' of the Council membership. Certain decisions on vitally important questions require consensus, meaning 'the absence of any formal objection'.[32] The Council is the ISA's executive body, and is empowered to formulate policies for the ISA. In addition, the Council

1 supervises and co-ordinates the implementation of all matters within the competence of the ISA, such as who shall be permitted to mine;
2 proposes candidates to the Assembly for election to the posts of Secretary-General, the Governing Board and the Director-General of the Enterprise;
3 establishes additional subsidiary organs within the ISA;
4 enters, with the Assembly's approval, into agreements with the UN or other international organisations.[33]

Other Council functions include the appointment of fifteen members 'of the highest standards of competence and integrity' to both the Legal and Technical Commission (LTC) and the Economic Planning Commission (EPC). The LTC reviews all mining-company work plans for activities in the Area, and provides recommendations to the Council. It assesses environmental consequences of activities in the Area, and if requested by the Council, it supervises activities in the Area. The LTC recommends actions for environmental protection and monitoring and can suspend work within the Area, if an activity seems harmful. Additional and highly significant functions it has are to make production authorisations and to calculate production ceilings. The EPC also reviews and analyses supply, demand, and price trends for metals produced in the Area, while attempting to view these conditions as they relate to both importing and exporting states, especially the developing States Parties. The concern is that certain metals produced – nickel, copper, cobalt, or manganese – could flood the world market, causing damage to the economies of major metal producers, such as Canada, Zambia, and Zaïre. If developing States Parties suffer economic damage from activities in the Area, the EPC may propose to the Council methods of compensating them or providing economic adjustment assistance. Because of these problems, two EPC members must come from developing States Parties whose economies depend on mineral exports that are the same as those to be produced in deep seabed mining. The EPC membership, also, includes experts in seabed mining, in mineral resource management, and in international trade and international economics.[34]

The Secretariat

The Secretariat is headed by a Secretary-General, who is elected by the Assembly from among candidates suggested by the Council. The term of office is four years and an incumbent may be re-elected. The Secretary-General's responsibilities include the submitting of an annual report to the Assembly and the appointing of administrative and technical personnel within the ISA. Allowing for the constraints of expertise needed, these appointments must reflect an equitable geographic distribution. The Secretary-General and staff within the Secretariat are not to have financial associations with ocean mining activities, are expected to remain autonomous from outside political influences, and are responsible only to the ISA.[35] Meeting this last expectation may be difficult, considering the highly politicised nature of the ISA's formation. On the other hand, there are many international civil servants (as in the UN) whose allegiance to their agency affiliation is stronger than their allegiance to their home countries.

The Enterprise

Under the guidance of a Director-General and a fifteen-member Governing Board (elected by the Assembly with the recommendation of the Council). The Enterprise acts for the Authority. The Enterprise will mine for the Authority in its own right (functions of the Enterprise are discussed in more detail below).

Preparatory Commission

To facilitate the implementation and operation of the ISA (including its seabed mining functions) and the Tribunal, UNCLOS III participants incorporated into the Final Act a provision (Resolution I) for establishing a Preparatory Commission (PREPCOM), when the fiftieth state signed the Convention. Having received the necessary signatures to the Convention and with the adoption of the Final Act on 9 December 1982, PREPCOM was officially convened. PREPCOM's membership includes all Convention signatories. Those not signing the Convention but who signed the Final Act are eligible to participate as observers in the deliberations of the Commission but cannot vote.[36] The US signed the Final Act but has refused to participate in PREPCOM's proceedings, even though other non-signatories of the Convention do attend. To emphasise the American position, the Reagan Administration on 30 December 1982 withheld its 1983 *pro rata* share (US $500,000–$700,000) of the UN budget for funding PREPCOM's work,[37] and it has since continued this policy. The decision by the US not to participate as an observer in PREPCOM is regrettable for two reasons. Its presence could have a significant impact in guiding the work of PREPCOM. More important, however, is that the UNCLOS III delegations have learned that they can function without the US.

PREPCOM, composed of a plenary body with four special subcommissions, has been

1 drafting administrative and financial management regulations to enable the Authority to begin operating;
2 preparing provisional agendas and rules for the first session of the Assembly and council;
3 making recommendations for relationships between the ISA and the UN and other international organisations;
4 studying potential problems of land-based producer states of minerals that will be mined in the area and finding ways to help these states make economic adjustments, such as establishing a compensation fund;
5 drafting rules and procedure to permit the ISA to begin operating the Enterprise.[38]

Pioneer Investors

In addition to Resolution I, which established PREPCOM, UNCLOS III adopted a second resolution titled 'Governing Preparatory Investment in Pioneer Activities Relating to Polymetallic Nodules', usually referred to as Pioneer Investor Protection (PIP). This resolution attempts to reassure states with large investments in seabed mining research that they will not be overlooked when the ISA allocates first-round licences for prime manganese nodule mining sites.

Under Resolution II, France, India, Japan, and the USSR, their state companies, or one of their natural or juridical persons were designated as eligible to apply for one priority claim as pioneer investors if, by 1 January 1983, they had invested at least $30 million (constant 1982 US$) in seabed exploration and research; 10 per cent of this expenditure must have been in a specifically identified ocean area. Four other entities (mining consortia) whose components possess the nationality of or are effectively controlled by one or more of the following states (Belgium, Canada, Italy, Japan, the FRG, the Netherlands, the UK, and the US) are considered eligible to register as pioneer investors. In addition to the eight PIP candidates, LDC States Parties could become eligible, if they met the same total-dollar-investment criterion. The investment cut-off date, however, for the LDCs was 1 January 1985. Several LDCs subsequently demanded an extension of this closure date, and in 1986 the rule was changed to allow any one or group of LDCs to apply until the Convention comes into force, with no limitation on the number of sites they may claim. These claims, though, must not conflict with the state pioneer investors' claims. This concession to the LDCs revived earlier demands by the Eastern European Bloc for a second site; PREPCOM, also, met the Eastern European Bloc's demand, and it may now apply for a second site until the Convention becomes effective.[39] To date (mid-1987), only the state entities, France, India, Japan, and the USSR have applied for registration as pioneer investors. None of the consortia has done so.

Overlapping claims

PREPCOM has had five sessions and resumed sessions since it began functioning in 1983. Progress has been slow within its four special commissions (I – Land-Based Producers, II – The Enterprise, III – Seabed Mining Code, IV – International Tribunal for the Law of the Sea). Indirectly part of the difficulty came from the unresolved problem of overlapping claims by three of the four state pioneer investors – France, Japan, the USSR, and India; the latter had no overlapping claim. The US, the UK, and FRG added to the complexity of the overlapping claims problems in the Clarion-Clipperton zone (CCZ) by independently licensing consortia to operate there. Some of these licence areas overlapped with the USSR's claim.[40] Although the special commissions had no direct involvement in the overlapping claims issue, Commission II could not perform its operational responsibilities until the claims issue was resolved, because all overlapping claims disputes had to be settled before PREPCOM could honour pioneer registrations.[41]

The last and most intractable dispute involved France and the USSR, but this was settled in late 1987, with registration taking place on 17 December. PREPCOM (under Resolution II) approved deep seabed mining operating sites for the four states signing the UNCLOS Convention, including France, Japan, India, and the USSR. Mine sites for the ISA were designated and the USSR also relinquished additional areas for future mining activities of non-PIP entities. These claims lie between the Clarion-Clipperton Fracture Zones (7°–16°N) and the Hawaiian Islands and Mexico (124°–154°W). This area is the world's best explored and prime nodule region (Figure 2.3). Elisabeth Mann Borgese sees the successful resolution of these overlapping claims problems and the agreement of the consortia (as well as the US Government) as a 'new phase for the Prep Com' and a move 'in the direction of a universally recognised regime'.[42]

Mining prescriptions under the Seabed Authority

For any entity to prospect (prior to exploration or exploitation), it must identify the approximate area(s) where it desires to work and have submitted to the Authority a satisfactorily written statement that it will meet all regulations and required procedures such as protecting the environment and, eventually, providing training programmes for developing states' seabed mining personnel. Before full-scale exploration and exploitation may begin, all contracting parties must submit a work plan to the Council, which acts with the advice of the Legal and Technical Commission. The work plan must include a description of the methods and technology to be used and a statement about where this technology may be purchased. The Council must evaluate work plans in the order received and begin its evaluation within six months after the Convention goes into force. Applicants must comply with all Convention requirements for (1) protecting the environment; (2) assisting with the training of ISA and developing States Parties personnel in the techniques and applications of scientific knowledge 'relating to activities in the Area so that the Enterprise and all State Parties may benefit therefrom'.[43]

After prospecting work is finished, a contract application has been made and a work plan submitted, the mining company or state firm must divide the area investigated into two parts of nearly equal size and value, and must provide the Authority with data on the location, abundance, and metal content of any nodules found there. The original Convention states that, within forty-five days, the Authority must choose one of the two units as a reserve area,[44] but in September 1986 (under an agreement titled the 'Arusha Understanding') PREPCOM moved to allow the first four pioneer investors – France, India, Japan, USSR – to pre-select $52,300km^2$ of their site (which they have done). This provision, in effect, predetermined the area designated for the Enterprise and allowed for more easily negotiated adjustments in overlapping claims.[45] The four pioneer investor states have also agreed to help the ISA 'in the exploration of a mine site ... for the Enterprise's first operation and to assist in ... preparing a plan of work for that mine site'.[46]

The Enterprise may opt to mine or not to mine its reserve areas. It can mine alone, form joint-ventures, or make production-sharing agreements with developing States Parties. A developing State Party 'or any natural or juridical person sponsored ... and effectively controlled by it' may submit a work plan to mine in a given reserve area.[47]

At the time when applicants submit a work plan and ask for a contract,

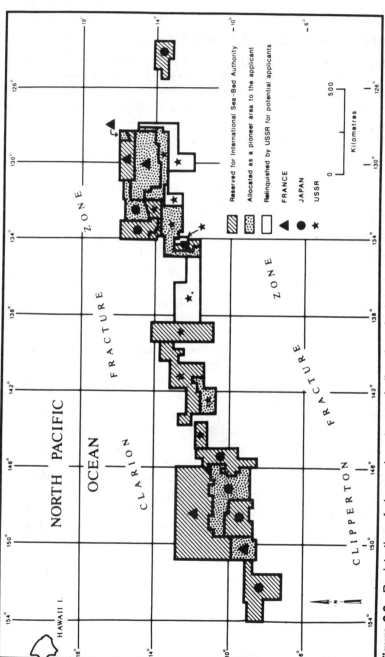

Figure 2.3 Registration of pioneer investors and allocation of mine sites for the International Seabed Authority

Sources: 'Registration of Pioneer Investors in the International Sea-Bed Area in Accordance with Resolution II of the Third United Nations Conference on the Law of the Sea', *Law of the Sea Bulletin*, Special Issue II (April 1988), pp. 6, 25, 82, 116, and maps (drafted by C. H. Harrington) provided by J. Flanagan, Letter, National Oceanic and Atmospheric Administration, DOC, Washington, DC, 11 Oct. 1988.

Note: Resolution II established an upper limit of 150,000 km^2 for each mine site. 'Potential applicants' refers to the western multinational consortia that possess the nationality of or are controlled by the nationals of Belgium, Canada, the FRG, Italy, Japan, the Netherlands, the UK, and the US.

they must pay US$500,000. If the Authority's costs in processing the application are less than this sum, the difference will be returned to the applicant. From the date of entry into force of the contract, the mining firm must also pay an annual fixed fee of $1 million. Within a year after the start of commercial production, the mining firm will begin to pay either the fixed fee or a production charge, whichever is greater. They may also pay a combination of production charge and a share of net proceeds. Should the contractor decide to pay a production charge only, the rate is set 'at a percentage of the market value of the processed metals produced from the polymetallic nodules recovered from the area covered by the contract'. During the first ten years of commercial production, the rate is fixed at 5 per cent and from the eleventh year until the end of mining operations, the rate is 12 per cent. If the contractor wants to pay the Authority by a combination of production charge and a share of the net proceeds, the rate of the production charge is a percentage of the market value (total production × average price) of the metals produced from the nodules mined. The percentage charged will be 2 per cent during a first-period of commercial production and 4 per cent in a second-period of commercial production. The first-period of commercial production begins 'in the first accounting year of commercial production' and terminates 'in the accounting year in which the contractor's development costs with interest on the unrecovered portion thereof are fully recovered by his cash surplus'. The second-period of commercial production begins in the accounting year succeeding the end of the first-period of commercial production and continues to the termination of the contract.[48] If the Enterprise operates mine sites for its own use, the necessary funding (other than sales of minerals mined) will come from application fees, production charges and profit charges paid by the mining consortia.

Dispute settlement

The Convention provides for several dispute settlement fora in matters of ocean use and abuse, including an International Tribunal for the Law of the Sea (ITLS), conciliation commissions, and arbitration tribunals.

International Tribunal for the Law of the Sea

The Tribunal has twenty-one members elected through a secret ballot by States Parties at a meeting to be convened by the UN Secretary-General. Each State Party will have had an opportunity to nominate two candidates to stand for election. Its members may not

exercise any political or administrative function, or associate actively with or be financially interested in any of the operations of any enterprise concerned with the exploration of the resources of the sea or the seabed or other commercial use of the sea or thc seabed.

Members serve nine years and may be re-elected. For the first term of office, however, seven members will serve only three years and seven others only for six years, to be chosen by lot by the UN Secretary-General. Each State Party may have only one of its nationals sitting on the Tribunal at any one time.[49]

Ordinarily the Tribunal will sit in Hamburg in the FRG, but it may meet in other places when it seems appropriate.[50] Some people question the appropriateness of the Tribunal's sitting in Hamburg, because the FRG did not sign the Convention and it has not ratified it. There is a general agreement that should the FRG not have acceded within one year of the Convention's coming into force, another host state will be selected. There is a vigorous effort within the FRG's Bundestag in favour of accession.[51]

An important feature of the Tribunal is its special Sea-Bed Disputes Chamber (TDC) composed of eleven members (from the TDC) chosen by a majority of the Tribunal's elected members. The TDC shall hear disputes concerning the terms of contracts for mining in the area. Its decisions are final and will be enforced 'in the territories of the States Parties in the same manner as judgments or orders of the highest court of the State Party in whose territory the enforcement is sought'. Members will sit for three years and are eligible for a second term.[52]

Conciliation

When a dispute occurs between Parties over an interpretation or application of the Convention, they may take their case before a five-member conciliation commission. The UN Secretary-General will maintain a list of conciliators nominated by all States parties, who may make no more than four nominations. The disputants, unless they otherwise agree, will appoint two conciliators each, preferably from the Secretary-General's list. Within thirty days after all four conciliators have been appointed, the disuputants will appoint (from the Secretary-General's list) a fifth conciliator (who serves as chair). Within twelve months after the commission's constitution, it must submit a report to the UN Secretary-General who then reports the findings to the Parties. Although the commission's findings are not binding on the Parties,[53] this avenue of dispute settlement should help maintain harmony.

Arbitration

If States parties fail to agree on a dispute settlement forum, the dispute must go to arbitration. As in conciliation, the UN Secretary-General maintains a list of arbitrators, who are nominated by the States Parties. Both disputants appoint one member to a five-member tribunal, preferably from the Secretary-General's list. The other members are selected by mutual agreement between the States Parties. If they cannot agree on these appointments, the President of the ITLS may appoint. An arbitration tribunal's decisions are by majority vote of its members. Awards must be limited to the precise subject matter under dispute and premised on specific reasons. Unless the States Parties have agreed 'to an appellate procedure' in advance, 'the award shall be final and without appeal'.[54]

When disputes arise concerning the 'interpretation or application' of Convention articles that deal with (1) navigation-associated pollution and dumping from vessels, (2) fisheries, (3) preservation and protection of the marine environment, and (4) marine research, special arbitration procedures may be used. Lists of experts, nominated by States Parties, are maintained by various international bodies, as the International Maritime Organization, the Food and Agricultural Organization, the Inter-governmental Oceanographic Commission, and the United Nations Environmental Programme. Again, disputants select experts (two in this case) from the States Parties' nominated lists. Appointment of the president of a special arbitral tribunal comes via mutual agreement. A special arbitration tribunal's functions, voting and awards procedures are the same as those used in regular arbitration tribunals.[55]

Conclusions

The complexity of the overlapping-claims disputes exemplifies the many profound problems inherent in implementing the prescriptions of a document as wide ranging as the UNCLOS III convention, which encompasses all resources and uses of the oceans. Although the Convention will probably receive the sixty ratifications needed to bring it into force, it will be a long process. Because the US and several other states with a seabed mining capability have chosen not to sign or to accede to the UNCLOS III Convention, prospects for its future are problematic. Many observers feel that, although the Convention has not been ratified by the necessary sixty states, it has already had a major impact on the law of the sea – not on formal law but on customary law.

In contrast to formal agreements or treaties, customary law develops when a community of world states, through reciprocity, recognises and conforms to uniform actions.[56] Numerous elements contained within the Convention are already becoming a functional part of the operating procedure and national laws of many states, as for example, the 12-nmi (22km) territorial sea and the 200-nmi EEZ. Because ratification of the Convention by the sixtieth state is still several years in the future, time will help parts of it to become customary law.

Is the US violating the etiquette of recognised customary international law?[57] Defenders of US policies would say:

No, we have signed nothing, and our President, Ronald Reagan, has stated repeatedly that we will neither sign nor ratify the Convention in its present form and furthermore, deep seabed mining is not prohibited by the 1958 Geneva Convention on the High Seas, which has not yet been superseded.

The less truculent might point out that, in many ways, the US is paralleling the Convention's mandates. For example, in March 1983 the US unilaterally declared an EEZ, an action allowed under the

Convention. Other actions, such as the issuing of deep seabed mining licences, have marked the US as an international renegade. Many astute students of the issue would defend the thesis that the US *is* in violation of international customary law. Despite worldwide criticism, the US will likely continue to pursue its ocean-affairs objectives, alone or in concert with other like-minded states.

Notes

1 C. C. Joyner, 'Normative evolution and policy process in the Law of the Sea', *Ocean Development and International Law Journal*, vol. 15, no. 1 (1985), p. 61.
2 M. C. W. Pinto, 'Emerging concepts of the Law of the Sea: some social and cultural impacts', in J. G. Richardson (ed.), *Managing the Ocean: Resources, Research, Law* (Lomond Publications, Mt Airy, MD, 1985), p. 298.
3 Joyner, 'Normative evolution and policy process', p. 6.
4 Pinto, 'Emerging concepts of the Law of the Sea', p. 298.
5 Joyner, 'Normative evolution and policy process', pp. 62–4.
6 W. Ullman, *A Short History of the Papacy in the Middle Ages* (Methuen, London, 1972), pp. 323–4; D. E. Worcester and W. G. Schaeffer, *The Growth and Culture of Latin America* (Oxford University Press, NY, 1956), p. 13.
7 Pinto, 'Emerging concepts of the Law of the Sea', p. 298.
8 R. P. Anand, *Origin and Development of the Law of the Sea: History of International Law Revisited* (Martinus Nijhoff, The Hague, 1983), pp. 78–9.
9 A. L. Hollick, 'Managing the oceans', *Wilson Quarterly*, vol. 8, no. 3 (1984), p. 77.
10 ibid., p. 78.
11 S. P. Jagota, 'The United Nations Convention on the Law of the Sea, 1982', in E. M. Borgese and N. Ginsburg (eds), *Ocean Yearbook 5* (University of Chicago, Chicago, IL, 1985), p. 12; Pinto, 'Emerging concepts of the Law of the Sea', p. 299.
12 Jagota, 'The United Nations Convention', p. 13.
13 F. C. F. Earney, 'Law of the Sea, resource use, and international understanding', *Journal of Geography*, vol. 84, no. 3 (1985), p. 105.
14 L. Juda, 'The Exclusive Economic Zone: compatibility of national claims and the UN Convention on the Law of the Sea', *Ocean Development and International Law Journal*, vol. 16, no. 1 (1986), p. 2; see also R. D. Eckert, *The Enclosure of Ocean Resources: Economics and the Law of the Sea* (Hoover Institution Press, Stanford, CA, 1979), pp. 142–5.
15 F. C. F. Earney, 'Law of the Sea', p. 105.
16 F. C. F. Earney, 'Ocean space and seabed mining', *Journal of geography*, vol. 74, no. 9 (1975), p. 541.
17 'Natural resources of the subsoil and sea bed of the continental shelf', Presidential Proclamation no. 2,667, *Federal Register*, vol. 10 (28 Sept. 1945), p. 12,303.
18 'Sea Law – "a rendezvous with history"', *UN Chronicle*, vol. 19, no. 6 (1982), p. 15.
19 F. M. Auburn, 'The international seabed area', *International and Comparative Law Quarterly*, 4th series, vol. 20, pt 2 (1971), p. 177.
20 'Sea Law', p. 15.
21 Pinto, 'Emerging concepts of the Law of the Sea', p. 300.
22 United Nations, *The Law of the Sea: Official Text of the United Nations Convention on the Law of the Sea* (UN, NY, 1983), pp. 27, 42.
23 A. V. Lowe, 'The development of the concept of the Contiguous Zone', *British Yearbook of International Law*, vol. 52 (1981), pp. 109–69.

24 M. H. Nordquist, 'Foreword', *San Diego Law Review*, vol. 22, no. 4 (1985), p. 726.
25 B. M. Taitt, 'The exclusive Economic Zone: A Caribbean perspective, part I – Evolution of a concept', *West Indian Law Journal*, vol. 7, no. 1 (1983), p. 42; see also E. Ferraro Costa, 'Peru and the Law of the Sea Convention', *Marine Policy*, vol. 11, no. 1 (1987), pp. 45–57.
26 ibid., pp. 43–9.
27 K. Ramakrishna, R. E. Bowen, and J. H. Archer, 'Outer limits of continental shelf: a legal analysis of Chilean and Ecuadorian Island claims and US response', *Marine Policy*, vol. 11, no. 1 (1987), pp. 58–68.
28 F. C. F. Earney, 'The United States Exclusive Economic Zone: mineral resources' in G. Blake (ed.), *Maritime Boundaries and Ocean Resources* (Croom Helm, London, 1987), pp. 162–81.
29 J. B. Smith, 'Managing nonenergy marine mineral development – genesis of a program', paper presented at Oceans 85 Conference, 12–14 Nov. 1985, San Diego, CA, p. 7.
30 F. C. F. Earney, *Ocean Mining: Geographic Perspectives*, Meddelelser fra Geografisk Institutt ved Norges Handelshøyskole og Universitetet i Bergen, no. 70 (Geografisk Institutt, Bergen, 1982), p. 138.
31 United Nations, *Law of the Sea*, pp. 55–6.
32 ibid., p. 56.
33 Earney, *Ocean Mining*, p. 139.
34 United Nations, *Law of the Sea*, pp. 59–62.
35 ibid., pp. 50, 64.
36 ibid., p. 175.
37 J. L. Malone, 'Who needs the sea treaty?', *Foreign Policy*, no. 54 (Spring 1984), pp. 58–9.
38 C. Murphy, 'LOS Preparatory Commission begins work', *Soundings*, vol. 8, no. 2 (1983), p. 1; United Nations, *Law of the Sea*, pp. 175–6; for an excellent critique of the PREPCOM's functions, see E. M. Borgese, 'The Preparatory Commission for the International Sea-Bed Authority and for the International Tribunal for the Law of the Sea: third session', in E. M. Borgese and N. Ginsburg (eds), *Ocean Yearbook 6* (University of Chicago Press, Chicago, IL, 1986), pp. 1–14.
39 'Preparatory Commission', p. 4.
40 ibid., p. 2. The CCZ is a prime manganese nodule mining area in the north-eastern Equatorial Pacific Ocean.
41 'Implementation of Resolution II', *Oceans Policy News* (May 1987), p. 4.
42 Letters: E. M. Borgese, Chairman, International Ocean Institute, Halifax, Nova Scotia, Canada, 31 Dec. 1987 and Valletta, Malta, 26 Aug. 1987.
43 United Nations, *The Law of the Sea*, pp. 44–5, 50, 113.
44 ibid., pp. 44–5, 119.
45 'Arusha understanding', *Oceans Policy News* (May 1986), pp. 5–7.
46 'Preparatory Commission', pp. 2–4.
47 United Nations, *Law of the Sea*, pp. 118–20.
48 ibid., pp. 121–2; for more detailed discussions, refer to Annex III, Article 13, 'Financial terms of contracts', pp. 120–6.
49 ibid., pp. 141–2.
50 ibid., p. 140.
51 Borgese, Letter, 26 Aug. 1987.
52 ibid., pp. 147–9.
53 ibid., pp. 98, 137–9.
54 ibid., pp. 149–52.
55 ibid., pp. 152–4.

56 J. L. Jacobson, 'Law of the Sea – what now?' *Naval War College Review*, vol. 37, no. 2/seq. 302 (1984), p. 85.
57 For a good discussion of this question, see B. Shingleton, 'UNCLOS III and the struggle for law: the elusive customary law of seabed mining', *Ocean Development and International Law Journal*, vol. 13, no. 1 (1983), pp. 33–63; see also J. K. Gamble, Jr. and M. Frankowska, 'The significance of signature to the 1982 Montego Bay Convention on the Law of the Sea', *Ocean Development and International Law Journal*, vol. 14, no. 2 (1984), pp. 121–60.

Chapter three

UNCLOS III Convention and alternatives: the view of governments and industry

States within the world community, depending on their technological capabilities and level of economic development, have differing perceptions of the UNCLOS III Convention. Most Third World states want seabed mining to proceed. On the other hand, Third World and developed states highly dependent on income from minerals like those to be extracted from the seabed would prefer a limited production. Some states with significant concerns for maintaining adequate supplies of strategic minerals fear that the ISA, to be established under the UNCLOS Convention umbrella, may not protect their interests, and they have sought to establish an alternative regime. Because of poor international mineral market conditions and the yet unsettled LOS regime situation, several seabed mining consortia – with considerable time and money already invested in seabed mining research – are hesitant to proceed.

The United States' view

During the administrations of Presidents Richard Nixon, Gerald Ford, and Jimmy Carter, US negotiators played a strong, positive role in shaping the emerging UNCLOS III Convention. But many in government and the general public felt it would be unwise for the US to agree to what they saw as an effort to 'collectivise' oceanic resources. During the decade of the 1970s, while negotiators were at work in New York and in Geneva, the US Congress made numerous efforts to pass unilateral legislation to allow consortia to mine the deep seabed. Not until 1978 did the State Department lend direct support to these deep seabed mining bills. Year after year (1971–9), deep seabed mining legislation failed to pass or died on the calendar. Finally in June 1980, the Congress passed PL–96–283, the Deep Seabed Hard Mineral Resources Act (DSHMRA).[1] Allegedly part of the motivation for passage of the act was to hasten the UNCLOS III negotiation process, and according to William Jones, a now retired US Foreign Service ambassador, it was intended as an interim arrangement.[2] The Act authorised the head of the National oceanic and Atmospheric Administration (NOAA) to issue deep seabed exploration licences and exploitation permits to eligible citizens 'pending conclusion of an acceptable Law of the Sea Treaty' which addresses the same issue.[3] Passage of the Act foretold policy changes to come.

The Reagan reassessment

In the 1980 presidential campaign the Republican Party's candidate, Ronald Reagan, focused a significant part of his platform on a re-examination of US foreign policy, including UNCLOS III negotiations. Shortly before the November 1980 election, Elliot Richardson, ambassador and head of the US UNCLOS delegation, resigned. His deputy, George Aldrich, took the vacated post of head of delegation, until he was replaced by James Malone, Assistant Secretary of State for Oceans and International Environmental and Scientific Affairs. Soon after Malone took up his duties in March 1981, he announced that President Reagan was suspending negotiations at the UNCLOS III meetings.

Malone explained that the US would not negotiate until it had carefully examined the structure of the Draft Convention, a document the Reagan Administration viewed as badly flawed. This action pleased long-time critics of the UNCLOS negotiations, as for example Northcutt Ely, a noted Washington, DC, lawyer, who was also a lobbyist for the US Steel Corporation and its associated seabed mining consortium. Another notable critic, Robert Goldwin, a constitutional studies specialist at the American Enterprise Institute, called the 'common heritage of mankind' concept a mistaken notion of property law. Although the withdrawal by the US from negotiations may have pleased some, its action was deeply resented by many US citizens who had, in good faith, worked with the LOS delegations. Third World states and many of the developed states that had so often supported the negotiating positions of the US felt betrayed and abandoned.[4]

Return to the negotiating table

President Reagan, on 29 January 1982, decided to return the US to the negotiating table. He publicly announced US objectives for reshaping the Convention, objectives that Leigh Ratiner, a negotiator and advisor to US negotiators at several UNCLOS sessions, felt were reasonable.[5] In February the US delegation returned to an informal UNCLOS meeting called specifically to consider questions raised by the US during its year of absence.[6] Inexcusably the US negotiators arrived without instructions. These did not come until after the session had begun. This situation arose primarily because of major differences of opinion within the Reagan Administration about whether instructions should be so strict as to tie the hands of the negotiators or to allow them bargaining room. Furthermore, some felt it was both hopeless and unwise for the US to continue negotiating, and if the US wanted to have its consortia mine the deep seabed, they could do so with the US adhering to a 'mini-treaty', outside an LOS Convention.[7] Finally, armed with instructions, the Malone team presented a forty-three page document that outlined Washington, DC's, misgivings about the proposed Draft Convention; the text made only a few specific suggestions about how the Convention might be improved. Collectively the document did not present a viable, definable position, but was rather a call for wholesale renegotiation.[8]

When the session formally reopened in March, the US delegation still had nothing concrete to put on the negotiating table. Finally, at the end of the first week, it presented what came to be called the 'Green Book'. This document contained material William Wertenbaker aptly described as 'an anthology of virtually everything that anyone in the Administration had ever wanted in regard to seabed mining'. The 'Green Book' proposal was so nebulous that many conference delegates quit taking the US seriously. As a result, the G–77 rejected it. They had become convinced that the US did not really want to negotiate, but only desired to gut the product of eight years of labour.[9]

Led by Jens Evenson of Norway, ten other ambassadors from among the developed states (Australia, Austria, Canada, Denmark, Finland, Iceland, Ireland, New Zealand, Sweden, and Switzerland) attempted to work out compromises. These focused on guaranteeing the US a permanent seat on the Council and simplifying procedures for acquiring mining contracts from the Authority. When the Group of Eleven's suggestions went before the G–77 and US negotiators, many thought that, if the US accepted them, the G–77 states would also approve. But the US did not accept the proposals. During the next several weeks, negotiations centred on PIP problems, with little progress being made. Throughout this period US negotiators were receiving mixed signals from Washington, DC, because of the opposed factions within the State Department and strong lobbying from outside interests. One faction demanded hard-line negotiations without compromise, whereas another pushed for compromise.[10] The US remained concerned about many issues, especially those dealing with forced transfers of mining technology[11] to the Enterprise or to LDCs and with the possible adoption of amendments to the seabed mining provisions (Part XI) of the Convention, without a concurrence of the US Senate.[12]

By the third week in April, negotiations were still in process, but the Conference President, Tommy Koh, was determined that debate should end. After calling for final amendments (more than 250 were presented) and then asking those offering them not to call for individual votes (only a few were called), on 28 April 1982, Koh closed debate. Some observers of the UNCLOS negotiations suggest that if Koh had not forced the issue at that time, the overall consensus may have disintegrated, given the US pressure on many delegations to vote against adoption. Other observers contend that if only more time had been available, the reservations of the US might have been overcome. But time had run out, and on 30 April, Koh called the question, with the US (to the dismay of all delegations) insisting on a formal vote. The US, along with Israel, Turkey, and Venezuela, voted against adoption, although the latter three did so for reasons unrelated to the seabed mining issue.[13] On 9 July President Reagan reinforced this decision when he announced that the US would not be a signatory in the coming December.[14] Then in November he sent Donald Rumsfeld to several European states and to Japan to persuade them not to sign the Convention and to encourage them to join with the US in a mini-treaty.

Reciprocating states agreements

In reality, long before the Rumsfeld mission, the US had begun an effort to design a multilateral alternative to the UNCLOS Convention. This effort focused on organising seabed mining 'reciprocity agreements', that is to recognise mutually the exclusivity of mining licences issued by the reciprocating states and not to grant conflicting licence areas. When the US Congress passed the DSHMRA in 1980, which reflected concern by industry and US Congressmen that long-established deep seabed mining consortia might not be allowed to exploit the oceans, discussions with other concerned states (UK, FRG, and France) were already under way. The 1980 Act formally authorised these negotiations by directing

> NOAA in consultation with the Secretary of State and the heads of other appropriate departments and agencies, to designate as reciprocating states those other nations which establish seabed mining programs which are compatible with and recognize the U.S. program.[15]

These states continued discussions throughout the following year,[16] and by December 1981 the FRG, the UK, and France had adopted legislation fairly similar to that of the US,[17] all of which contained clauses allowing reciprocation. Their main concerns were to harmonise (1) time frames for exploration (not before 1 January 1981) and exploitation (not before 1 January 1988), (2) regulations for environmental protection, and (3) allowable dimensions for exploration areas.[18] In early 1982 consensus was nearly reached on a negotiating text for a reciprocating states agreement. The agreement might have been signed at that time, but France was worried that the agreement would allow such large mine site areas that space for mining (in prime areas) would be inadequate for other states or for the Enterprise. Then the UK and FRG balked, fearing that signing an agreement immediately before the opening (in March) of a new round of UNCLOS meetings might adversely affect negotiations.[19] Others felt such an agreement might stimulate progress at the UNCLOS sessions. That pressure was being exerted on UNCLOS members became clear when President Koh made a personal appeal for the four states not to proceed in their efforts to establish a parallel legal framework for mining the deep seabed.

On 2 September 1982 – four months after the UNCLOS delegates had adopted the Draft Convention – the US, UK,[20] FRG, and France signed a document titled 'Seabeds: Polymetallic Nodules Agreement Between the United States of America and Other Governments', which incorporated the concerns of the nearly agreed upon text of January 1982. It also (1) established criteria and procedures to resolve conflicts that might occur from overlapping claims made during the filing and processing of applications by consortia to mine under each state's legislation, and (2) ensured that joining in this agreement would not prejudice a state's decision to become a Party to the UNCLOS III Convention.[21]

Given that a primary objective of the reciprocating states in developing interim legislation was to create compatible regulations among themselves and with the strictures and objectives of the Convention, how successful were they? One researcher, Richard Luoma, attempted to answer this question. Writing in the *Journal of Maritime Law and Commerce*, he compared deep seabed mining Acts passed by the US, the FRG, the UK, and France, as well as the USSR. Early on, the reciprocating states considered the USSR eligible to become a Party, if its seabed mining legislation were compatible. The USSR has not become a Party; instead the USSR claims its seabed mining law is a protective response against the actions of the reciprocating states.

Luoma analyses how well the legislation of the four reciprocating states and the USSR meets three main concerns. He asks whether their laws serve the interest of (1) the general mining industry; (2) the individual nodule-mining consortia; and (3) the international community.[22]

Concerns of the general mining industry include whether the law provides for 'an exclusive and comprehensive right to explore by all reasonable means for an adequate period' and a complete 'right of development and an absolute right to acquire ... an agreed portion of any find'. Rental charges, taxes, licence fees and royalties cannot be too burdensome so there 'can be an early return of capital or the generation of a reasonable profit'. Individual mining consortia concerns encompass security of tenure, protection of mining claims, and a good return on investments. Finally, international community concerns involve provisions for sharing profits with Third World and with land-locked and geographically disadvantaged states and for protecting the marine environment and human life on the sea. The US's law, perhaps because it has been discussed since 1971, is the most detailed and adequate in meeting each category of concern, whereas the UK's law is the least.[23]

In Luoma's view – albeit a highly subjective one – under the first category (the general mining community), the US and FRG laws are adequate, but those of the UK and France are not. The UK's provisions fail to identify a specific time span allowed for exploring and developing a mining site, and the taxes imposed are too high. France provides no time frame, and the size allowed for mine sites may be inadequate. The USSR's legislation is not applicable here, because the state will undertake all capital-investment risks and administer all mining operations.

Concerns within category 2 (individual mining consortia) are favourably met by the US, the FRG, and the USSR. Each state specifically protects individual mining site claims. In most ways, France protects individual mining firms, but is not specific enough in defining a firm's tenure of operation. Similarly the UK's law does only a fair job in this category, because the mining enterprise's mine-site tenure can be guaranteed only if the government happens to allow it enough time to operate its claim. Finally, in category 3 (international community), all states other than the US fail either to provide adequate revenue sharing or to protect the marine environment. The UK, FRG, USSR, and France are especially lacking in arrangements for environmental protection. The FRG, however, has excellent provisions for sharing revenues with other states; the UK and USSR's provisions for revenue sharing are weak and France does not even mention the topic (Table 3.1).[24]

Table 3.1 Adequacy of deep seabed mining laws for the general mining industry, individual mining firms, and international community

	United States	Federal Republic of Germany	United Kingdom	USSR
General mining industry	good	good	fair	NA[a]
Individual mining firms	good	good	fair	good
International community	good	poor	poor	poor

Source: Compiled from R. T. Luoma, 'A comparative study of national legislation concerning the deep sea mining of manganese nodules', *Journal of Maritime Law and Commerce*, vol. 14, no. 2 (1983), pp. 243–68.
Note: a. Not applicable.

Co-operation among the four original reciprocating states has continued since the 2 September 1982 Agreement. And on 3 August 1984 the number of reciprocating states grew to eight when a new accord, 'Provisional Understanding Regarding Deep Seabed Matters', was signed by Belgium, Italy, the Netherlands, and Japan, as well as the US, UK, FRG, and France.[25] The Parties, all with deep seabed mining investments and technical capabilities, agreed not to allow mining before 1 January 1988 and to settle disputes amicably. The understanding provided for avoiding registration and mining operations conflicts by requiring notification and consultation (which they have done) before (1) applying for the registration of claims, (2) issuing authorisations to mine, or (3) conducting mining activities. Those states that have signed the UNCLOS Convention are in no way compromised. Japan, France, Italy, and the Netherlands have signed, but not ratified, the Convention.[26] If they ratify the Convention and it comes into force (without their abrogating their reciprocity agreements), they will be in violation of the Convention. Article 137 'Legal Status of the Area and Its Resources' says:

No state shall claim or exercise sovereignty or sovereign rights over any part of the Area or its resources, nor shall any State or natural or juridical person appropriate any part thereof. No such claim or exercise of sovereignty or sovereign rights nor such appropriation shall be recognised.

All rights in the resources of the Area are vested in mankind as a whole, on whose behalf the Authority shall act. These resources are not subject to alienation. The minerals recovered from the Area ... may only be alienated in accordance with this Part and the rules, regulations and procedures of the Authority.

No State or natural or juridical person shall claim, acquire or exercise rights with respect to the minerals recovered from the Area except in accordance with this Part. Otherwise, no such claim, acquisition or exercise of such rights shall be recognized.[27]

The reciprocating states' actions have prompted accusations that they are acting illegally. Some G–77 critics suggest that international financial institutions should not make loans to deep seabed mining ventures not sanctioned by the ISA.[28] On 30 August 1985 PREPCOM declared that any claim, action, or agreement taken outside the PREPCOM which deals with the Area and its resources is illegal if it is not compatible with the UNCLOS III Convention. Malone has long rejected this notion. He contends that

1 the seabed portions of the Convention are very new and contractual and are not part of customary international law,
2 provisions of the Convention will be binding only on those ratifying it,
3 agreements among parties with mutual oceanic interests are consistent with international law.

Both the government of Italy (which signed the Convention on 7 December 1984) and Italian LOS specialists have echoed Malone's position. Italy's PREPCOM representative, in the interests of harmony, voted for passage of the PREPCOM declaration, but before doing so, he spoke against it. Individuals in Italy with a deep concern for the success of UNCLOS's objectives also have spoken against the PREPCOM declaration; they noted that:

The Preparatory Commission is not a court of law and ... when the Convention ... enters into force for a number of states, it will, obviously, not be binding on those states that will not have ratified it. ... Only when a vast majority of States, including all major groups and shades of interests in deep seabed mining ... become party to the Convention, will ... the 'treaty be of an universal character generally agreed upon ... ' Only then will it become meaningful to talk of the regime 'established' by the Convention for seabed mining as the 'only regime'.[29]

Alternate (mini-) treaty

Some specialists in international law suggest that the reciprocating states' approach to co-operation outside the UNCLOS Convention is too narrow.[30] Rather than focusing only on seabed mining, what may be needed is a more inclusive treaty which contains provisions parallel to those of the Convention, that is in matters concerning the EEZ, the territorial sea, the continental shelf, the marine environment, and the navigation of international straits. Anthony D'Amato, in an editorial published in the *American Journal of International Law*, suggested that one substantive provision which would lend credibility to the treaty as something other than a cover for selfish exploitation would be provisions for sharing seabed mining income with the have-nots of the world. He feels that the amount shared (which could be increased through time) should be at least as large as the tax established by the US under its 1980 DSHMRA. The percentage should not be so high, however, as to discourage mining consortia from investing in mining ventures. Should mining begin under the reciprocating-states agreement, the funds generated by the royalty will be deposited in a 'Deep Seabed Revenue Sharing Trust Fund'. If the US ratifies the Convention before 28 June 1990, the moneys will go to the ISA.[31] This approach could provide an alternate treaty with more political clout (recognition) in customary international law.

As alluded to in the previous chapter, 'customary law develops when a community of world states, through reciprocity, recognizes and conforms to uniform actions'.[32] This description of how customary law develops implies the element of time. Thus, for an alternate treaty to the UNCLOS Convention to have weight in customary law, the sooner it is negotiated and put into force the sooner it will gain that weight, for only through time will a body of accepted principles emerge.

D'Amato stressed that this situation could be important to the parties in an alternate treaty, because if future disputes were to come before the International Court of Justice, the Court would likely weigh the merits of a case by looking at those elements of customary law that have evolved within each of the two treaty structures. Whichever system has developed the stronger (most adhered to) customary law relating to the specifics of the case will probably receive the positive judgement.[33] Kathryn Surace-Smith, writing for the *Columbia Law Review*, supported this perspective when she said:

The deep seabed mining ... provisions of the Convention ... must be ... evaluated to determine the probability that they will generate customary law and to ascertain their legal implications for the United

States. In addition, an analysis of existing customary norms is crucial to an understanding of the legal parameters governing future United States activity in Seabed mining.[34]

Her admonition could be extended to all non-signatories of an alternate treaty. Although treaties cannot bind non-parties, under some circumstances treaties do create customary law that becomes binding on all states, including non-parties. This process seems already under way in the case of the UNCLOS Convention (as with the nearly universal acceptance of the 200-nmi EEZ concept), even though it is not yet in force.

The land-locked and geographically disadvantaged states' view

One mission of the UNCLOS III negotiations was to develop a Convention that contributes to a 'new international economic order' to help financially and technologically impoverished states. A close look at the Convention shows little evidence of the mission's having been accomplished for the more than fifty land-locked and geographically disadvantaged (LGD) states, many of which have weak economies and burgeoning populations. They need help, but how much can they expect from the Convention?[35]

A fundamental problem of the LGD states is that the acceptance of the 200-nmi EEZ concept has reaffirmed their tacit exclusion (under the 1958 GCCS) from the best fishing waters and the richest petroliferous areas of the oceans, as well as many other nearshore minerals. Convention Articles 69 and 70 proclaim the right of these Parties to 'participate on an equitable basis, in the exploitation of an appropriate part of the surplus of the living resources of the exclusive economic zones of coastal states of the same subregion or region', to be arranged 'through bilateral, subregional or regional agreements'.[36] The developed LGD states may share surpluses only in the EEZs of other developed states within the subregion or region. The reference to 'surpluses' and to 'agreements' means that the have-not states will be at the mercy of coastal states, many of whom are desperately poor themselves. In sum, they cannot expect much help from this provision of the Convention.

As for mineral resources within the coastal states' 200-nmi EEZs, the LGD states receive nothing. In situations where a state's continental shelf extends beyond the 200-nmi limit, there is a special provision. The provision requires that in waters lying within the 200 to 350-nmi zone – as measured from the baseline of its territorial sea – the coastal state must (after the fifth year of production) make an annual payment to the ISA of 1 per cent of the value of the minerals produced. The rate increases annually by 1 per cent for the next six years, then remains the same (7 per cent) thereafter. These funds are paid to the ISA and will be distributed to all Parties based on 'equitable sharing criteria, taking into account the interests and needs of developing states, particularly the least developed and the land-locked among them'. However, those developing coastal states that are net importers of the mineral(s) produced are exempt from this requirement.[37]

In oceanic regions designated as the 'Area', the ISA must equitably share financial and other economic benefits with all the less developed states, especially those not having gained 'full independence or other self-governing status'.[38] Theoretically the less developed LGD states can participate in deep seabed ventures jointly with other Third World states or with the Enterprise, but this prospect seems unlikely for most land-locked states.

If viewed in total, the Convention makes only limited provisions (via the Area) for the LGD states to share in a potential ocean mineral harvest. It must seem to many of them, especially the land-locked states, that they might be invited to dinner but would be served only the crumbs.

The Soviets' view

Two-thirds (47,000km) of the USSR's borders is coast-line. For this reason alone, one would expect the USSR to have an interest in the UNCLOS III negotiations. Its interests are diverse and in many ways parallel those of the US, including fishing, navigation, and seabed mining.

When the Soviets attended UNCLOS I in 1958 in Geneva, their perspective was narrow, focusing almost exclusively on territorial-sea issues. They still viewed themselves as a continental bastion, with their surrounding waters serving as an isolating moat. UNCLOS II in 1960 achieved nothing, and by the mid-1960s, with expanding international interests and growing naval, research, fishing, and merchant fleets, the Soviets had become more outward-looking, more concerned about ocean space far from their own shores. Consequently they felt uneasy as they watched territorial-sea jurisdictions creeping seaward throughout the world.

In 1966 the Soviets circulated a petition among some sixty states, asking if they were interested in an UNCLOS III whose focus would be to set territorial limits at 12nmi (22km). They were determined not to be excluded from their traditional areas of oceanic access. The US, receptive to the USSR's overture, suggested that problems of international straits be added to the agenda.[39] The Soviets agreed and preliminary discussions began.

Then, in 1967, came Arvid Pardo's electrifying proposal to the UN that seabed resources be designated a common heritage of mankind and that a treaty be established to supervise their use. Soon, international straits and territorial seas issues were absorbed into this broader agenda, despite efforts by the USSR and the US to keep them separate.[40] By 1977 several of the two superpowers' original concerns were in part settled, resulting in most subsequent debate and political heat being focused on seabed mining problems.

The Soviets have used seabed mining issues to denigrate the motives of the US's participation in UNCLOS III. They have usually favoured proposals of the G–77 and have castigated the US for its withdrawal from negotiations. Yet, because of their displeasure with the provisions of Resolution II of the Convention that deals with preparatory investment in polymetallic nodule mining pioneering activities, the USSR, along with its Eastern European satellites (except Romania) abstained during the 30 April 1982 vote on the Draft Convention. The USSR, however, signed the Convention later that year (10 December). To date (late 1987) it has not ratified the Convention. Even though a state has not ratified the Convention, it may file for a claim as a Pioneer Investor, and the USSR did so. After it settled an overlapping-boundary claim with France, the USSR's claim (along with those of Japan and France) was registered on 17 December 1987. The three State Pioneer Investors' claims lie within the Clarion-Clipperton zone (CCZ) of the Pacific Ocean. The USSR has also applied for a claim in a 300,000-km^2 area in the Indian Ocean.[41]

Industry's view

For two decades US-based marine mining consortia have investigated deep seabed geological and mineralogical conditions. They also have made detailed metallurgical studies and are now ready for full-scale testing of mining and processing operations, but they are reluctant to push ahead until assured of a stable seabed mining regime and firmly established mining rights; until recently many managers in the industry felt that the only truly viable regime would come from outside the UNCLOS III Convention,[42] and they long considered the constraints and demands of the Convention so burdensome that free-market mining operations would be impossible.[43]

Some entrepreneurs continue to look upon the ISA as nothing more than an international monopoly, that has a stacked deck or, at the least, holds all the trumps. The ISA's most controversial functions include its right to

1 cancel mining permits;
2 grant Third World states permission to mine the deep seabed either individually, within partnerships, or jointly with the Enterprise;
3 set ceilings on the annual production of nickel, cobalt, copper, and manganese, all important strategic minerals;
4 demand mining and processing technology transfers to the Enterprise or Third World states.[44]

Technology transfers

Probably the major complaint focuses on Article 5 in Annex III of the Convention. Under this article, if the Enterprise or designated developing states cannot obtain 'on fair and reasonable terms and conditions', the technology they desire, then a contractor must transfer (at a fair market value) all legally held technology that is being used for mining in the Area. If a contractor uses technology they do not legally own, then they must acquire assurance from the owner that the owner will transfer to the Enterprise (or designated parties) any technology provided to the contractor for work in a seabed mining claim, otherwise the contractor may not use the technology in question in their seabed claim.[45]

Transferrable technology includes all specialised equipment, manuals, and operating instructions associated with contractually undertaken mining in the Area.[46] No distinction is made between patented or 'trade secret' technology.[47] The technology transfer provisions also require industry to train Enterprise and/or Third World states' seabed mining personnel.[48] Disputes arising during technology transfer negotiations are subject to compulsory settlement and binding commercial arbitration.[49]

To assist the Enterprise in technology purchases, contractors – at the time they submit mine-site work plans – must have indicated if and where the technology they plan to use is available on the open market. Although the Enterprise must purchase a mining firm's transferred technology on 'fair and reasonable commercial terms and conditions, to the same extent as made available to [that firm]',[50] what is fair and reasonable will be debatable. Given the prospect of arbitrary technology transfers (despite the existence of some loopholes), industry's fears might be justified. In most instances, developing this technology took decades of labour and millions in capital investments. Besides the mining consortia, numerous major equipment suppliers for mining companies have said in hearings before the US Congress that they will refuse to provide technical materials to buyers under these conditions.[51]

Some in industry suggest that government assistance, such as insurance programmes guaranteeing compensation if technology transfers are demanded, might encourage industry to begin mining when metal markets improve. Critics of this approach insist that the US Government should not go into the insurance business, seeming to overlook federal insurance programmes in many other sectors of industry and society. In sum, if the US should ratify the Convention, its mining consortia might refuse to proceed with plans to mine the seabed. They may insist on mining under the US flag or not at all.

A US Constitutionalist could argue that, if the US ratifies the Convention, it will be violating Article VI of its Constitution which protects citizens from having their property taken without 'just' compensation. It is ironic that many Convention signatories have constitutions guaranteeing their citizens the right to own property without an arbitrary transfer, unless they are justly compensated.[52] Although the desire of the have-not states to share in a more equitable distribution of income and technical know-how is understandable, causing individual business enterprises to subsidise the world community creates another inequity.

Are there alternatives to the Convention's present technology-transfer provisions? Several investigators suggest that the US Government and those of other developed states could pay the mining companies or supply firms for the transferred technology.[53] Marsteller and Tucker feel that an 'International Investment Guarantee' system might offer a solution. This system, to be operated by the International Investment Agency of the International Bank for Reconstruction and Development, would work much like automobile collision insurance programmes, whereby a premium would be paid on the value of a specific technology. If the technology were expropriated, they could make a claim.[54] Still another suggestion is that the ISA could provide specific exclusivity periods for the mining firm's use of its technology, after which – upon payment of a royalty – the less developed states or the Enterprise could use the technology. In the view of the mining consortia, however, none of these suggestions solves their problem, that is how to remain competitive when they must transfer their proprietary technology to competitors.[54] As is suggested by the work of Porter Hoagland, at Woods Hole Oceanographic Institution, the problem of international and domestic patent laws will surely become an issue under coerced transfers of technology.[56]

Production ceilings

Another major criticism many in industry have for the present structure of the UNCLOS Convention centres upon limitations set for the production of several minerals, such as nickel, cobalt, copper, and manganese – the Big Four – contained in manganese nodules. This issue is especially sensitive because both developing and developed states have much at stake. Numerous national governments and existing onshore producers of minerals to be mined from the seabed fear they will experience severe economic crises if large quantities of these minerals become competitively available on the world market.[57] On the other hand, deep seabed mining consortia personnel worry that production ceilings might reduce their efficiency, and those officials of states highly dependent on imports of these strategic metals fear that seabed production limitations might jeopardise their sources of supply for the Big Four metals.[58]

Canada and nickel

The world's onshore sulphide nickel reserves (those deposits currently most economically exploitable) are estimated at approximately 53 million tonne.[59] In addition, large supplies of nickel are potentially available in poorer-grade laterite and sulphide deposits. Among the Big Four strategic minerals contained in deep seabed polymetallic nodules, nickel (an important component of superalloys) is presently the most important to potential producers. Investigators estimate that the world's oceans may contain 73,000 million tonnes of polymetallic nodules. Pacific Ocean nodules contain an average of approximately 1.3 per cent nickel and, in some instances, as much as 1.9 per cent.[60] The 1.3 per cent value applied to the total estimated world nodule tonnage indicates a potential nickel content of about 900 million tonnes.[61] Such a large nickel resource is of vital interest to major world nickel producers (Canada) and consumers (US).

Table 3.2 World nickel mine production and capacity, 1983 (thousand tonnes)[a]

Area	Production	Capacity[b]
North America		
Canada	121.8	250
United States	——	15
Total	121.8	265
Latin America		
Brazil	7.1	13
Colombia	13.6	27
Cuba	37.3	38
Dominican Republic	20.2	34
Guatemala[c]	——	11
Total	78.2	123
Africa		
Botswana	17.5	18
Republic of South Africa	20.5	32
Zimbabwe	10.9	16
Total	47.9	66
Europe		
Albania	5.8	9
Finland	6.0	9
Greece	15.0	23
USSR	169.6	163
Yugoslavia [c]	12.0	27
Other[d]	4.7	4
Total	213.1	235
Asia		
Indonesia	46.6	59
People's Republic of China	8.0	12
Philippines[c]	19.0	41
Total	73.6	112
Oceania		
Australia	90.0	109
New Caledonia	62.9	109
Total	152.9	218
Other[e]	0.2	9
World total	683.7	1,028

Source: S. F. Sibley, 'Nickel', in *Mineral Facts and Problems 1985*, USBM Bulletin 675 (USGPO, Washington, DC, 1985), p. 536.
Notes: a. Converted from thousands of short tons; b. Rounded to nearest thousand; c. Standby or partially standby capacity; d. Includes the German Democratic Republic, Poland, and Norway; e. Includes Burma, Mexico, and Morocco.

The world's nickel mining industry is highly competitive, because twenty-five countries have nickel mines and because of present-day market conditions. Owing to a weak demand, mine production in 1983 was only 66.5 per cent of capacity (Table 3.2).[62] In 1985, however, demand for nickel was on the increase, if only slightly.[63] Leading world producers in the mid-1980s included the USSR (25 per cent), Canada (18 per cent), Australia (13 per cent), New Caledonia (9 per cent), Indonesia

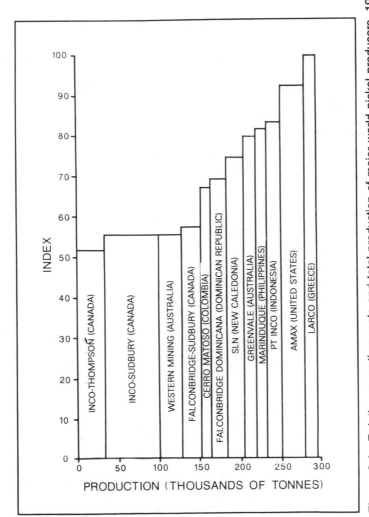

Figure 3.1 Relative operating costs and total production of major world nickel producers, 1983

Source: After Mineral Policy Sector, Energy, Mines and Resources Canada, *The Canadian Minerals Sector: A Framework for Discussion and Consultation* (EMRC, Ottawa, Feb. 1985), p. 18. With permission.

Note: Excellent ores and up-to-date mining and processing methods provide Canada with a production cost advantage.

(7 per cent), and Cuba (5 per cent).[64] Canada is one of the world's lowest-cost nickel producers. In 1983 Canada produced nickel at a cost of 30 to 50 per cent less than many other major producers, such as the Philippines, Indonesia, Greece, and the US (Figure 3.1).[65] Regardless of Canada's relatively low production costs, its government and industry recognise the potential problems polymetallic nodules might present to their domestic industry. In 1976 the Canadian Government decided to determine more precisely the impacts deep seabed mining could have on Canada's mineral industries, especially nickel – despite its competitive edge. The Department of Energy, Mines and Resources initiated a project titled 'Deep Ocean Mining Study'. Between 1976 and 1982 the Study Group developed an extensive data base collection programme and it did detailed investigations of (1) the types and potential reserves of seabed minerals, (2) the involvement and motivation of the national government and corporations in seabed mining, (3) the status of seabed mining technology, and (4) the likely timing and competitiveness of commercial production. Implementation of this research programme helped the Canadian UNCLOS delegation to be one of the best informed during negotiations[66] and to have had a major impact in establishing the production criteria and output ceilings for nickel, as well as for cobalt.

Under Article 151, 'Production Policies', of the Convention, the negotiators attempted to establish a system 'to promote the growth, efficiency and stability of markets for ... minerals derived from the Area, at prices remunerative to producers and fair to consumers'. Applicants for production authorisations must provide the ISA with an estimate of their anticipated annual production of nickel. The production total applied for will be granted' unless the sum of that level and the levels already authorised exceeds the nickel production ceiling' set by the ISA. Production ceilings for a given year will be established by using a set of trend-line values for nickel consumption, 'derived from a linear regression of the logarithms of actual nickel consumption for the most recent 15-year period for which such data are available, time being the independent variable'. The ISA reserves for the Enterprise an initial production quantity of 38,000 tonnes of nickel (to be calculated as part of the annual production ceiling), with an absolute production of 46,500 tonnes allowed for any individual work plan.[67]

Individual producer production ceilings for cobalt, copper, and manganese are determined by how much nickel production is allowed under their contract.[68] The ceilings should help protect those states whose mining industries are highly dependent on these minerals.

Production ceilings could be significant for the future of Canada's mineral industries, given that it produces not only nickel but also copper and cobalt, the latter as a by-product of nickel refining. Nickel by-product producers (as in South Africa's platinum mines) and government-subsidised producers (as in Cuba, the Philippines, and the USSR) could make competition difficult for nickel firms in a free-market economy like

Canada's. Because of a severe recession in the world's mineral economy, the nickel industry since 1980 has been cutting production. Despite an international over-capacity, Cuba in 1986 put a new nickel-oxide plant into operation, and the People's Republic of China (PRC) expects to have a new facility on stream by 1990.[69]

In 1982 nickel prices (in US $) dropped to a low of $3.50/kg, 50 cents below the price considered necessary for laterite nickel miners (Canada produces from sulphides) to make a profit, resulting in closures of unsubsidised mines in Guatemala, the Philippines, and the US.[70] Because prices in the 1980s have remained near or below production costs, even the world's lowest-cost producers (Canada and Australia) are suffering. Inco, Canada's premier producer, returned to profit only in 1985, after four years of losses. In 1986 Inco managers were predicting an increase in demand.[71]

World nickel consumption during the last decade (1976–86) experienced little growth.[72] But according to projections by the US Bureau of Mines, world demand for primary nickel (refined from ores) will likely increase at an annual rate of 3.1 per cent from 1983 to the year 2000. This increase translates into a quantity demanded of 816,000 tonnes of primary nickel in 1990 and 1.2 million tonnes in the year 2000.[73]

Nickel industry analysts expect that the demand for primary nickel in the US in 2000 will be between 220,000 and 255,000 tonnes – most likely about 245,000 tonnes. Mineral economists derived this figure by applying a trend value of 2.5 per cent growth annually, using 1983 as the base year. Consumption of secondary nickel (recovered from scrap) in 2000 should be approximately 73,000 tonnes. Secondary nickel in 1983 supplied about one-quarter (46,000 tonnes) of the total US industrial demand of (185,000 tonnes) for nickel. Although the US has large tonnages of low-grade nickel-bearing deposits, these are presently uneconomic; the US has only one nickel mine, in Oregon. Consequently imports supply nearly all of its primary nickel consumed, with more than half coming from Canada. Because of Canada's dependability and proximity as a supplier of nickel, the US must be concerned for what happens to Canada as a nickel producer,[74] perhaps even to the point of paying a premium, if LDC-subsidised producers become competitive.

Zaïre and cobalt

Cobalt has been used for thousands of years as a colouring additive; today it has a diversity of uses because of its multiple physical properties. Cobalt, as an additive, provides high strength, heat and wear resistance, and superior magnetic qualities. Producers of jet-engine components, permanent magnets, electrical equipment, cutting tools, catalysts and paint pigments and dryers are among its major consumers. Although cobalt has a variety of uses, world demand is relatively small when compared with many other minerals.[75]

Like nickel, only a few world states account for the major share of cobalt mine ore production – Zaïre (48 per cent) Zambia (10 per cent), USSR (10 per cent), Australia (8 per cent), Cuba (7 per cent), and Canada (7 per cent). Four of the major cobalt producers (USSR, Australia, Cuba, and Canada) are among the top nickel producers. As noted earlier, this situation occurs mainly because cobalt is produced as a by-product in both copper and nickel smelting. Thus cobalt mine production capacity exceeds actual production by 48 per cent (Table 3.3).

Table 3.3 World cobalt mine production and Capacity, 1983 (tonnes)[a]

Region	Production[b]	Capacity
North America		
Canada	1,582	4,081
Cuba	1,650	1,814
Total	3,232	5,895
Europe		
Finland	907	1,361
USSR	2,358	2,721
Total	3,265	4,082
Africa		
Botswana	223	317
Republic of South Africa	159	227
Zaïre	11,302	15,873
Zambia	2,404	4,535
Zimbabwe	73	91
Total	14,161	21,043
Oceania		
Australia	1,814	2,041
New Caledonia	272	371
Philippines	599	1,270
Total	2,685	3,628
World total	23,343	34,648

Source: W. S. Kirk, 'Cobalt', in *Mineral Facts and Problems 1985*, USBM Bulletin 675 (USGPO, Washington, DC, 1985), p. 172.
Notes; a. Converted from thousands of 1bs; b. Estimated on the basis of recovered cobalt content.

In Canada copper-cobalt-nickel ores are mined in the Sudbury and Abitibi districts of Ontario and the Thompson district of Manitoba. The Sudbury ores contain about 0.07 per cent cobalt. Zaïre and Zambia's copper ores also contain significant amounts of cobalt. Of the world's total (3.6 million tonnes) economic cobalt reserves, Zaïre has 39 per cent (1.4 million tonnes), the largest of any country. Cuba has 1.0 million tonnes of economic reserves and Zambia 0.4 million tonnes. If Zaïre's marginally economic and some subeconomic reserves are added in, that country's reserve total is 2.1 million tonnes.[76]

Zaïre in 1983 produced 29 per cent of the world's primary refined cobalt metal. The USSR, a larger consumer and the second-most important producer of primary refined cobalt metal, produced over 25 per cent of the world output. Worldwide, refined cobalt metal production in 1983 exceeded 18,100 tonnes (Table 3.4), slightly short of world demand (20,200 tonnes);[77] this deficit resulted from the very depressed production situation in the nickel and copper industries. Given Zaïre's heavy dependence on cobalt as a by-product of copper production (which because of fluctuating demand is highly volatile), its economy can suffer severe stress.[78] This point does not, however, imply that copper is less important to Zaïre than is cobalt, because the amount of refined copper is much larger (227,000 tonnes in 1983).[79] On a world scale, seabed mining will have little effect on overall copper output; seabed copper production would meet less than 1 per cent of world demand.[80]

Table 3.4 World primary refined cobalt metal production, 1983[a] (tonnes)[b]

Country	Production[c]
Zaïre	5,297
USSR	4,535
Zambia	2,407
Finland	1,455
Japan	1,371
Canada[d]	1,059
Norway	998
United Kingdom	726
Federal Republic of Germany	145
United States	93
Zimbabwe	73
Total	18,159

Source: W. S. Kirk, 'Cobalt', in *Mineral Facts and Problems 1985*, USBM Bulletin 675 (USGPO, Washington, DC, 1985), p. 172.
Notes: a. Estimated contained cobalt. Belgium is not listed because it is mainly a processor of refined metal; b. Converted from thousands of 1bs; c. Preliminary data; d. Data represent the output within Canada of metallic cobalt of both Canadian and non-Canadian origin.

Because manganese nodules contain important amounts of cobalt – deposits in prime areas of the Pacific Ocean may contain 0.26 per cent cobalt – oceanic sources could contribute a large amount of cobalt to the world market. As noted earlier, geologists estimate the worldwide gross-weight tonnage of nodules at 73,000 million tonnes. Of this total an estimated 2,100 million tonnes are now potentially recoverable in the north-east equatorial Pacific Ocean; these nodules could contain some 4.1 million tonnes of cobalt, approximately 38 per cent of the world's total known onshore resources.[81] Another potential cobalt source is ferromanganese crusts that lie at relatively shallow depths and cover large areas of the Pacific Basin, especially in association with seamounts and slopes of volcanic islands. The thickness of these crusts (up to 15cm in a few areas) varies greatly (average 2cm) as does their cobalt content (0.3 per cent to 1.2 per cent).[82]

Considering the availability of cobalt in manganese nodules, as well as in ferromanganese crusts, the concern of the cobalt industry and national governments for setting production ceilings within the Convention is understandable. A US Government study indicates that seabed mining of cobalt could depress cobalt prices by 25–30 per cent. If seabed mining does begin, industry and governments should have little fear of running out of cobalt.[83]

International consortia – membership, licensing, and investments

Unlike the majority of mineral producing firms, which are usually domestically constituted, most seabed mining consortia have a broad membership of important international corporations, which reduces the financial risk for individual firms and provides a variety of technical and managerial expertise. This broad membership, however, reduces the cohesion within the consortia, because the participants' current objectives (and reasons for joining) vary. Furthermore, nearly all the consortia have some participation by government-affiliated companies or are subsidised by national governments (Table 3.5). The national governments, in most cases, desire a wider supply base for mineral materials and want to help their firms to compete in what may well become a more highly competitive industry, a situation viewed with considerable apprehension by those in the incumbent industry.

Despite industry's rather pessimistic view of the future for seabed mining, the three US-based consortia and the UK-based Kennecott Consortium (KCON) have applied for and been issued (1984) exploration licences under the 1980 DSHMRA (Figure 3.2). The US-based consortia are Ocean Management Incorporated (OMI), Ocean Minerals Company (OMCO), and Ocean Mining Associates (OMA).

OMI, formed in 1975, has member companies from four countries (Canada, US, FRG, and Japan), and has vigorously pursued manganese

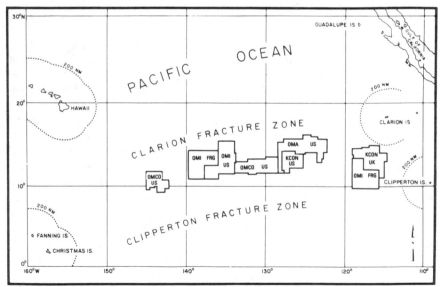

Figure 3.2 Licensed deep seabed mining consortia exploration areas

Sources: After C. E. Harrington (comp.), National Ocean Service, National Oceanic and Atmospheric Administration, Nov. 1987; P. Hoagland, III, 'Performance requirements in ocean mineral development', *Marine Policy Reports*, vol. 9, no. 3 (1987), p. 6. with permission.
Notes; KCON – Kennecott Consortium: OMA – Ocean Mining Associates; OMCO – Ocean Minerals Company; OMI – Ocean Management Incorporated. Licensing governments are the FRG, US, and UK.

nodule research and pilot-testing programmes in the Pacific Ocean. One of its members, Preussag AG, a large FRG extractive resources company, also has other extensive experience in deep seabed research such as that associated with polymetallic muds in the Red Sea. OMI's Japanese member (DOMCO) has been supported by loans from Japan's national government.[84] Another member (Sedco-Forex) is now using a diluting

Table 3.5 Ocean mining consortia, members and subsidiaries, as of March 1987

Name	Ownership (%)	Country
OCEAN MANAGEMENT INCORPORATED (OMI)[a]		
Inco Ltd	25	Canada
Arbeitsgemeinschaft Meerestechnisch Gewinnbare Rohstoffe (AMR)[b]	25	FRG
Metallgesellschaft AG		
Preussag AG		
Sedco-Forex (Schlumberger Group)	25	US
Deep Ocean Mining Co. Ltd (DOMCO)	25	Japan
Composed of 19 Japanese companies		
ARBEITSGEMEINSCHAFT MEERESTECHNISCH GEWINNBARE ROHSTOFFE (AMR)[b]		
Metallgesellschaft AG	50	FRG
Preussag AG	50	FRG
OCEAN MINERALS COMPANY (OMCO)[c]		
Cyprus Minerals Co.	50	US
Lockheed Corp.	50	US
Lockheed Systems Co., Inc.		
Lockheed Missiles & Space Co., Inc.		
OCEAN MINING ASSOCIATES (OMA)[d]		
Essex Minerals Co. (USX Corp.)	25	US
Sun Ocean Ventures, Inc. (Sun Co.)	25	US
Union Seas, Inc. (Union Miniere SA)	25	Belgium
Deep Sea Systems, Inc.[a] (Ente Nazionale Idrocarburi)	25	Italy
KENNECOTT CONSORTIUM (KCON)[e]		
Kennecott Corp. (Standard Oil of Ohio/British Petroleum Co.)	40	UK
British Petroleum Development Ltd (British Petroleum Co.)	12	UK
Rio Tinto Zinc Deep Sea Mining Enterprises Ltd.	12	UK
Consolidated Gold Fields	12	UK
Noranda Exploration, Inc.[g] (Noranda Mines Ltd)	12	Canada
Mitsubishi Corp.	12	Japan
GEMONOD[f]		
Institut Francais de Recherches pour l'Exploitation de la Mer (IFREMER)	50	France
Commissariat à l'Energie Atomique (CEA)	35	France
Technicatom (subsidiary of CEA)	15	France
DEEP OCEAN RESEARCH AND DEVELOPMENT COMPANY LIMITED (DORD)		
composed of about fifty Japanese companies	—	Japan

Sources: Letters or telephone interviews: a. L. Messalum, Vice President, OMI, NY, 9 Jan. 1987; b. R. Fellerer, Managing Director, AMR, Hannover, FRG, 8 and 28 Jan. 1987; c. C. G. Welling, Senior Vice President, OMCO, Santa Clara, CA, 30 Jan. 1987; d. R. J. Greenwald, OMA, Gloucester Pt, VA, 7 Dec. 1986; e. L. Mercando, Director of Process Metallurgy, KCON, Salt Lake City, UT, 30. Jan. 1987; f. J. P. Lenoble, President, Afernod, Ifremer, Paris, France, 30 Jan. 1987; see Bibliography for complete entries. Note: g. A US subsidiary corporation.

clause of OMI's joint-venture agreement and is thus not contributing financially. In August 1984 OMI received a US exploration licence for a 136,000km^2 area in the Pacific Ocean's Clarion-Clipperton zone (CCZ), and in late 1985 it also acquired one from the FRG.[85] In June 1986 OMI applied for a revision of its US licence. It asked permission to reduce its projected expenditures to one-third of the original figures presented with its licence application. In addition, it wants to reduce ship-time research to only a little more than half the original amount. OMI still plans to file for a mining permit by 1994, that is within the ten-year licence period. The revised exploration plan came about because OMI, as a result of overlapping claims negotiations by France and the USSR, obtained information that allowed them to reduce projected expenditures and time allocations. Basing its judgement on this new information, OMI will concentrate on identifying outstanding nodule areas. Detailed exploration work in other areas of its claim will be delayed until commercial recovery of nodules in prime areas begins.[86]

In 1977 four firms – two from the US and two from the Netherlands – formed OMCO, a consortium that has focused especially on nodule mining systems and techniques for recovering manganese (as well as copper, nickel, and cobalt) from manganese nodules. According to James Broadus and Porter Hoagland, of the Marine Policy Center at Woods Hole Oceanographic Institution, US-based Lockheed Corporation organised OMCO as 'an outlet for the company's high technology deep sea operation capabilities', with Lockheed earning revenues by selling equipment and technology to the other members of the consortium.[87] But, according to a spokesman for Lockheed, the company initiated formation of the consortium 'to perform research and engineering in ocean mining, and other members of the consortium were recruited to help with the work and the funding'.[88] OMCO is the only fully privately owned and financed ocean-mining consortium.

OMCO's Netherlands firms, Billiton International Metals BV and Royal Bos Kalis Westminster BV, gradually became inactive and then withdrew from the consortium (12 December 1985).[89] Lockheed's only partner is Cyprus Minerals Company, an independent spin-off from Amoco Corporation (previously Standard Oil of Indiana).[90] OMCO, like OMI, has applied for and received permission to reduce its number of cruises and to cut back on planned programmes for at-sea survey work and systems testing. Although cutting back in its exploration and testing programmes, OMCO says it plans to file for a mining licence by 1994.[91]

KCON, the first group formed (1974), is an unincorporated consortium whose namesake, Kennecott, was recently absorbed by British Petroleum through its ownership of Standard Oil of Ohio, a firm that purchased Kennecott Corporation. Broadus and Hoagland have noted that the Kennecott Group's members seem to have been truly 'motivated by the prospect of eventual seabed mining profits, rather than, as may have been the case in the Lockheed Group, by more immediate revenues from the sale of research and development services'.[92] KCON, in October 1984, received from NOAA an exploration licence for an area encompassing 65,000km^2. In late 1984 KCON also received an exploration licence from the UK,[93] under that country's 1981 Deep Sea Mining (Temporary Provisions) Act.

OMA, formed in 1974, originally had two countries represented in its corporate membership, but now has four firms from three countries – the US, Belgium, and Italy. The entry of Ente Nazionale Idrocarburi (ENI), Italy's national petroleum company, in late 1980 has injected significant amounts of capital into OMA. More recently the ENI ocean mining subsidiary, SAMIN Ocean Ventures, has been dissolved, with its shares being held inactively by the ENI subsidiary holding company. The US granted OMA an exploration licence in August 1984. This consortium is the only one planning to extract manganese from nodules, but because OMA has no corporate members from the States Party to the Convention, it is not eligible as a pioneer investor unless it is specifically named by PIP and a parent state ratifies the Convention. Broadus and Hoagland have suggested that if the USSR had had 'knowledge of the OMA area claim coordinates', it could have taken advantage of OMA's surveying and prospecting work by applying to the PREPCOM for the 'identical area claim under PIP'. This action would have provided the Soviets 'a low-cost public relations benefit by permitting them to assume the role of "enforcer" of the 1982 Convention against the "outlaw" claims of the United States partnership'. Broadus and Hoagland foresaw, however, that the Soviets' efforts might be foiled, because through conflicting-claims-resolution provisions under a Private Industry Arbitration Agreement, 'OMA and the other consortia have agreed to exchange portions of previously claimed areas, so the original OMA site will have become part of the applications of others'. A Soviet effort to claim-jump OMA through PIP, therefore, would involve the Soviets in conflicts with other groups represented in the PREPCOM that are prospectively PIP participants.[94]

In addition to the four multinational US-based consortia, there arc several others with memberships from within only one state. For example, AMR (a member of OMI) functions as a separate seabed mining consortium in the FRG. Originally AMR had three members – the two current partners and Salzgitter AG. The latter suspended its membership in 1984. If Salzgitter re-enters the consortium, AMR's ownership structure will revert to a three-way division. Because of its OMI membership, AMR has an interest in three separate deep seabed exploration locations.[95] The FRG, in late 1985, under its 1980 Act on the Interim Regulation of Deep Seabed Mining, granted both OMI and AMR exploration licences. The FRG was to have acted in the previous April, but delayed doing so because of the delicate situation associated with the FRG's being the future home of the International Tribunal for the Law of the Sea.[96]

Two other consortia with members from only one country are Groupment pour la Mise au Point Moyens Necessaires à l'Exploitation Development de Technologies Préliminaires (GEMONOD) of France and Deep Ocean Research and Development Company Limited (DORD) of Japan. GEMONOD, formed in 1984 by a joint decision of the Ministry of the Economy, Finances and Budget and the Ministry of Industry and Research, is the successor of AFERNOD (Association Française d'Etude et de Recherche des Nodules). AFERNOD vigorously pursued exploration and testing programmes, especially for remotely controlled ore-retrieval shuttle units. No French firms are involved with consortia outside of France.[97] DORD is the successor to a Japanese group of about forty companies that were members of Deep Ocean Mining Association (DOMA) and the Continuous Line Bucket Group (CLB), whose main focus is identified by the title. DORD has a very broad industrial membership; it includes about fifty Japanese companies, including producers of ships, ferrous and non-ferrous metals, chemicals, electrical appliances and cables, as well as firms involved in mining. The national government finances about 80 per cent of DORD's research budget. In 1981 DORD began a multi-million dollar research programme to develop a manganese nodule mining system, an effort scheduled for completion by 1989, but which is now delayed by two years. In the mid-1980s DORD was the most active of the various consortia.[98]

From 1962 to 1984 six of the major consortia (including originals and successors) invested an estimated $635 million in seabed mining research programmes. Among the consortia OMCO led the way in investments (26.6 per cent of the total), followed, in order, by OMA, KCON, DORD, AFERNOD, and OMI (Table 3.6). Investments declined sharply after 1980 among the US-based consortia and for AFERNOD, GEMONOD's predecessor (Figure 3.3). DORD is the only consortium to increase expenditure during the 1980s; it may spend a total of $90 million between 1982 and 1990.[99]

In contrast to the three US-based consortia, which have held back during the last several years, the Japanese and French consortia have pushed forward in their exploration and testing programmes. US consortia may be squeezed out by their French and Japanese counterparts. The agency responsible, NOAA of the Department of Commerce (DOC), did not issue proposed deep seabed hard mineral commercial recovery regulations until 25 July 1986.[100] Many in industry feel that before the consortia can fully plan for future exploration and mining operations and obtain financing, they need these regulations spelled out, although some observers doubt this claim.

As is necessary in obtaining an exploration licence, acquiring a mining licence will require two steps. First, NOAA must establish an applicant's general eligibility and then make a detailed critique of the mining plan. Even though the consortia now have a better understanding of the proposed regulations, they may be unable to obtain financing, because many major financial institutions that underwrite seabed mining activities

Table 3.6 Estimated seabed mining expenditure of consortia and predecessors (1982 US $ Millions)

Year	OMA	KCON	OMI	OMCO	AFERNOD	DORD	Total
(1962–4)	(11.8)	(2.5)	NA				(14.3)
1965	3.8	0.8	NA	0.6			5.2
1966	3.7	0.8	NA	0.6			5.1
1967	3.6	1.9	NA	0.6			6.1
1968	3.4	1.8	NA	0.6			5.8
1969	3.3	1.7	NA	0.5			5.5
1970	9.9	6.6	NA	1.2			17.7
1971	4.7	1.6	NA	9.5	0.5	0.2	16.5
1972	4.6	7.7	NA	9.2	1.9	1.1	24.5
1973	2.2	7.3	NA	8.6	4.7	1.1	23.9
1974	1.9	7.8	NA	7.8	4.3	1.9	23.9
1975	17.8	8.0	1.8	7.1	4.7	2.1	41.5
1976	8.4	10.9	8.4	7.1	7.0	2.5	44.3
1977	15.8	11.4[a]	15.8	6.7	6.8	2.7	59.2
1978	14.7	9.2	35.3[a]	30.1	4.6	3.4	97.3[a]
1979	18.5[a]	7.9	6.9	46.9[a]	4.9	5.3	90.4
1980	5.2	NA	0.6	24.4	4.9	9.3	44.4
1981	8.4	NA	0.2	5.3	9.4[a]	10.4	33.7
1982	12.0	NA	0.1	2.0	6.9	11.1	32.1
1983	2.9	NA	0.1	0.2	6.0	13.1	22.3
1984	1.4	NA	NA	NA	4.2	15.9[a]	21.5
Totals	158.0	87.9	69.2	169.0	70.8	80.1	635.0
% of Total	24.9	13.9	10.9	26.6	11.1	12.6	100.0

Source: *The Ocean and the Future*, 'Hearings', testimony of J. M. Broadus before the Subcommittee on Oceanography of the Committee on Merchant Marine and Fisheries, House of Representatives, 99th Cong., 1st. sess., 24 Oct. 1985 (USGPO, Washington, DC 1986), p. 154.
Notes: a. Peak year of investment. NA. Not available.

have said they will not lend the consortia money until the UNCLOS Convention is revised to reduce what they see as political risks, production risks, and market risks.[101] Because nickel and copper markets are saturated (with cobalt markets nearly so), the first one or two consortia to begin full-scale ocean mining, if competitive enough, may capture the few available markets and may force out new entrants.[102] On the other hand, some firms may bide their time, only to take advantage of the mistakes (and failures) of early entrants.

In addition to the likely competition among the private and mixed-form consortia, several state mining operations may add to the saturated metal market; India, South Korea, the PRC, and the USSR have state enterprises or state research programmes. India's state firm works within the Department of Ocean Development and has invested several millions of dollars, but precise information is unavailable. India has focused much research on manganese nodule deposits in the Indian Ocean, where it plans to mine. South Korea, within its Korean Ocean Research and Development Institute (KORDI), has only recently begun deep-seabed mining research, but claims to have spent $30 million by 1984.[103] Its researchers organised one three-week cruise (16 November–7 December 1983), employing the *Kana Keoki*, a research ship based in the Hawaiian

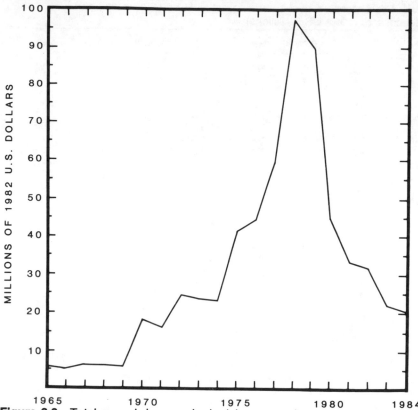

Figure 3.3 Total annual deep seabed mining consortia research expenditure, 1965–84.

Source: From data contained in testimony by J. M. Broadus, in *The Ocean and the Future*, 'Hearings', before the Subcommittee on Oceanography of the Committee on Merchant Marine and Fisheries, House of Representatives, 99th Cong., 1st sess., 24 Oct. 1985 (USGPO, Washington, DC, 1986), p. 154. Notes; Includes expenditures by KCON; OMA; OMCO; OMI and predecessor companies; AFERNOD (succeeded by GEMONOD); and DOMA and CLB Group.

Islands. A recent KORDI study recommended that Korea move forward in programmes to assess further seabed mining's potential and to establish strategies and an administrative agency to prepare for possible future deep seabed mining activities.[104] The final form of organisation (state-owned, mixed-form, or private) that South Korea's deep seabed mining programme will take is unclear.[105] Since 1976 the PRC has pursued a manganese nodule recovery programme which is administered by the National Bureau of Oceanography. The PRC reportedly has spent about $40 million in seabed-mining technical research and $8 million in seabed surveys.[106] The USSR has explored the deep seabed's manganese nodule resources since the early 1960s,[107] and in the mid-1970s began detailed prospecting in areas such as the CCZ. No expenditure estimates are available for the USSR.[108]

The most recently formed seabed mining group emerged in April 1987.

The USSR, Bulgaria, Cuba, Czechoslovakia, the German Democratic Republic, and Vietnam signed an agreement establishing an organisation named Interkeanmetall. The group has announced its intention to request the PREPCOM to award it deep seabed mine sites.[109]

Conclusions

For those who had placed much hope in the future of the UNCLOS III Convention, it was unfortunate that Ronald Reagan was elected to the US Presidency. His administration shattered what might have been a nearly unanimous decision to establish a new regime for the use and management of one of the world's last resource frontiers. The Reagan Administration's refusal to sign or to seek ratificaton of the Convention has damaged the credibility of the US. In the end, as Ratiner has so cogently put it, the US opted for 'the pursuit of principle over pragmatism'.[110]

Although the US has not signed or ratified the Convention, a future administration may reverse this policy. Whatever the US does, it seems likely that the Convention eventually will receive the sixty ratifications needed for it to enter into force. How the Convention's entry into force will affect parallel efforts to control deep seabed mining is difficult to predict. But it is clear that the UNCLOS III Convention has

1 support from the majority of states,
2 several of its concepts already incorporated into customary
   international maritime law, which will weaken an alternate regime,
3 a majority of adherents with few economic and technological strengths.

International mining consortia and lending institutions will continue their cautious policies for entering into commercial deep seabed ventures in the Area. They are mainly concerned about depressed mineral markets but are also wary of uncertainites associated with attempting to mine without an effective LOS regime or an alternate treaty being in place. Some observers (although it is unclear why) suggest that, if government subsidies were available to US consortia, they might go ahead with mining. But many in industry do not want government assistance, even though they must shoulder large investments. For example, one industry representative in 1983 stated that his consortium (on top of the $120 million already spent) would need to invest, over a period of five years, another $250 million to $300 million to perfect its mining and processing technology. Another five years would be needed to gear up commercial production and would require $1,000 million to $1,500 million. It would take another five to ten years before the firm could begin to recoup its investment.[111] This example illustrates why large-scale commercial mining of the deep seabed is unlikely until well after the turn of the century, certainly by US consortia and probably by others as well.

When deep seabed mining does begin, onshore producers of metals such as cobalt, nickel, and copper will experience added competitive stress. It is possible, also, that it will not be manganese nodules harvested from the seabed but rather the cobalt-rich ferromanganese crusts and hydrothermal polymetallic sulphides. Significantly the Convention does not specifically address these resources by name.[112] Ironically Arvid Pardo is among those who have forcefully pointed out this weakness in the Convention.[113] A cynic might suggest that this gap in the Convention's language makes little difference, because most of the best areas of ferromanganese crusts and polymetallic sulphides have already been given away to coastal and archipelagic states through an international acceptance of the 200-nmi EEZ concept.

Notes

1 'Deep Seabed Hard Mineral Resources Act', Public Law 96–283, 96th Cong., 28 June 1980, 94 Stat., pp. 553–86.
2 W. B. Jones, 'Risk assessment: corporate ventures in deep seabed mining outside the framework of the UN Convention on the Law of the Sea', *Ocean Development and International Law*, vol. 16, no. 4 (1986), pp. 341–2.
3 *Federal Register*, vol. 46, no. 56 (24 March 1981), p. 18,448.
4 W. Wertenbaker, 'A Reporter at Large: the Law of the Sea – part I', *New Yorker* (1 Aug. 1983), p. 45.
5 L. S. Ratiner, 'The Law of the Sea: a crossroads for American foreign policy', *Foreign Affairs*, vol. 60, no. 5 (1982), p. 1,010.
6 W. Wertenbaker, 'A reporter at large: The Law of the Sea – part II', *New Yorker* (8 Aug. 1983), p. 71.
7 Ratiner, 'The Law of the Sea', pp. 1,007–12.
8 Wertenbaker, 'A reporter at large: part II', p. 71.
9 ibid., pp. 71–2.
10 ibid., pp. 74–7.
11 See T. F. Marsteller, Jr. and R. L. Tucker, 'Problems of the technology transfer provisions in the Law of the Sea Treaty, *IDEA – The Journal of Law and Technology*, vol. 24, no. 4 (1983), pp. 167–80.
12 D. W. Arrow, 'Seabeds, sovereignty and objective regions', *Fordham International Law Journal*, vol. 7, no. 2 (1983–4), pp. 198–201. Amendments could originate in a Review Conference to be held fifteen years after 1 January of the year when commercial mining first begins (under an approved work plan) in the Area.
13 For a useful critique of the varied agendas of those voting for and against the Draft Convention, consult 'Sea Law: their reasons why', *UN Chronicle*, vol. 19, no. 6 (1982), pp. 16–22.
14 Wertenbaker, 'A reporter at large: part II', pp. 75–7.
15 *Federal Register*, vol. 46, no. 56, p. 18,448.
16 Consortia management felt these discussions were progressing so slowly that, in 1981, they began conferring among themselves in an effort to develop their own reciprocal agreement. This effort resulted in an agreement by industry that provided for the exchanging of mine site co-ordinates to avoid overlaps and the submission of disputes to compulsory and binding arbitration, if the parties cannot reach agreement; see R. L. Brooke, 'The current status of deep seabed mining', *Virginia Journal of International Law*, vol. 24, no. 2 (1984), pp. 398–9.
17 Comptroller General of the United States, *Uncertainties Surround Future of U.S. Ocean Mining: Report to the Congress of the United States*, GAO/

NSIAD–83–41 (General Accounting Office, Washington, DC, 6 Sept. 1983), p. 10. The FRG's legislation is titled 'Act of Interim Regulation of Deep Sea Mining' (Aug. 1980); UK – 'Deep Sea Mining (Temporary Provisions) Act of 1981' (July 1981); France – 'Exploration and Mining of Major Seabed Resources' (Dec. 1981).

18 L. Kimball, 'Is there a mini-treaty? Will there be one?' *Neptune*, no. 19 (March 1982), p. 7.

19 Comptroller General of the United States, '*Uncertainties Surround Future of U.S. Ocean Mining*', p. 12.

20 For a useful discussion of the UK's position on the Convention, see E. D. Brown, 'The United Nations Convention on the Law of the Sea 1982: the British Government's dilemma', *Current Legal Problems*, vol. 37 (1984), pp. 259–93.

21 US Department of State, *Seabeds: Polymetallic Nodules Agreement Between the United States of America and Other Governments*, Treaties and Other International Acts Series 10562 (DOS, Washington, DC, 2 Sept. 1982), pp. 2–17.

22 R. T. Luoma, 'A comparative study of national legislation concerning the deep sea mining of manganese nodules', *Journal of Maritime Law and Commerce*, vol. 14, no. 2 (1983), pp. 243–68.

23 ibid., p. 252.

24 ibid., pp. 251–68.

25 J. A. Clemons, 'Recent developments in the Law of the Sea, 1983–1984', *San Diego Law Review*, vol. 22, no. 4 (1985), p. 826.

26 J. L. Malone, 'Freedom and opportunity: foundation for a dynamic oceans policy', *Department of State Bulletin*, vol. 84, no. 2,093 (1984), pp. 78–9.

27 United Nations, *The Law of the Sea: Official Text of the United Nations Convention on the Law of the Sea* (UN, NY, 1983), pp. 42–3.

28 A. McDonald, 'Mines in a lawless sea', *Geographical Magazine*, vol. 54, no. 9 (1982), p. 503. .

29 'Italian position on the Law of the Sea Convention', *Italy and the Law of the Sea Newsletter*, no. 15 (Jan. 1986), p. 3.

30 K. Surace-Smith, 'United States activity outside of the Law of the Sea Convention: deep seabed mining and transit passage', *Columbia Law Review*, vol. 84, no. 4 (1984), pp. 1,049–50.

31 A. D'Amato, 'An alternative to the Law of the Sea Convention', *American Journal of International Law*, vol. 77, no. 2 (1983), pp. 281, 284–5. The tax imposed under the US's DSHMRA is calculated as 3.75 per cent of 20 per cent of the imputed 'fair market value of the commercially recoverable metals and minerals' extracted from the seabed, that is at a rate of 0.75 per cent; see 'Deep Seabed Hard Mineral Resources Act', pp. 582–3.

32 J. L. Jacobson, 'Law of the Sea – what now?' *Naval War College Review*, vol. 37, no. 2/seq. 302 (1984), p. 85.

33 D'Amato, 'An alternative to the Law of the Sea Convention', p. 283.

34 Surace-Smith, 'United States activity outside of the Law of the Sea Convention', p. 1,035.

35 For a good critique of the land-locked states' problems relative to the EEZ, see M. I. Glassner, 'Land-locked states and 1982 Law of the Sea Convention', *Marine Policy Reports*, vol. 9, no. 1 (1986), pp. 8–14.

36 United Nations, *Law of the Sea*, pp. 24–5.

37 ibid., p. 29.

38 ibid., p. 43.

39 A. P. Allison, 'The Soviet Union and UNCLOS III: pragmatism and policy evolution', *Ocean Development and International Law*, vol. 16, no. 2 (1986), pp. 110–14.

40 ibid., p. 114.
41 'Seabed mining: Soviet bloc forms seabed mining venture', *Minerals and Materials: A Bimonthly Survey* (April/May 1987), p. 17.
42 'Declaration of policy of the American Mining Congress', *Mining Congress Journal*, vol. 64, no. 11 (1978), pp. 89–90; 'American Mining Congress declaration of policy', *Mining Congress Journal*, vol. 73, no. 10 (1987), p. 15.
43 Comptroller General of the United States, *Uncertainties Surround Future of U.S. Ocean Mining*, p. 28. According to Martin Glassner, a participant in many UNCLOS III sessions, by the tenth session several of the consortia had stated that they could accept the negotiated seabed provisions; Letter: M. Glassner, Professor and Chairman, Department of Geography, Southern Connecticut State University, New Haven, CT, 25 Nov. 1987.
44 Recent critics say processing technology was not meant to be included, but a close reading of the Convention text (Annex III, Article 5, Paragraph 5) shows that it was specifically indicated; see United Nations, *Law of the Sea*, p. 116.
45 ibid., pp. 115–16.
46 ibid.,
47 Marsteller and Tucker, 'Problems of the technology transfer provisions in the Law of the Sea Treaty, p. 169.
48 United Nations, *Law of the Sea*, pp. 115–16.
49 Marsteller and Tucker, 'Problems of the technology transfer provisions in the Law of the Sea Treaty', pp. 171–3.
50 United Nations, *Law of the Sea*, p. 115.
51 Marsteller and Tucker, 'Problems of the technology transfer provisions in the Law of the Sea Treaty', p. 173.
52 ibid., pp. 177–8.
53 D. Silverstein, 'Proprietary protection for deepsea mining technology transfer; new approach to seabeds controversy', *Journal of the Patent Office Society*, vol. 60, no. 3 (1978), pp. 143, 145–6, 169.
54 Marsteller and Tucker, 'Problems of the technology transfer provisions in the Law of the Sea Treaty', pp. 178–80.
55 A study done by the Comptroller General's Office in the US federal government contends that with appropriate modifications (such as precisely defined terms for 'fair and reasonable' and the use of Article 302 in the Convention which excludes defence-sensitive technologies) no great damage would result from technology transfers. Furthermore, the technology transfer requirement expires ten years from the date the Enterprise begins mining; see Comptroller General of the United States, *Impediments to US Involvement in Deep Ocean Mining Can Be Overcome: Report to the Congress of the United States*, EMD–82–31 (General Accounting Office, Washington, DC, 3 Feb. 1982), p. 28.
56 See P. Hoagland III, 'Seabed mining patent activity: some first steps toward an understanding of Strategic Behavior', *Journal of Resource Management and Technology*, vol. 14, no. 3 (1986), pp. 211–22.
57 For a detailed analysis of the potential impact of deep seabed mining on world metal markets, see J. B. Donges (ed.), *The Economics of Deep-Sea Mining* (Springer-Verlag, Berlin, FRG, 1985), especially pp. 62–252.
58 For useful analyses of the US's strategic minerals problems, see L. H. Bullis and J. E. Mielke, *Strategic and Critical Materials* (Westview Press, Boulder, CO, 1985); G. J. Mangone (ed.), *American Strategic Minerals* (Crane Russak, New York, 1984), especially Chapter 4 by J. R. Moore, 'Alternative sources of strategic minerals from the seabed', pp. 85–108.
59 S. F. Sibley, 'Nickel', in *Mineral Facts and Problems 1985*, USBM Bulletin 675 (USGPO, Washington, DC, 1985), pp. 537–8.

60 J. H. DeYoung Jr. *et al.*, *International Strategic Minerals Inventory Summary Report – Nickel*, USGS Circular 930–D (DOI, Washington, DC, 1985), p. 8.
61 Sibley, 'Nickel', pp. 537–8.
62 Because of the nickel industry's economic problems, all world producers (except the PRC) have worked since 1980 to establish a body called the International Nickel Discussion Group; see 'International issues and actions – nickel', *Minerals and Materials: A Bimonthly Survey* (Oct./Nov. 1985), pp. 5–6.
63 'Nickel', *Mining Annual Review, 1986* (Mining Journal, London, 1986), p. 66.
64 Sibley, 'Nickel', p. 536.
65 Mineral Policy Sector, Energy, Mines and Resources Canada, *The Canadian Minerals and Metals Sector: A Framework for Discussion and Consultation* (EMRC, Ottawa, Feb. 1985), p. 18.
66 Deep Ocean Working Group, Departmental Co-ordinating Committee on Ocean Mining, *Deep Ocean Mining Study: Final Report*, Division Document no. 1983–2 (Department of Energy, Mines and Resources Canada, Ottawa, Canada, 1983), pp. 8–9.
67 United Nations, *Law of the Sea*, pp. 48–9.
68 ibid., p. 49.
69 'International issues and actions – nickel', *Minerals and Materials; A Bimonthly Survey* (Feb./March 1986), p. 5.
70 Sibley, 'Nickel', p. 547.
71 R. Regan, 'Nickel keeps looking for a better price', *Iron Age Metals Producer*, vol. 229, no. 9 (1986), p. 63.
72 J. P. Schade, 'Current and future uses of nickel: defending existing markets and searching for new ones', *Minerals and Materials: A Bimonthly Survey* (April/May 1986), p. 6.
73 Sibley, 'Nickel', p. 548.
74 ibid., pp. 543–4, 547.
75 W. S. Kirk, 'Cobalt' in *Mineral Facts and Problems 1985*, USBM Bulletin 675 (USGPO, Washington, DC, 1985), pp. 171–2, 177.
76 ibid., p. 174.
77 ibid., pp. 172, 182.
78 For several years, Zaïre and Zambia have had an understanding to maintain the price of cobalt. Since November 1986 the price has been set at US$15.43/kg. In the past, Zambia is said to have undercut the set price. See 'International issues and actions: cobalt', *Minerals and Materials: A Bimonthly Survey* (Dec. 1985/Jan. 1986), p. 11; 'Cobalt: Zaïre and Zambia reaffirm price ceiling', *Minerals and Materials: A Bimonthly Survey* (April/May 1987), p. 15.
79 J. L. W. Jolly, 'Copper', in *Mineral Facts and Problems 1985*, USBM Bulletin 675 (USGPO, Washington, DC, 1985) p. 201.
80 Comptroller General of the United States, *Impediments to US Involvement in Deep Ocean Mining Can be Overcome*, p. 22.
81 J. M. Broadus, 'Seabed materials', *Science*, vol. 235, no. 4,791 (1987), p. 855.
82 Kirk, 'Cobalt', p. 174.
83 Department of International Economic and Social Affairs, UN, *Methodologies for Assessing the Impact of Deep Sea-Bed Minerals on the World Economy*, ST/ESA/168 (UN, NY, 1986), p. 43.
84 J. M. Broadus and P. Hoagland III, 'Conflict resolution in the assignment of area entitlements for seabed mining', *San Diego Law Review*, vol. 21, no. 3 (1984), p. 559.
85 'Declaration on national licenses', *Oceans Policy News* (May 1984), p. 4.

86 'Proposed revision of OMI deep seabed mining license', *Oceans Policy News* (June 1986), p. 5.
87 Broadus and Hoagland, 'Conflict resolution', p. 557.
88 Letter: R. Beall, Public Relations Co-ordinator, Lockheed Missiles and Space Co., Sunnyvale, CA, 21 April 1987.
89 Letter: M. B. Fisk, Law of the Sea Officer, UN, NY, 8 Jan. 1987.
90 Letter: M. J. Fraser, Vice President of Technology, Cyprus Minerals Company, Englewood, CO, 30 Jan. 1987.
91 'Deep seabed mining license amendment', *Oceans Policy News* (Nov. 1985), p. 7.
92 Broadus and Hoagland, 'Conflict resolution', p. 559.
93 'Declaration on national licenses', p. 4.
94 Broadus and Hoagland, 'Conflict resolution', p. 572.
95 Letter: R. Fellerer, Managing Director, Arbeitsgemeinschaft meertechnisch gewinnbare Rohstoffe, Hannover, FRG, 28 Jan. 1987.
96 'Declaration on national licenses', p. 4; 'National mining licenses', *Oceans Policy News* (March–April 1985), p. 4.
97 Letter: J–P. Lenoble, President, Afernod, Ifremer, Paris, France, 30 Jan. 1987.
98 Letter: T. Oyama, General Manager, Technical Department, Technology Research Association of Manganese Nodules Mining System, Tokyo, Japan, 25 June 1987; 'Developments concerning the international area (as of 9 July 1986)', p. 3.
99 'Corporate interests and activities in seabed mining', mimeographed data supplied by Letter: M. B. Fisk, Law of the Sea Officer, UN, NY, 8 Jan. 1987.
100 *Federal Register*, vol. 51, no. 143 (25 July 1986), pp. 26,794–824.
101 Comptroller General of the United States', *Impediments to US Involvement in Deep Ocean Mining Can Be Overcome*, p. 31.
102 Comptroller General of the United States, *Uncertainties Surround Future of U.S. Ocean Mining*, p. 29.
103 Broadus and Hoagland, 'Conflict resolution', p. 564.
104 'Developments concerning the international area', pp. 3–4.
105 Broadus and Hoagland, 'Conflict resolution', p. 564.
106 'Developments concerning the international area', pp. 3–4.
107 Broadus and Hoagland, 'Conflict resolution', p. 563.
108 'Developments concerning the international area', pp. 3–4.
109 'Seabed mining: Soviet bloc forms seabed mining venture', p. 17.
110 Ratiner, 'The Law of the Sea', p. 1,018.
111 Comptroller General of the United States, *Uncertainties Surround Future of U.S. Ocean Mining*, p. 30.
112 Article 133 of the Convention defines resources as 'all solid, liquid or gaseous mineral resources *in situ* in the Area at or beneath the sea-bed, including polymetallic nodules'. Article 151, paragraph 9, also, contains a brief reference to minerals other than manganese nodules: 'The Authority shall have the power to limit the level of production of minerals from the Area, other than minerals from polymetallic nodules, under such conditions and applying such methods as may be appropriate by adopting regulations in accordance with Article 161, paragraph 8', which puts the decision-making process in the hands of the Council; see United Nations, *Law of the Sea*, pp. 42, 49, 56–7.
113 A. Pardo, 'Ocean space and mankind', *Third World Quarterly*, vol. 6, no. 3 (1984), p. 569.

Minerals of the deep seabed

Minerals of the deep oceans have long held a fascination for both laymen and scientists. Marine researchers have been intensively exploring the seabed and sampling and analysing its rocks and sediments. From these studies they are developing hypotheses about the processes responsible for seabed mineral formation and are considering the potential significance of these minerals to humankind.

Ferromanganese nodules and crusts

Ferromanganese nodules (manganese nodules) and ferromanganese crusts (crusts) have a wide distribution and have in common a composition dominated by manganese and iron oxides. Both the manganese nodules and crusts may be exploited in future, but a gargantuan task remains ahead to map and sample them more carefully and to develop and test machines and methods for extracting and processing the metals contained. These deposits of 'black gold' will not be won easily from the oceans.

Manganese nodules

Lacustrine ferromanganese concretions have been known to northern Europeans for centuries, but marine nodules were not discovered until 1868, during the *Sofia* Expedition led by A. E. Nordenskiöld; the nodules were recovered in the Kara Sea, east of Novaya Zemlya. This discovery was soon followed by the more celebrated work of Sir John Murray and A. Renard (1873–6) on *HMS Challenger*, which acquired numerous manganese nodule samples.[1] At first, these dark often potato-like concretions were scientific curiosities. Not until the 1950s with the publication of the seminal work of John Mero (who did basic research in manganese nodule analyses) did the scientific and business communities begin to consider manganese nodules as an exploitable source for minerals, such as copper, nickel, cobalt, and manganese.[2] Although these four metals are presently of greatest interest to mining consortia, others – zinc, vanadium, and molybdenum – may become important as well.

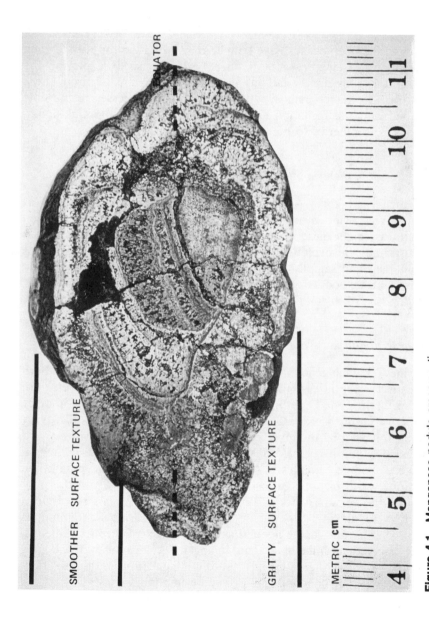

Figure 4.1 Manganese nodule cross-section
Source: Courtesy Kennecott Corporation. With permission.

Morphology

Nodules vary in size from that of a small pea to a large cantaloup, and occur in various shapes – spheroidal, ellipsoidal, botryoidal (grape-clusters), and irregular. Their surface may be smooth and/or granular, with both textures occurring on the same nodule. The top surface is usually smoother than the buried surface which is often rough and irregular. Internal fractures are common and usually filled with clay. Variations in the mineral content of the growth layers give many nodules a banded appearance when sectioned; their surface colour is usually black to dark reddish-brown (Figure 4.1).[3]

Distribution

Manganese nodules occur in all the oceans, although they are not common in the Arctic Ocean (Figure 4.2). True manganese nodules lie on or partly or wholly buried in soft sediments, usually in the deep sea, in contrast to crusts, which occur at shallower depths and on hard substrates largely devoid of permanent sediments. Nodules' chemical associations and layering have been known for 'many years, but the means by which nodules are maintained at the sediment-water interface are still controversial and not fully understood'. It is generally accepted that nodules form through accretion from a dual source of mineral elements – seawater and seabed sediments, especially the latter.[4] Their characteristics vary greatly from one oceanic area and environment to another,[5] even within a given dredge sample.

Nodules are most abundant where (1) sedimentation rates are relatively low; (2) nuclei are present; (3) water depths are between 3,000 and 6,000m; (4) benthic activity helps keep the nodules from becoming buried;[6] (5) strong bottom currents occur; (6) a high oxidising potential exists in the depositional environment; and (7) the sea-floor's sediment

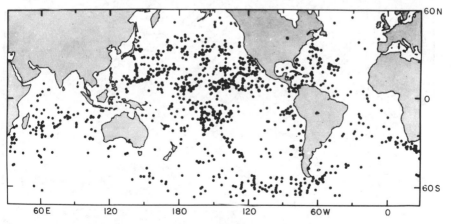

Figure 4.2 Sites where deep seabed manganese nodules have been collected

Source: G. R. Heath 'Manganese nodules: unanswered questions', *Oceanus*, vol. 25, no. 3 (1982), p. 37. With permision.

surface does not have a reducing condition,[7] that is an absence of oxygen. One determinant for the abundance and size of nodules is the time they have been exposed to formation processes. Their accretion rate (except where metals of hydrothermal origin are available) is very slow, only a few mm in 1 million years.[8] Thus, in general, the farther a seabed area is from a sea-floor spreading centre, the larger the nodules are likely to be,[9] a consequence of increasing age of the sea-floor with distance from a spreading centre.

Nuclei and sedimentation

For nodule formation to begin, some type of nucleus must be present, as for example, volcanoclastics, nodule fragments or biogenic materials such as a whale's earbone or a shark's tooth, around which accretion can occur. If potential nuclei become buried under sediments, the formation of manganese oxides is inhibited. Thus the seabed sedimentation rate is a primary factor in manganese nodule distributions. The Atlantic Ocean has a relatively high amount of sediment carried into it by continental rivers, whereas the Indian and Pacific Oceans – with fewer rivers debouching into them – have considerably less sedimentation (Figure

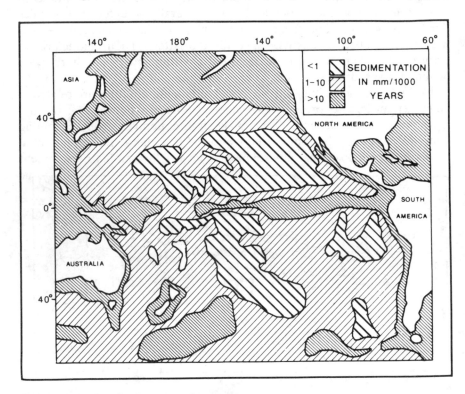

Figure 4.3 Pacific Ocean sedimentation rates

Source: After D. Z. Piper and M. E. Williamson, 'Composition of Pacific Ocean ferromanganese nodules', *Marine Geology*, vol. 23, no. 4 (1977), p. 293. With permission.

4.3).[10] The Pacific Ocean also has great depth and areas with strong bottom currents that help keep potential nodule nuclei free of sediments. These conditions may contribute to the Pacific's primacy in nodule populations and perhaps to their relatively rich metal content.

Accretion and mineral content

Beginning soon after the *Challenger* voyages of Murray and Renard, scientists have offered various explanations for the origin of the nodules' metal content and their formation process. In 1891 Murray and Renard published a report on their *Challenger* expeditions.[11] They suggested that the manganese might originate from oceanic sources – weathering marine basalts and sea-water. Renard favoured sea-water as the source and Murray the marine basalts.

In his early papers Murray focused on the correlation between evidence of deep sea marine volcanic debris and manganese accumulation. He felt that the manganese formation might be accentuated by volcanic carbon dioxide.[12] Later (in 1894) he recognised the importance of manganese derived from rivers and streams, as well as shallow marine sediments. His statements appear limited to shallow marine oxides, whereas he seemed to adhere to his earlier hypothesis for the origin of abyssal nodules:

> Indeed, all observations go to show that the quantity of manganese dioxide in these abyssal deposits is in direct relation to the abundance and basic character of the erupted rocks and minerals associated with them, and extent to which these minerals and rock particles have undergone alteration.[13]

In the late 1950s Edward Goldberg and Gustav Arrhenius attributed the nodules' minerals to continental sources that were dissolved in and later precipitated from the sea-water,[14] and in the 1960s Enrico Bonatti and Y. Rammonhanroy Nayudu (as well as others) suggested that the nodules' minerals originate from marine volcanic effusions, with strong interactions between the sea-water and hot lavas, whereby large amounts 'of iron, manganese and other elements are leached out of the lava' with the manganese being subsequently partially separated from the iron and 'then precpitated on the ocean floor in the vicinity of the effusion, thus forming the nodules'.[15] Some investigators attributed nodule accretion, in part, to biogenic absorption and subsequent secretion by foraminifera,[16] or to oxidising microbial flora within the manganese nodules that catalyse Mn^{+2} to Mn^{+4}. The Mn^{+4} proceeds to absorb additional Mn^{+2}, thus causing a progressive growth.[17]

The source of the minerals (sediments or sea-water) is thought to contribute to variations in nodule morphology and mineralogy, even within a given nodule. Work by Werner Rabb showed that many nodules' lower surfaces contain high values for nickel, copper, zinc, molybdenum, and lead, whereas upper surfaces are reversed.[18] Peter Halbach and Rainer Fellerer (among many others) have suggested that the lower surfaces reflect diagenetic precipitation processes associated with leaching of the sediments, whereas the upper surfaces represent precipitates acquired from the sea-water.[19]

Minerals in manganese nodules occur in three general phases: manganese minerals, iron oxide minerals, and accessory minerals. The main manganese minerals are todorokite $(Mn^{+2}, Ca, Mg) Mn^{+4} O_7 \bullet H_2O$; birnessite $Na_4Mn_{14}O_{27} \bullet H_2O$; and vernadite $(Mn, Co^{+3}) Mn_6O_{13} \bullet H_2O$. According to Benjamin Haynes, Stephen Law, and David Barron, 'copper, nickel, molybdenum, zinc and some cobalt generally associate with the vernadite phase by substituting in Mn^2+ ...', thereby helping to 'stabilize manganese crystal structure in the marine environment'.[20] David Piper, J. R. Basler, and James Bischoff, however, observed very little divalent Mn in a recent study of manganese nodules.[21] And Frank Manheim does not believe that Co is associated with Fe phases in nodules, rather Co and Fe are both associated with hydrogenetic phases (where the tops of nodules are exposed to sea-water), which accounts for a correlation between the two minerals in nodules.[22] The most important iron oxide minerals include akaganeite, feroxyhyte, goethite, and lepidocrocite. Titanium and lead also associate with the iron oxide minerals in nodules. Accessory minerals, that is those occurring in very small amounts, include zeolites and sheet silicates, clastic volcanics and biogenics, with the latter including minerals such as aragonite, apatite, and calcite.[23]

Researchers have identified more than seventy elements in manganese nodules.[24] The variety stems, in part, from the excellent sea-water scavenging qualities (by adsorption) of iron and manganese oxides, especially the latter. Those nodules containing large amounts of todorokite are usually rich in copper and nickel.[25] Nodules are richest in copper and nickel where the oceans support the most abundant populations of planktonic organisms, as in the equatorial region of the central Pacific. As the rain of Plankton accumulates on the seabed, the mineral elements are recycled into the nodules.[26] According to David Piper and Michael Williamson, molting products and faecal matter of oceanic organisms, also, could contribute to transfers of metals out of the water column to seabed sediments and on into the nodules.[27]

Analyses of data contained in the Scripps Institution of Oceanography's Sediment Data Bank (based on the pioneering work of Jane Frazer and Mary Fisk),[28] demonstrate that the depth at which nodules form contributes to variations in the types and quantities of metals contained.

> Manganese shows little correlation with depth, although the data suggest a decrease in the maximum values and possibly an increase in the minimum values with increasing depth. No nodule analysis shows more than 32 per cent Mn below depths of 5,300m, and only one reports less than about 13 per cent Mn below 6,000m [Figure 4.4].[29]

When nodules form at depths greater than 3,000m, they usually have a higher copper and nickel content than at shallower depths. A combined

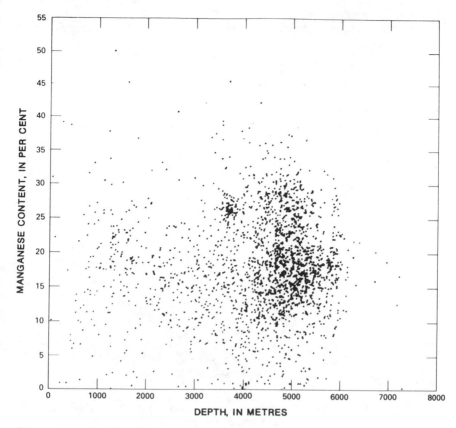

Figure 4.4 Relationship of depth and manganese content in ferromanganese nodules (as represented by data in the Scripps Institution of Oceanography's Sediment Data Bank)

Source: V. E. McKelvey, N. A. Wright, and R. W. Bowen, *Analysis of the World Distribution of Metal-Rich Subsea Manganese Nodules*, USGS Circular 886 (USGS, Arlington, VA, 1983), p. 21.

nickel and copper content maximum occurs at around 5,000m. The depth relationship, however, is not linear. At about 2,900m to 3,000m, there is a depth threshold, above which nickel and copper (combined) rarely measure more than 1 per cent (Figure 4.5).[30] Although the enrichment of copper and nickel in the equatorial productivity zone was already well established in the early 1970s,[31] David Cronan offered a hypothesis to explain why the high enrichments of these metals do not occur above 3,000m. He suggested that only from about 3,000m downward to the lysocline[32] do biotically concentrated inorganic carbonate materials begin to dissolve – thus liberating metals (especially copper) which are then available for nodule formation processes.[33]

Several hypotheses have been posed to explain the varying amounts of cobalt. Piper and Williamson in the late 1970s suggested that high cobalt values are a function of the presence of oxidising seabed currents.[34] More recently, however, Halbach and Manheim pointed out that nodules (and crusts) with high amounts of cobalt form in areas where the oxygen level

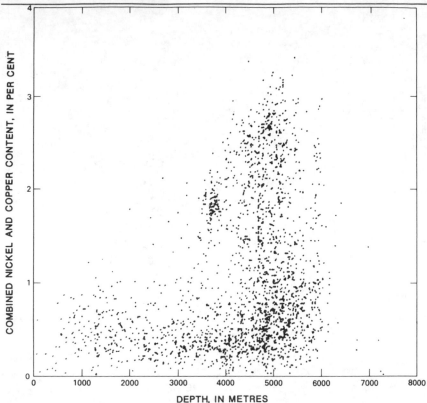

Figure 4.5 Relationship of depth and combined nickel and copper content of Pacific Ocean ferromanganese nodules (as represented by data from 1,770 sampling sites contained in Scripps Institution of Oceanography's Sediment Data Bank).

Source: V. E. McKelvey, N. A. Wright, and R. W. Bowen, *Analysis of the World Distribution of Metal-Rich Subsea Manganese Nodules*, USGS Circular 886 (USGS, Arlington, VA, 1983), p. 14.

is low;[35] Manheim attributed these high values to cobalt enrichment with respect to manganese in the water. Biological organisms (as the phytoplankton) have not been established as an important source of cobalt, because cobalt 'is less effectively removed from sea-water by phytoplankton than other transitional metals' – copper and nickel.[36] Cobalt's relationship to depth is more complex than for nickel and copper. Value concentrations in crusts are at a maximum at about 1,000m to 2,000m; they then decline rapidly to 3,000m where they decrease slowly with increasing depth.[37] Below a depth of 3,000m, cobalt rarely measures more than 0.6 per cent (Figure 4.6).[38]

Average metal contents in Pacific Ocean nodules are appreciably higher than those of the Atlantic and Indian Oceans. Pacific Ocean nodules, excluding samples from the CCZ (see Figure 3.2) have about 1.5 times as much manganese as Atlantic Ocean nodules. The value for nickel in Pacific nodules is aproximately 2.4 times that in the Atlantic and

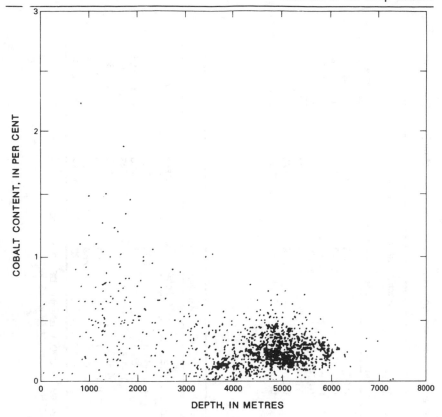

Figure 4.6 Relationship of depth and cobalt in Pacific Ocean ferromanganese nodules (as represented by data from 1,710 sampling sites contained in Scripps Institution of Oceanography's Sediment Data Bank).

Source: V. E. McKelvey, N. A. Wright, and R. W. Bowen, *Analysis of the World Distribution of Metal-Rich Subsea Manganese Nodules*, USGS Circular 886 (USGS, Arlington, VA, 1983), p. 18.

1.8 times that in the Indian Ocean. The copper value measures about 4.2 and 2.2 times more than those of the Atlantic and Indian Oceans, respectively. The average cobalt content of 0.27 per cent in Pacific Ocean nodules is the same as in the Atlantic Ocean; the average value for cobalt in the Indian Ocean is 0.21 per cent (Table 4.1)

Because of the variation in nodule-abundance and metal-content averages on a world scale, as well as locally, caution should be used in making specific judgements about specific areas. Both Jane Frazer and Vincent McKelvey strongly emphasised that, because the preponderance of sampling has been in the north-eastern Equatorial Pacific (the CCZ region), the data reflect a sampling bias.[39] On the other hand, some researchers feel that, because of the large manganese nodule population in the Pacific's CCZ (relative to other areas), the sampling bias is not as great as might appear.

Table 4.1 Sampling stations, metal content and ranges (in %) and nodule concentration (in kg/m^2) for all stations in the Pacific, Atlantic, and Indian Oceans

Metal	All Pacific stations			Pacific stations outside CCZ[a]		
	Number of Stations	Mean	Range	Number of Stations	Mean	Range
Manganese	1,777	20.1	0.07–50.3	1,378	18.84	0.07–50.3
Iron	1,772	11.4	0.30–41.9	1,377	12.77	0.42–41.90
Nickel	1,784	0.76	0.01–1.95	1,378	0.63	0.01–1.80
Copper	1,771	0.54	0.01–1.90	1,365	0.41	0.01–1.90
Nickel and copper	1,770	1.30	0.02–3.44	1,364	1.04	0.02–3.44
Cobalt	1,710	0.27	0.01–2.23	1,314	0.29	0.01–2.33
Nodule concentration	321	10.89	0.12–70.80	268	11.61	0.18–70.80

Metal	All Atlantic stations			All Indian stations		
	Number of stations	Mean	Range	Number of stations	Mean	Range
Manganese	298	13.25	0.04–40.9	303	15.25	0.60–32.30
Iron	299	16.97	1.54–50.0	303	14.23	1.33–39.63
Nickel	297	0.32	0.01–1.56	302	0.43	0.01–1.58
Copper	297	0.13	0.01–0.88	300	0.25	0.01–1.66
Nickel and copper	296	0.44	0.03–2.30	300	0.69	0.03–3.24
Cobalt	285	0.27	0.01–1.44	289	0.21	0.01–0.94
Nodule concentration	0	—	—	23	10.82	0.40–43.2

Source: V. E. McKelvey, *Subsea Mineral Resources*, Chapter A of USGS Bulletin 1689 (USGPO, Washington, DC, 1986), p. 32.
Note: a. The CCZ (Clarion-Clipperton Zone) is here defined as latitude 7° N to 15° N, longitude 114° W to 153° W.

Ferromanganese crusts

Recently marine scientists have been less interested in deep seabed nodules and more involved with studies of ferromanganese crusts, especially those rich in cobalt. Many in industry and the scientific community think some cobalt-rich ferromanganese crusts may be mined before commercial nodule production begins.[40] Several reasons account for this view:

1 crusts usually hold more cobalt and, also, contain other minerals found in nodules (nickel, copper, zinc, and platinum) which gives them a greater overall metal content;
2 they occur in much shallower waters than do the best nodules, most of which lie within very deep abyssal regions;
3 they are often located within the 200-mile EEZs of the coastal states, making them both more accessible to markets and politically safe.

Distribution

On hard substrates and where soft sediments are absent, iron and manganese minerals occur as crusts rather than as nodules, although some 'nodules' may form around discrete nuclei in these environments.[41] Cobalt enrichment in manganese crusts has been recognised since the early 1950s when researchers found them associated with Pacific Ocean islands and seamounts. In the late 1960s and early 1970s scientists from the Hawaii Institute of Geophysics examined ferromanganese crusts in the Hawaiian Archipelago.[42] In 1981 an FRG-sponsored research cruise focused for the first time on the cobalt-rich crusts in the mid-Pacific region; data collected indicated the presence of major deposits with a probable economic potential. The USGS in 1983 and 1984 followed up on these findings, collecting data which showed

> that significant portions of the upper slopes of certain seamounts and ridges may be covered with from 2 to more than 4cm of black oxide dominated by manganese and iron but containing cobalt, nickel, lead, cerium, molybdenum, vanadium and many other minor metals.[43]

A few Pacific Ocean deposits examined in 1986 during a cruise south of Johnston Atoll were found to have a thickness of up to 15cm.[44] The western Atlantic Ocean, also, has important deposits, especially on the Blake Planteau, off the south-eastern coast of the US. Some areas on the Blake Plateau have a thickness of 5cm.[45]

Accretion and metal content

As in nodule formation, time is a key factor in ferromanganese crust development. According to Manheim, only those seamount substrates with an age of at least 15 million years have had enough time to acquire large accumulations of crusts, and seamounts should be 60 million to 80 million years of age to have truly significant crustal development.[46]

Cobalt crusts form more slowly than manganese nodules. The slower their formation rate, the higher their cobalt content. A study of ferromanganese crusts in the mid-Pacific and Line Islands region indicates growth rates of only 1–2.8mm per million years.[47] Manheim noted: 'cobalt-rich crusts appear to be the slowest forming mineral and rock deposits known, accumulating at the rate of a molecular layer (unit cell thickness of 4.7 Å) per 1 to 3 months allowing for porosity'. But he cautioned that

> crust accumulation rates from dating by uranium daughters must be interpreted with care because they can be applied only to the top millimeter or so of cobalt-rich crusts. They do not yield information about earlier hiatuses and changes in growth rate.[48]

Examination of seamounts (as well as submerged portions of islands) shows that their sides have the greatest crustal development, especially between depths of 500m and 2,500m.[49] With increasing depth, deposits become thinner and the coverage more irregular, probably a function of downslope movements of talus and sediments caused by slumping and sliding, as well as 'Taylor Column' currents not yet fully understood by oceanographers. Crust accretion in relatively shallow waters may also be inhibited by sediments originating from coral detritus (Table 4.2)[50]

Just as depth is important to the distribution and growth rate of crusts, it is similarly related to metal content. Hein *et al.* have demonstrated that crusts with the highest cobalt content occur in the central and south-central Pacific Ocean, as in the Kingman Reef and Palmyra Atoll and Marshall Island groups. Table 4.3 presents a generalised qualitative ranking for various island and atoll groups, based on grade and permissive area calculations.[51] Crusts, overall, contain from 1.4 to 2.7 times more cobalt than nodules, but their nickel, copper, and zinc values are usually less (Table 4.4).[52]

Cobalt is removed from ambient sea-water by oxides through selective precipitation. Beyond this statement, according to Manheim, how cobalt content is affected by various physical conditions in the marine environment is still not fully understood. He has, however, succinctly reviewed several hypotheses offered to account for variations of cobalt content in crusts, many of which parallel those presented to explain the mineral content of nodules. For example, vulcanism has been offered as an explanation, given that seamounts, island slopes, and volcanically formed ridges often have significant amounts of cobalt in their associated crusts and nodules. But analyses of discharges from hydrothermal vents and their associated sulphide and oxide accretions have shown that their cobalt content is relatively low. Cold water alteration of basalts has been suggested as a source of the cobalt, but this hypothesis does not explain why crusts at higher elevations have more cobalt than do those near deep seabed basalts. Another hypothesis holds that pressure-sensitive minerals (vernadite and todorokite) influence cobalt accretion, but vernadite occurs

Table 4.2 Depth distribution for cobalt in ocean ferromanganese crusts

Depth (m)	HR-MSA[a] (177)[i] Mean[j]	MI-WMPM[b] (71)[i] Mean[j]	NLI-MPM[c] (192)[i] Mean[j]	SLI-FP[d] (97)[i] Mean[j]	TCP[e] (537)[i] Mean[j]	TA[f] (77)[i] Mean[j]	I[g] (102)[i] Mean[j]	TWO[h] (679)[i] Mean[j]
0–1,000	0.91	1.64	—	1.98	1.31	0.85	—	0.97
1,000–1,500	0.94	0.84	1.05	1.25	1.00	0.80	0.76	0.83
1,500–2,000	0.66	0.74	0.93	1.41	0.86	0.58	0.39	0.76
2,000–2,500	0.70	0.87	0.59	—	0.67	0.55	0.40	0.61
All depths	0.76	0.77	0.77	0.78	0.76	0.50	0.42	0.63

Source: After F. T. Manheim, 'Marine cobalt resources', Science, vol. 232, no. 4,750 (1986), p. 604. With permission. Copyright 1986 by the AAAS.
Notes: a. Hawaiian Ridge-Musicians Seamount Area; b. Marshall Islands-West Mid-Pacific Mountains; c. Northern Line Islands-Mid-Pacific Mountains; d. Southern Line Islands-French Polynesia; e. Total Central Pacific; f. Total Atlantic; g. Total Indian; h. Total World Oceans; i. Number of samples used to calculate the mean; j. Concentrations are in per cent of oven-dried matter weight.

Table 4.3 Estimated resource potential of crusts within the United States EEZ of Hawaii and former and current trust and affiliated territories

Pacific Area	Relative ranking	Potential
Marshall Islands[a]	1	high
Micronesia[a]	2	high
Johnston Island	3	high
Kingman-Palmyra	4	high
Hawaii-Midway	5	medium
Wake	6	medium
Howland-Baker	7	medium
Northern Mariana Islands	8	low
Jarvis	9	low
Samoa	10	low
Belau/Palau[a]	11	low
Guam	12	low

Source: Modified from J. R. Hein, F. T. Manheim, and W. C. Schwab, *Cobalt-Rich Ferromanganese Crusts from the Central Pacific*, OTC 5234, Offshore Technology Conference, May 1986, pp. 119–26, as cited in Office of Technology Assessment, US Congress, *Marine Minerals: Exploring Our New Ocean Frontier* (USGPO, Washington, DC, 1987), p. 74.
Note: a. Former US Trust Territories; information on the status of the former US Trust Territories was provided by Letter: C. E. Harrington, Chief Geographer, Nautical Charting Division, NOAA, DOC, 3 Nov. 1987. As of 15 July 1987, the Republic of Palau was still administered by the US; although a compact of free association has been concluded and approved by the US Congress, the process has not been completed in Palau. See US Department of State, 'New country codes for the Federated States of Micronesia, Marshall Islands, and Palau', *Geograhic Notes*, issue 6 (15 July 1987), p. 22.

Table 4.4 Concentrations of principal metals of nodules and crusts in world oceans at all depths (oven dried weight in %)[a]

Metal	Nodules		Crusts			
	World Oceans[b] (Mean)[d]	Clarion-Clipperton Zone[c] (Mean)[e]	World Oceans (Mean)[f]	Pacific Ocean (Mean)[g]	Atlantic Ocean (Mean)[h]	Indian Ocean (Mean)[i]
Al	2.7	2.9	1.19	1.06	1.27	1.31
Fe	13.6	6.9	16.48	16.09	18.56	16.46
Mn	17.4	25.4	21.62	23.06	20.07	18.04
Co	0.27	0.24	0.63	0.73	0.53	0.38
Ni	0.55	1.28	0.45	0.47	0.40	0.39
Cu	0.34	1.02	0.14	0.16	0.11	0.13
Pb	0.093	0.040	0.158	0.163	0.163	0.150
Zn	0.120	0.140	0.068	0.073	0.080	0.056

Source: a. After F. T. Manheim 'Marine cobalt resources', *Science*, vol. 232, no. 4750 (1986), p. 603. With permission. Copyright 1986 by the AAAS; b. V. E. McKelvey, N. A. Wright, and R. W. Bowen, *Analysis of the World Distribution of Metal-Rich Subsea Manganese Nodules*, USGS Circular 886 (USGS, Arlington, VA, 1983), various pages; c. M. R. Scott *et al.*, 'Rapidly accumulating manganese deposits from the Median Valley of the Mid-Atlantic Ridge', *Geophysical Research Letters*, vol. 1, no. 8 (1974), p. 355.
Notes:d. N=1979(Mg and Fe); e. N=234(Al) and 2237 (Ni); f. N=1005 (Mn); g. N=803(Mn); h. N=75(Mn, Fe, Co, Ni, and Cu); i. N=127(Mn, Co, and Cu).

at all depths, so the difference in cobalt content should not be appreciable. Unlike their nickel and copper affinity, the phytoplankton seem not to extract enough cobalt to supply the seabed sediments with major amounts of cyclable cobalt. A final hypothesis suggests that cobalt crusts form best where the water is richest in oxygen (providing the greatest potential for oxidation). This suggestion is countered by calling attention to the cobalt-rich crusts that occur in relatively oxygen-deficient zones in the Equatorial Pacific region.[53]

Adding to the complexity of the debate over explanations for crusts' cobalt content (as well as other minerals) is the problem of seamount and island formation, migration, and change of elevation associated with hot spots and plate tectonics. A good example of this process is the Hawaiian Archipelago, where a hot spot in the vicinity of the island of Hawaii creates islands and seamounts that then migrate slowly to the northwest. The changes in elevation and latitudinal location create a mixed mineral-isation and accretion rate.[54]

Polymetallic sulphides

Earth has more than a dozen oceanic plates that are in nearly constant tectonic flux, diverging, converging, and transforming (Figure 4.7). Marine geologists are especially interested in the boundaries of these plates, for it is here that a major part of seabed mineral genesis occurs.[55] Yet only about 1 per cent of these boundaries has been adequately explored. There are three types of plate boundaries – transform, convergent, and divergent. Each type has associated with it physical processes that lend themselves to the genesis of certain mineral groups. The emphasis here will be upon divergent boundaries, the location of recent discoveries of polymetallic sulphides, ores that may have major significance to the future of seabed mining.

Divergent boundaries

As was briefly described in Chapter 1, the oceanic plates are driven by internal physical forces within the earth not yet completely understood. These forces are thought to depend on the dynamics of convectional heat transfer, whereby deep-lying magma from the asthenosphere rises toward the seabed in mid-ocean areas and then accretes to adjacent and already cooled basaltic lithosphere. The process is relatively continuous so that the newly formed lithosphere is displaced away from the spreading centre. These spreading centres are classed as divergent boundaries and extend for about 54,000km around the world.[56]

As magma moves upward from the asthenosphere, it comes into contact with lithospheric rocks containing sea-water. Through contraction cracks and tectonically induced faults, the sea-water percolates downward

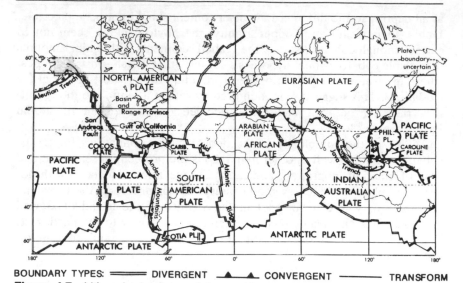

BOUNDARY TYPES: ═══════ DIVERGENT ▲▲ CONVERGENT ──────── TRANSFORM

Figure 4.7 Lithospheric plates of the world

Source: B. A. McGregor and M. Lockwood, *Mapping and Research in the Exclusive Economic Zone* (USGS, Reston, VA, 1985), p. 14.

and laterally from the seabed to become part of the circulating convection system. As the water circulates through the seabed, it becomes heated by and makes contact with the magma below. During this process, it takes into solution metallic minerals. These hydrothermal solutions are then carried upward to be disharged through active vents ('black [or white] smokers') where iron and heavy-metal sulphides (as well as native sulphur) precipitate from solution (Figure 4.8). Over time, these precipitates build platforms and mounds surmounted by chimney-like vents, containing sulphides of zinc, copper, iron, silver, cobalt, and gold, among others. Some chimneys, as in the East Pacific Rise area located just south of Baja California (at 21° N), attain heights of 20m and widths of several metres. The vented plumes (Figure 4.9) may rise several hundred metres to form hydrous oxides 'that slowly settle out to accumulate as ... metalliferous sediments' which often occur along the flanks of mid-ocean ridges.[57]

Evidence, to date, indicates that the faster a spreading centre is diverging, the greater is the chance for hydrothermal black-smoker venting.[58] Until recently, no vents had been discovered in areas with a spreading rate (half-rate) of < 2cm/yr, whereas those with a half-rate > 2cm/yr have some vents and those with a half rate > 6cm/yr often have many. Peter Rona, writing for *Marine Mining* in 1985, stressed, however, that slow-spreading centres (which compose more than 50 per cent of the divergent boundaries) should not be dismissed as lacking in hydrothermal mineralisation potential.[59] That same year, a team of government and university scientists led by Rona discovered 'the first black-smoker-type

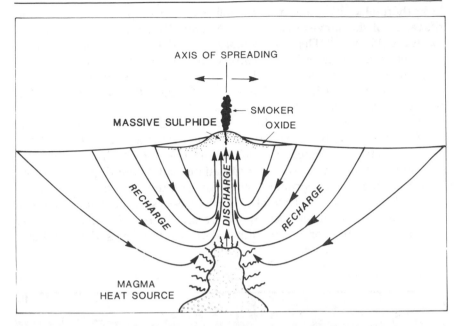

Figure 4.8 Spreading centre sea-water circulation model (illustrating discharging, recharging, and massive sulphide deposition)

Source: After R. A. Koski, W. R. Normark, J. L. Morton, and J. R. Delaney, 'Metal sulphide deposits on the Juan de Fuca Ridge', *Oceanus*, vol. 25, no. 3 (1982), p. 47. With permission.

Figure 4.9 Black smoker on the East Pacific Rise (between 2,700m and 3,000m, taken from Woods Hole Oceanograhic Institution's submersible research vehicle, *Alvin*)

Source: Photo by Dudley Foster. Courtesy WHOI. With permission.

hydrothermal venting and massive sulfide mineral deposits at a site in the rift valley of the slow-spreading Mid-Atlantic Ridge, near latitude 26° N, longitude 45° W.'[60] This find demonstrated that slow-spreading centres can have 'a complete series of hydrothermal mineral deposit types', which has significance for future exploration in other slow- to intermediate-spreading centres such as the Gorda Ridge lying within the 200-nmi EEZ off the US Oregon and California coasts.[61]

An excellent example of a highly mineralised slow-spreading centre is the Red Sea Rift, lying between the Saudi Arabian Peninsula and the African Continent. Whereas much of the emerging hydrothermal solutions entering waters above most spreading centres are dispersed, in the Red Sea they have become highly concentrated, making the area a potentially rich source of metals. The Red Sea spreading zone contains thick, marine evaporites that overlie emerging hydrothermal solutions. Hot, dense, metal-containing brines form as they pass through these

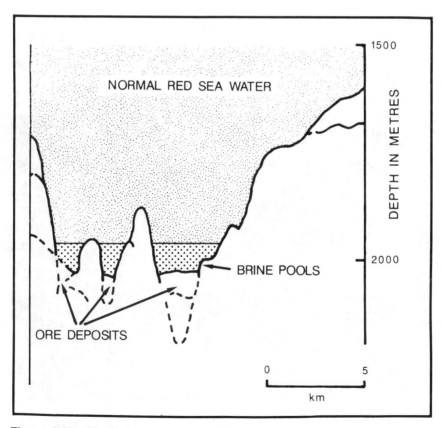

Figure 4.10 Metal-rich precipitates lying in Red Sea deeps

Source: After J. M. Edmond, 'The geochemistry of Ridge Crest Hot Springs', *Oceanus*, vol. 27, no. 3 (1984), p. 19. With permission.

evaporites, becoming trapped (because of their density) in numerous basins as deep as 2,000m. When emerging brines cool in their upper portion and come into contact with overlying sea-water, precipitates form and settle to the bottom of the deeps to become metalliferous muds (Figure 4.10).[62]

In addition to metal genesis, divergent boundaries, under certain conditions, can be the site of hydrocarbon formation, as evidenced by today's petroleum production in the Red Sea. Where the divergence is confined within very narrow margins, organic matter becomes trapped and then buried by adjacent continental sediments. With the proper metamorphosis, hydrocarbons may develop.[63]

Convergent boundaries

As one part of a plate is emerging from out of the asthenosphere, another part, where a continent lies in its path, is returning (via subduction) into the asthenosphere (see Figure 1.2). This process creates tremendous pressure and heat, causing a gradual melting of both the subducting plate and portions of the adjacent continental plate. As the subducting plate is forced downward, the oeanic crust containing metallic enrichments such as chromium, silver, gold, nickel, copper, zinc, and the platinum group metals may melt, with the metallic minerals reconcentrating in eruptive volcanic rocks on an adjacent continent such as the Andes Mountains of western South America or an adjacent volcanic island chain such as Japan, the Philippines, and Indonesia. At times, portions of these mineral-containing basalts of oceanic crusts may be shaved off and pushed up on to the land to form ophiolites, as on the island of Cyrpus where copper ores have long been extracted. Finally, where organic sediments become trapped, buried, and altered in deep trenches formed during subduction, hydrocarbons can develop.[64] Offshore Peru, for example, has several petroleum fields adjacent to its subduction zone.[65]

Transform boundaries

When plates move parallel to one another, mineralisation similar to that occurring along divergent plate boundaries may develop. This mineralisation, however, is usually not of the same magnitude.[66]

Conclusions

Much progress has been made during the last two decades in understanding the mineralisation processes responsible for forming the ferromanganese nodules and crusts and the polymetallic sulphides. Paradoxically, however, the more we learn about oceanic geology and metal genesis, the more we understand (by analogy) the continents' geology and metal distributions.[67] This knowledge can help us identify the best places to seek onshore minerals. Because new onshore discoveries will probably be more accessible and cheaper to extract than those of the deep seabed, mineral producers may delay going offshore to mine.

Notes

1 F. T. Manheim, Book Review of G. P. Glasby, *Marine Manganese Deposits*, in *Geochemica et Cosmochimica Acta*, vol. 42, no. 5 (1978), p. 541.
2 See, for example, J. L. Mero, *The Mining and Processing of Deep-Sea Manganese Nodules* (Institute of Marine Resources, Berkeley, CA, 1959); J. L. Mero, *The Mineral Resources of the Sea* (Elsevier, Amsterdam, 1965).
3 B. W. Haynes, S. L. Law, and D. C. Barron, 'An elemental description of Pacific manganese nodules', *Marine Mining*, vol. 5, no. 3 (1986), pp. 244–6; W. Rabb, 'Physical and chemical features of Pacific deep sea manganese nodules and their implications to the genesis of nodules', in D. R. Horn (ed.), *Papers from a Conference on Ferromanganese Deposits on the Ocean Floor* (National Science Foundation, Washington, DC, 1972), p. 32.
4 F. T. Manheim, 'Marine cobalt resources', *Science*, vol. 232, no. 4,750 (1986), p. 605.
5 J. E. Andrews and G. H. W. Friedrich, 'Distribution patterns of manganese nodule deposits in the Northeast Equatorial Pacific', *Marine Mining*, vol. 2, nos 1/2 (1979), pp. 1–2.
6 U. von Stackelberg, 'Influence of hiatuses and volcanic ash rains on the origin of manganese nodules of the Equatorial North Pacific (*Valdivia* Cruises VA-13/2 and VA-18)', *Marine Mining*, vol. 3, nos 3/4 (1982), pp. 302–4, 309.
7 T. J. Rowland, 'Non-energy marine mineral resources of the world oceans', *Marine Technology Society Journal*, vol. 19, no. 4 (1985), p. 14.
8 T. L. Ku and W. S. Broeker, 'Radiochemical studies on manganese nodules of deep-sea origin', *Deep Sea Research*, vol. 16 (1969), pp. 628–30; D. Heye and V. Marchig, 'Relationship between the growth rate of manganese nodules from the Central Pacific and their chemical constitution', *Marine Geology*, vol. 23, nos 1–2 (1977), p. M22.
9 G. P. Glasby, 'The role of submarine volcanism in controlling the genesis of marine manganese nodules', *Oceanography and Marine Biology: An Annual Review*, vol. 11 (1973), p. 37.
10 Rowland, 'Non-energy marine mineral resources', p. 14.
11 J. Murray and A. F. Renard, 'Report on deep-sea deposits based on specimens collected during the voyage of H.M.S. *Challenger*, 1872–1876', in C. W. Thomson (ed.), *Report on the Scientific Results of the Voyage of H.M.S. 'Challenger'*, vol. 5 (Eyre & Spottiswoode, London, 1891), pp. 1–525.
12 J. Murray, 'On the distribution of volcanic debris over the floor of the ocean and its character, source and some of the products of its disintegration and decomposition', *Proceedings of the Royal Society of Edinburgh*, vol. 9 (Royal Society, Edinburgh, 1876), pp. 247–61.
13 See J. Murray and R. Irvine, 'On the manganese oxides and manganese nodules in marine deposits', *Transactions: Royal Society of Edinburgh*, vol. 37 (Royal Society, Edinburgh, 1894), pp. 721–42.
14 E. D. Goldberg and G. O. S. Arrhenius, 'Chemistry of Pacific Ocean pelagic sediments', *Geochimica et Cosmochimica Acta*, vol. 13, nos 2–3 (1958), see especially pp. 194–201.

15 E. Bonatti and Y. R. Nayudu, 'The origin of manganese nodules on the Ocean Floor', *American Journal of Science*, vol. 263, no. 1 (1965), p. 17.

16 J. Greenslate, 'Micro-organisms in the construction of manganese nodules', *Nature*, vol. 249, no. 5,453 (1974), pp. 181–3.

17 H. L. Ehrlich, 'The role of microbes in manganese nodule genesis and degradation', in D. R. Horn (ed.) *Papers from a Conference on Ferromanganese Deposits on the Ocean Floor* (National Science Foundation, Washington, DC, 1972), pp. 63–70.

18 Rabb, 'Physical and chemical features of Pacific deep sea manganese nodules', p. 35.

19 P. Halbach and R. Fellerer, 'The metallic minerals of the Pacific seafloor', *GeoJournal*, vol. 4, no. 5 (1980), pp. 413–14.

20 Haynes, Law, and Barron, 'An elemental description of Pacific manganese nodules', pp. 242–3.

21 D. Z. Piper, J. R. Basler, and J. L. Bischoff, 'Oxidation state of marine manganese nodules', *Geochimica et Cosmochimica Acta*, vol. 48, no. 11 (1984), pp. 2,347–55.

22 Letter: F. T. Manheim, Geologist, Office of Energy and Marine Geology, Atlantic-Gulf Branch, USGS, Woods Hole, MA, 1 Nov. 1987.

23 Haynes, Law, and Barron, 'An elemental description of Pacific manganese nodules', p. 243.

24 See B. W. Haynes *et al.*, *Laboratory Processing and Characterization of Waste Materials from Manganese Nodules*, USBM Report of Investigations 8938 (USBM, Washington, DC, 1985).

25 V. E. McKelvey, *Subsea Mineral Resources*, Chapter A of USGS Bulletin 1689 (USGPO, Washington, DC, 1986) p. 56.

26 Manheim, 'Marine cobalt resources', p. 605.

27 D. Z. Piper and M. E. Williamson, 'Composition of Pacific Ocean ferromanganese nodules', *Marine Geology*, vol. 23, no. 4 (1977), p. 297.

28 See, for example, J. Z. Frazer and M. B. Fisk, 'Geological factors related to characteristics of sea-floor manganese nodule deposits', *Deep-Sea Research*, vol. 28a (1980), pp. 1,533–51.

29 V. E. McKelvey, N. A. Wright, and R. W. Bowen, *Analysis of the World Distribution of Metal-Rich Subsea Manganese Nodules*, USGS Circular 886 (USGS, Arlington, VA, 1983), p. 19.

30 ibid., p. 14.

31 D. R. Horn, B. M. Horn, and M. M. Delach, *Ferromanganese Deposits of the North Pacific Ocean*, Technical Report no. 1, NSF GX-33616 (National Science Foundation, Washington, DC, 1972), p. 6 and accompanying map no. 3.

32 The lysocline is that depth at which 'shells of Foraminifera exhibit a sharp increase in dissolution rate'; see Piper and Williamson, 'Composition of Pacific Ocean ferromanganese nodules', p. 297.

33 See D. S. Cronan, *Underwater Minerals*, (Academic Press, London, 1980), pp. 137–40.

34 Piper and Williamson, 'Composition of Pacific Ocean ferromanganese nodules', pp. 299–301.

35 P. Halbach and F. T. Manheim, 'Potential of cobalt and other metals of ferromanganese crusts on seamounts of the Central Pacific Basin', *Marine Mining*, vol. 4, no. 4 (1984), pp. 325, 331.

36 Manheim, 'Marine cobalt resources', p. 605.

37 Manheim, Letter, 1 Nov. 1987.

38 McKelvey, Wright, and Bowen, *Analysis of the World Distribution of Metal-Rich Subsea Manganese Nodules*, p. 18.

39 J. Z. Frazer, 'Manganese nodule reserves: an updated estimate', Marine

Mining, vol. 1, nos 1/2 (1977), p. 105; McKelvey, *Subsea Mineral Resources*, p. 30.
40 See, for example, T. Burroughs, 'Ocean mining: boom or bust?', *Technology Review*, vol. 87, no. 3 (1984), pp. 54–60.
41 Manheim, 'Marine cobalt resources', p. 608.
42 For a useful bibliography of these early studies, examine C. J. Johnson *et al.*, *Resource Assessment of Cobalt-Rich Ferromanganese Crusts in the Hawaiian Archipelago*, Report of the East-West Center and the Research Corporation of the University of Hawaii for the Minerals Management Service, DOI, Co-operative Agreement no. 14–12–0001–30177, 'Environmental studies in support of the proposed leasing for cobalt-rich crusts' (Resource Systems Institute, East-West Center, Honolulu, HA, May 1985), pp. 130–5.
43 Manheim, 'Marine cobalt resources', p. 601; see also J. R. Hein *et al*, 'Geological and Geochemical data for seamounts and associated ferromanganese crusts in and near the Hawaiian, Johnston Island and Palmyra Island Exclusive Economic Zones', USGS Open File Report 85-292 [assumed location only] (USGS, Washington, DC, 1985).
44 Manheim, Letter, 1 Nov. 1987.
45 Manheim, 'Marine cobalt resources', p. 602.
46 ibid., p. 601.
47 P. Halbach, M. Segl, D. Puteanus, and A. Mangini, 'Relationships between co-fluxes and growth rates in ferromanganese deposits from Central Pacific seamount areas', *Nature*, vol. 304, no. 5,928 (1983), pp. 716–19.
48 Manheim, 'Marine cobalt resources', p. 606; for a good discussion of hiatuses see von Stackelberg, 'Influence of hiatuses and volcanic ash rains on the origin of manganese nodules', especially pp. 302–4.
49 Rowland, 'Non-energy marine mineral resources', p. 15.
50 Manheim, 'Marine cobalt resources', pp. 601, 604.
51 J. R. Hein, L. A. Morgenson, D. A. Glague, and R. A. Koski, 'Cobalt-rich ferromanganese crusts from the Exclusive Economic Zone of the United States and nodules from the Oceanic Pacific', in D. Scholl, A. Grantz, and J. Vedder (eds), *Geology and Resource Potential of the Continental Margins of Western North America and Adjacent Ocean Basins*, American Association of Petroleum Geologists, Memoir 1986 (in press).
52 Manheim, 'Marine cobalt resources', p. 603.
53 ibid., p. 605
54 Johnson *et al.*, *Resource Assessment of Cobalt-Rich Ferromanganese Crusts*, pp. 14–16.
55 P. A. Rona, 'Potential mineral and energy resources at submerged plate boundaries', *Marine Technology Society Journal*, vol. 19, no. 4 (1985), p. 22.
56 ibid.
57 J. M. Edmond, 'The geochemistry of Ridge Crest Hot Springs', *Oceanus*, vol. 27, no. 3 (1984), p. 18.
58 R. D. Ballard, 'The exploits of *Alvin* and *Angus*: exploring the East Pacific Rise', *Oceans*, vol. 27, no. 3 (1984), p. 11.
59 P. A. Rona, 'Hydrothermal mineralization at slow-spreading centers: Red Sea, Atlantic Ocean, and Indian Ocean', *Marine Mining*, vol. 5, no. 2 (1985), pp. 117–18.
60 P. A. Rona *et al.* 'Black smokers, massive sulphides and vent biota at the Mid-Atlantic Ridge', *Nature*, vol. 321, no. 6,065 (1986), pp. 33–7.
61 P. A. Rona, 'Hydrothermal mineralization at slow-spreading centers: the Atlantic model', *Marine Mining*, vol. 6, no. 1 (1987), pp. 1, 6.
62 Edmond, 'The geochemistry of Ridge Crest Hot Springs', p. 19.
63 Rona, 'Potential mineral and energy resources', p. 18.
64 ibid., p. 19.

65 Interview: M. Chirinos Garcia, Director General, Ministry of Energy and Mines, Lima, Peru, 7 Aug. 1986.
66 Rona, 'Potential mineral and energy resources', pp. 20–1.
67 N. Wetzel, J. L. Ritchey, and S. A. Stebbins, 'A strategic mineral assessment of an onshore analog of a mid-oceanic polymetallic sulfide deposit', *Minerals and Materials: A Bimonthly Survey* (June/July 1987), pp. 6–13.

Chapter five

Technology and economics of deep seabed minerals

For more than two decades scientists and engineers have worked to locate and evaluate deep seabed minerals. They have also developed prototype systems for extracting these minerals and studied seabed mining's potential environmental consequences. Most efforts have focused on ferromanganese nodules, but in the late 1960s and early 1970s, researchers expanded their interests to include the Red Sea's polymetallic sulphide muds. And in the early 1980s they began to give attention to ferromanganese crusts and massive polymetallic sulphides. Many of the scientific findings and exploration and processing techniques developed have utility for each of these mineral deposit types (as well as hydrocarbons), but some are usable in only one mining type. Whatever the research effort – exploration, extraction, or beneficiation – progress is slow and costs are high. Because present metal markets are so poor, managers in industry face tough decisions about proceeding with research and development programmes. However, those consortia receiving governmental subsidies have the advantage of pushing forward while their counterparts in several other countries remain on hold.

Exploration

Because deep seabed minerals lie at great depths and vary significantly in areal density and mineral content, mapping and evaluating their economic potential is difficult. A total darkness, the great hydrostatic pressure, the resistance created by the viscosity of the sea-water overburden when towing objects, and an often irregular topography contribute to difficulties in deep seabed exploration. Manganese nodules present an added problem; they are a two-dimensional and usually a single-layer ore body.

For every unit of ore obtained in manganese nodules a much larger extraction area is needed than that required for most onshore deposits. This problem accounts for the very large areas contained in the exploration leases granted by the US, the UK, and the FRG; for example, OMI's lease (136,000km^2) is equivalent to the total area of Greece. The various physical constraints of the seabed and mineral deposits demand special exploration and sampling technologies. The technologies in use or under development vary from state-of-the-art to antique (Figure 5.1).

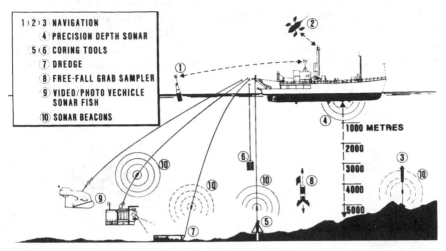

Figure 5.1 Marine survey and exploration technologies

Source: C. G. Welling, 'The Future of U. S. seabed mining: an industry view', *Mining Congress Journal*, vol. 68, no. 11 (1982), p. 21. With permission.

Acoustical devices

Because the seabed's topography and geology are important to seabed minerals' exploitability, careful surveys must be made. Acoustical methods such as echo-sounders operating at frequencies between 3 and 35 kHz provide seabed and subsurface profiles. A 3.5 kHz profiler can penetrate the upper 50m of sediments and log irregularities such as subsurface intrusions and surface extrusions of basalt.

Bottom-towed side-scan sonars provide information on obstacles and nodule coverage.[1] An echo-sounding system called Seabeam, different from most echo sounders, has sixteen narrow beams, each with an angle of 1.66°. This structure allows much more detailed topographic data collection than do conventional methods which use one broad beam of 30°.[2] Another system, GLORIA, towed 50m below the sea surface, maps the ocean floor in swaths of 14, 30, or 60km (Figure 5.2).[3] Other acoustical devices include pinger-transponder systems that – when lowered to the seabed – measure salinity, pressure, and temperature. Some acoustical systems can also activate various types of seabed equipment through coded signals from a surface craft.[4]

Optical systems

To make reasonably accurate estimates of seabed minerals analysts must have detailed sample information on their type, volume, density of distribution, and the amount of sediments covering them.[5] Various large, still cameras with wide-angle lenses and a capacity for hundreds of frames are used, as are small, flashlight cameras capable of only one shot; such cameras can be controlled via mechanical or sonar-electrical systems. These units, combined with 'boomerang' sampling devices, can take a photograph and then return automatically to the surface. In the last stages of prospecting surveys, special deep-sea television systems may be used.[6]

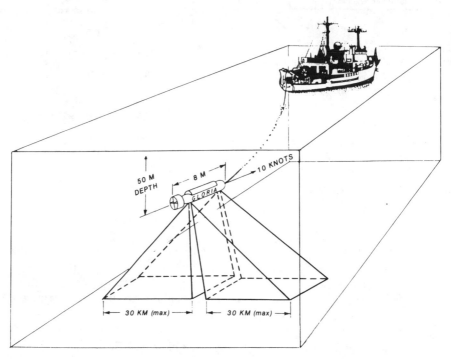

Figure 5.2 The GLORIA sidescan sonar 'fish' system

Source: B. A. McGregor, and M. Lockwood, *Mapping and Research in the Exclusive Economic Zone* (USGS, Reston, VA, 1985), p. 8.
Note: The GLORIA sidescan sonar 'fish' system has 30 transducers along each side, weighs 2 tonnes and is 8m long. The resulting sonographs, resembling radar images, are computer enhanced and presented in a mosaic. The final result is similar to an aerial photograph.

Navigation systems

Precise positioning capabilities are imperative, especially in the final stages of survey programmes. Satellite navigation (combined with Loran C and Doppler-sonar) help meet these needs. This technology is supplemented by radar-buoy systems, made up of radar and transponder units. When free-fall cameras and boomerang samplers are fitted with lamps and radios, surface craft can retrieve them.[7]

Other devices and techniques

Special instruments, such as the bathysonde, provide vertical profile measurements of the water column's pressure, temperature, and salinity. Anchored meters measure currents, and after completing their measurements, surface-craft personnel use sonar-transponders to trigger the current meters' release for retrieval at the surface. To obtain bottom samples, geologists use various grabs, dredge buckets, and corers. Seabed grab samples taken at intervals can be analysed by employing a Californium-252 radiation source. The process works by measuring gamma radiation given off by seabed materials after activation by neutrons.[8]

Engineers are now developing a device capable of working in water depths of 4,500m that can take deep core-drill samples (as deep as 53m), measure sediment strength and recover sediment samples.[9] Development of a dependable, portable seabed core drill is vitally important. Core drilling can be done from a surface drilling vessel, but it is very expensive.[10] Recently Canada successfully used a remotely controlled core drill in waters of 3,500m. Marine geologists employ these core drills to obtain mid-ocean ridge polymetallic sulphide samples.[11]

Remotely controlled and manned submersible vehicles provide significant help in deep ocean exploration. But most are designed to work in shallower waters.

Ferromanganese nodules

To date, the manganese nodules have been the focus of a large portion of deep seabed mining and processing research. The mining industry has made significant progress in designing usable manganese nodule extraction and processing methods and equipment but most systems are yet in the prototype stage.

Mining

Early efforts to develop nodule mining technology centred on continuous-line bucket (CLB) systems, using one or two vessels to tow the cable (Figure 5.3a). One now-defunct consortium, the CLB Group, had nearly twenty companies; its main purpose was to develop this one system of technology. But the system did not work well. Tests in 1973 showed that, at best, buckets contained only a 25 per cent load of nodules, whereas many buckets were filled with mud. Other problems included cable entanglements and difficulties in keeping the cable on the seabed.[12] One early nodule retrieval research programme focused on developing battery-powered and remotely controlled surface-to-seabed shuttle vehicles. AFERNOD, now a part of the contemporary French GEMONOD consortium, did this work, which currently seems to have been abandoned (Figure 5.3d)

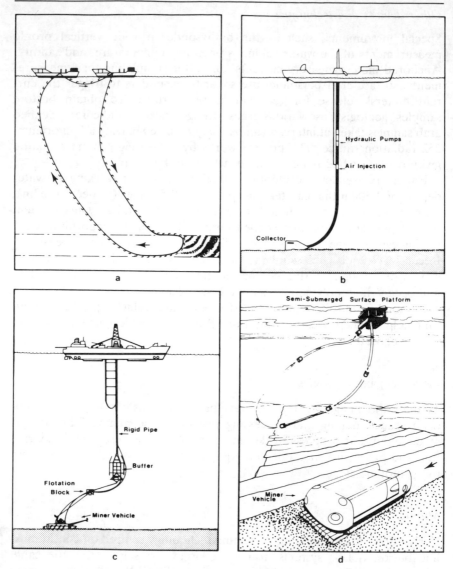

Figure 5.3 Seabed ferromanganese nodule mining systems

Source: R. W. Knecht, 'Deep seabed ocean mining', *Oceanus*, vol. 25, no. 3 (1982), p. 6. With permission.
Note: The systems include continuous-line-bucket dredging (3a), hydraulic and pneumatic lifting with a sweeper-collector or crawler-miner vehicle (3b) and (3c), and a futuristic, remotely controlled shuttle miner (3d).

As of 1987, most engineering efforts are focused on towed hydraulic and pneumatic dredging systems. These systems use a collector to sweep or syphon nodules into a lifting pipe that carries them to the surface by pumping (hydraulic) and bubbling (pneumatic) systems (Figures 5.3b and 5.3c). Problems with these methods include (1) the drag created by the

Figure 5.4 Ocean Minerals Company's test miner

Source: Courtesy Ocean Minerals Company. With permission.
Note: Ocean Minerals Company's one-tenth scale (9-m-wide/14-m-long) test miner required 17 years to develop. Propelled by Archimedian screws, the 100-tonne crawler-miner picks up nodules and pumps them as a slurry to a buffer which neutralises oscillations created by the miner (see Figure 5.3c), ship, and pipe's motions. The buffer sends the nodules on to the pipe-string for pumping to the mother ship. All moving, mining, and pumping operations are computerised.

Figure 5.5 Dynamically positioned *Glomar Explorer*

Source: Courtesy Ocean Minerals Company. With permission.
Note: From the 45,000-tonne/25,000-horse-power and dynamically positioned *Glomar Explorer*, crewmen lower a crawler-miner and a 5km pipe-string (see Figure 5.3c) to the seabed via a 'moonpool' below the derricc. The entire mining system cost about US$1,00 million to develop; a full-scale system will probably require US$1,000 million and have the dimensions of a super-tanker.

lifting pipe; (2) the motions of both ship and pipe that affect the collector's efficiency; and (3) the difficulty of maintaining a proper balance of nodule, water and air flows to the surface.[13] Pneumatic systems are simpler than the hydraulic systems, but require more energy to operate and a larger pipe to handle the nodules. One consortium, OMCO, has worked extensively with a self-propelled collector (Figure 5.4). This system has been tested (using a 5.5km-long pipe) from the dynamically positioned mining ship *Glomar Explorer* (Figure 5.5).[14]

Many once active nodule mining technology research programmes have been on hold since the early 1980s. In contrast, Japan is pushing rapidly ahead. The national government in 1982 passed legislation that established the Technology Research Association of Manganese Nodules Mining System. This organisation, with its function explicitly stated in its name, co-operates with twenty major Japanese corporations and also includes the Agency of Industrial Science and Technology (AIST), a unit of the national government.

Processing

Nodule beneficiation may be done on board ship or onshore. At-sea processing will probably be more expensive and difficult for handling nodule and beneficiation chemical wastes. Numerous consortia engineering programmes have developed nodule processing systems. The techniques depend on two basic methods – leaching (hydrometallurgy) and roasting/smelting (pyrometallurgy), or combinations of the two.[15] For example, in a system developed by OMI, electric furnace pyrometallurgy first smelts dried nodules, which produces an alloy of copper-cobalt-nickel (along with some iron), while rejecting most of the manganese, some iron and gange. The alloy is converted into a matte where more iron is removed from the converter slag. Some of the smelter slag, containing significant amounts of manganese, can be reduced into ferromanganese via another electric furnace. From this point on, hydrometallurgy (pressure sulphuric acid leaching, solvent extraction, and electro-extraction) processes remove the base metals of cobalt, copper, and nickel (Figure 5.6).[16] This method's feasibility has not been fully proven, because OMI's research programme has been shelved.[17]

Economics of nodule production

Because the main minerals (copper, nickel, cobalt, and manganese) contained in nodules are some of those most in demand by industry and for strategic needs, extraction technology, and economic analyses have focused on them. Numerous investigators have done detailed studies of hypothetical nodule-producing establishments to determine their economic competitiveness.[18] According to several studies, nodule enterprises are likely to be highly sensitive to metal prices and to energy and

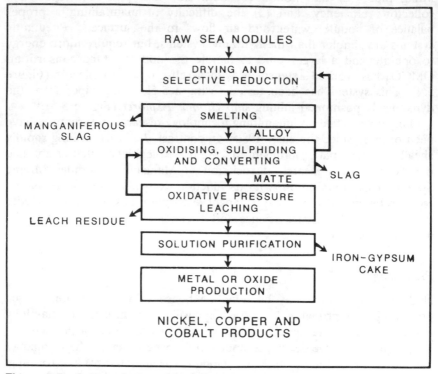

Figure 5.6 Schematic flow diagram for pyrometallurgical processing and recovery of copper, nickel, and cobalt from ferromanganese nodules

Source: R. Sridhar, W. E. Jones, and J. S. Warner, 'Extraction of copper, nickel, and cobalt from sea nodules', *Journal of Metals*, vol. 28, no. 4 (1976), p. 33. With permission.

capital costs. Political stability and investment climate will be important locational factors.[19]

Researchers at Texas A&M University (TA&M), under the leadership of John Flipse, have refined an ocean mining cost model, first developed at Massachusetts Institute of Technology. In 1982 Flipse completed a computer modelling study designed to (1) document the model's capacity to estimate seabed mining and operating costs and (2) provide NOAA's Marine Minerals Division with a capability of evaluating the effect of the DSHMRA on mining consortia profits. A later model refinement considered the effects of the US Economic Recovery Act of 1981 on mining costs. The analysis assumed a vertically integrated firm (including research, mining, and processing); a three-metal plant (copper, nickel, and cobalt) with a 3-million-tonne annual capacity; a gross investment of nearly $1,500 million; and a fixed capital investment of $1,000 million. The venture also was assumed to have

1 a lifetime of twenty-six years
2 a mine site in the Clarion-Clipperton zone;

3 an hydraulic mining system, requiring two ships;
4 three bulk carriers to move nodules from the mine site to the processing plant;
5 ore-unloading facility on the US West Coast, although Hawaii or Mexico could be used;
6 reduction/ammoniacal leaching process plant located some distance from the unloading site;
7 nodule feed with metal values of manganese (30–40 per cent); iron (8–10 per cent); copper (1.0–1.5 per cent); nickel (1.0–1.5 per cent); cobalt (0.1–0.5 per cent); and trace elements (10–20 per cent), as well as silica;
8 onshore waste disposal site at a considerable distance from the processing plant.[20]

The analysis projected that mining, transportation, processing, and waste dispoal operations would demand more than a $1,000 million investment in equipment, vehicles, processing plant and other facilities and would require an annual operating budget of over $200 million (Table 5.1), as well as $175 million in working capital. Anticipated annual revenues were put at nearly $423 million.[21]

Table 5.1 Estimated capital investment and annual operating costs for a three-metal/three-million-tonne nodule mining operation (millions of 1980 US $)

Expenditures	Capital Investment	Annual Operating Cost
Continuing preparations	—	6.0
Mining	294.7	68.6
Ore marine transportation	174.5	20.9
Ore marine terminal	30.0	2.7
Onshore transportation	39.7	7.5
Processing	458.2	100.1
Onshore waste disposal	22.8	6.9
Additional support	1.3	15.9
Regulatory	—	—
Total	1,021.2	228.6

Source: J. E. Flipse, *An Economic Analysis of a Pioneer Deep Ocean Mining Venture*, TAMU-SGT-82-201, COE Report no. 262 (Sea Grant College Program, Texas A&M University, College Station, TX, Aug. 1982), p. 42.

Those in industry expect processing to be the most costly phase of nodule production, with energy the paramount input. The TA&M study documents this expectation. Annual energy costs (coal and electricity), totalled nearly $46.2 million, slightly more than 46 per cent of the plant's total estimated direct operating costs (Table 5.2). The before-tax profit on the fixed capital investment of just over $1,000 million would be about $180 million. The after-tax profit would be less than $100 million, an internal rate of return (IRR) for twenty-six years of only 7 per cent. According to Flipse, at this rate of return (which must be viewed only as indicative), 'it is unlikely that ocean mining will be undertaken using the system defined ... unless a critical feedstock for a company's major product is produced or a national need for a strategic metal develops'.[22] A similar TA&M study done for a four-metal (including manganese) pioneer plant also showed an unfavourable IRR, in this case only 6.4 per cent after taxes.[23]

Table 5.2 Estimated annual direct operating cost for a three-metal/three-million-tonne nodule processing plant

Expenditures	Annual cost (thousands of 1980 US $)
Utilities and fuel	
Coal	34,875
Power	11,280
Other	1,370
Labour	16,530
Capital related charges	32,074
Materials and supplies	3,991
Total estimated cost	100,119

Source: J. E. Flipse, *An Economic Analysis of a Pioneer Deep Ocean Mining Venture*, TAMU-SG-82-201, COE Report no. 262 (Sea Grant College Program, Texas A&M University, College Station, TX, Aug. 1982), pp. 32–3.

Charles Johnson and James Otto, at the Resources Systems Institute of the East-West Center in Hawaii, have examined basic components in locational decision-making for processing plants. They believe these decisions will prove fundamental for a plant's ability to compete with onshore producers of the Big Four metals.

Because nodule processing plants are capital- and energy-intensive and not labour-intensive, advantages for locating in low-cost labour areas are limited. Johnson and Otto examined comparative advantages of locating a nodule processing plant in seven countries bordering the Pacific Ocean, including both developed countries and LDCs.[24] Using a base case for comparison and measuring only actual costs, Canada proved to have the best relative overall IRR and Hawaii the worst (Table 5.3). Canada's electrical energy costs, for example, are 2.5 times less than those for Hawaii and three times less than for Fiji (Table 5.4). But if Hawaii's geothermal energy could be harnessed to process nodules, its relative IRR position might be enhanced.[25] If adjustments for political and economic risks are injected into the IRR equation, some rankings could change considerably. The spring 1987 *coup d'état* in Fiji and the continuing unrest within the Philippines probably would make them less favoured than Hawaii, and could discourage decision-makers in considering them. If a given country has market incentives such as subsidies or if it presents little danger of expropriation, the minimum IRR necessary to attract investors might be lowered.

Table 5.3 Impact of different processing site locations on nodule project economics

Processing plant location	Internal rate of return (%)
Canada (Prince Rupert)	14.6
Colombia (Bahia Solano)	13.9
Australia (Gladstone)	13.1
Ecuador (Manta or Esmeraldas)	12.4
Philippines (Leyte or Mindanao)	11.2
Fiji (Nomosi Area or Savusavu)	11.1
Hawaii (Island of Hawaii)	10.8

Source: C. J. Johnson, *Economic and Business Investment Climates for Manganese Nodule Processing in Six Pacific Countries*, prepared for DOC Institutional Grant no. NO81AA-D-00070 (University of Hawaii Sea Grant Program, Honolulu, HA, 1985), pp. 50, 58, 74, 88, 97, 107, and as abstracted from the entire document.

Cobalt-rich crusts

It might appear that the cobalt-rich crusts are 'there for the taking'. In reality, the task is not so simple. Before exploitation may begin, the crusts' characteristics and distribution must be more carefully defined, and because much of the technology applicable to nodule mining will not work in crust extraction, new mining systems must be developed and tested. In addition, crusts are firmly attached to hard substrates, are often discontinuous and many times occur in areas with a relatively irregular topography. Once again, looming beyond the engineering difficulties, prospective producers face a currently depressed metal market. Cobalt's onshore 'resource life expectancy' has been calculated at 340 years and, as was noted earlier, in the mid-1980s world capacity to produce cobalt will exceed demand (see Table 1.1).[27] Thus, one might ask: 'Why bother?'

The answer to this question is fourfold:

1 Cobalt-rich crusts also contain numerous other minerals of strategic value;
2 several major cobalt supply areas are located in politically unstable countries;
3 crusts often occur within the 200-nmi EEZs of consuming countries;
4 for those countries dependent on imports, domestic production could help reduce balance-of-payments deficits.

Although some cobalt producers such as Canada and Zaïre may not welcome more cobalt on to the world market, the US (the world's largest consumer and importer) and others could benefit, especially if they have secure access to it, as in their EEZs.

The United States: a case study

With the establishment of its 200-nmi EEZ, the US acquired or gained access to a vast oceanic area – especially in the Pacific Ocean – surrounding

Table 5.4 Project costs for seven processing sites (millions of constant 1984 US $)[a]

Costing categories	Base case	Australia	Canada	Colombia	Ecuador	Fiji	Hawaii	Philippines
Capital Cost[b] + 10% working capital	1,485	1,485	1,485	1,635	1,635	1,635	1,560	1,560
Total operating cost/year	443	446	421	423	432	504	492	503
Mining	70	70	70	70	70	70	70	70
Transport (nodules)	40	45	28	30	30	36	25	50
Transport (metals)	30	35	30	20	20	35	35	30
Total processing plant/year	303	291	288	298	307	358	357	348
Energy: Coal	46	44	48	49	50	50	55	51
Coke	64	64	64	67	67	67	67	67
Oil	12	12	12	12	12	12	12	12
Electricity	54	48	40	60	68	118	100	111
Non-fuel materials	70	70	70	70	70	70	70	70
Water	2	2	2	2	2	2	2	2
Labour	25	26	27	13	13	14	26	10
Maintenance and insurance	30	30	30	30	30	30	30	30
Est. effective tax rate (%)	35	40	35	30	35	25	40	30
Loss carry forward	Yes	Yes	Yes	Yes	Yes	Yes	Yes	No
Depreciation (%)	20	20	20	20	10	20	20	20

Source: C. J. Johnson, *Economic and Business Investment Climates for Manganese Nodule Processing in Six Pacific Countries*, prepared for DOC Institutional Grant no. NO81AA-D-00070 (Sea Grant Program, University of Hawaii, Honolulu, HA, 1985) as abstracted from the entire document.
Notes: a. Cost estimates reflect data collected in each of the respective countries during 1984; b. Includes capital costs for mining, transport, and processing facilities.

its Hawaiian Islands, Johnston Atoll, and other island possessions. These waters have good potential for crusts with high values of cobalt and other minerals such as platinum (Table 5.5). One estimate notes that a 300-km^2 area with a 40 per cent surface coverage of accessible crusts could provide miners with about 3 million tonnes of crust for each 2cm of thickness.[27] The Hawaiian Islands and Johnston Atoll EEZs alone have an estimated 57,200km^2 of seamount target areas with a crust potential of 320 million tonnes. The US Pacific island areas have crusts containing enough cobalt (and manganese), at mid-1980s consumption rates, to last that country for cent ries. Most of its platinum needs could also be supplied.[28]

Table 5.5 Resource potential of cobalt, nickel, manganese, and platinum in seabed crusts of former and current US Trust and Affiliated Territories

Territory	Resource potential[a]			
	Cobalt	Nickel (Millions of Tonnes)	Manganese	Platinum (Thousands of Kilograms)[b]
Belau/Palau[c]	0.55	0.31	15.5	21.2
Guam	0.55	0.31	15.5	21.2
Howland-Baker	0.19	0.11	5.5	14.9
Jarvis	0.06	0.03	1.6	4.7
Johnston Atoll	1.38	0.69	41.6	108.9
Kingman-Palmyra	3.38	1.52	76.1	177.3
Marshall Islands[c]	10.55	5.49	281.3	668.7
Micronesia[c]	17.76	9.96	496.0	1,079.3
Northern Marianas	3.60	1.97	100.2	239.5
Samoa	0.03	0.01	0.8	1.2
Wake	0.98	0.51	26.8	62.2

Source: A. L. Clark, P. Humphrey, C. J. Johnson, and D. K. Pak, *Resource Assessment: Cobalt-Rich Manganese Crust Potential – Exclusive Economic Zones: US Trust and Affiliated Territories in the Pacific*, OCS Study MMS85-0006 (Minerals Management Service, DOI, Washington DC, 1985), p. 20.
Note: a. These are estimates of in-place resources and do not indicate either potential recoverability or mineable quantities; b. Converted and rounded from troy oz to kg, with 1kg equal to 32.15 troy oz; c. Former US Trust Territories; see Table 4.3, note a for explanation.

When it became clear that these seamount crusts represented a potentially rich mineral resource, the state of Hawaii and the US federal government organised a joint task force to prepare an environmental impact statement (EIS) for future crust extraction within the EEZ. US investigators, in a series of cruises, examined the entire axis of the Hawaiian Archipelago (HA), as well as Johnston Atoll (JA). FRG researchers also worked in waters surrounding Johnston Atoll (Figure 5.7). The researchers demonstrated that several crustal areas (I, G, and E) have high cobalt values (Table 5.6). Based on measurements of kg/m^2, the areas of greatest overall metals potential are in decreasing order – I, JA, G, H, C, B, A, F, E, and D (Figure 5.8).

Overall, the entire HA-JA study zone has considerable potential for metals, especially at depths between 800m and 2,400m. Those sub-areas designated for leasing contain an estimated 80.5 million tonnes of manganese, 1.6 million tonnes of nickel, and

Table 5.6 Resource potential summary of the Hawaiian Archipelago and Johnston Atoll Exclusive Economic Zones

Sub-area	Dry Weight (%)					Thousand Tonnes			
	Co	Ni	Mn	Fe	Crust[a]	Co	Ni	Mn	Fe
Included in initial offer									
A	0.471	0.287	18.01	19.55	26,854	126	77	4,836	5,250
B	0.551	0.280	19.27	20.01	37,413	206	105	7,209	7,486
C	0.630	0.330	18.65	18.70	14,110	89	47	2,631	2,639
G	0.892	0.504	24.14	15.20	10,834	97	55	2,615	1,647
H	0.648	0.332	20.16	18.91	17,487	113	58	3,525	3,307
I	0.964	0.483	25.90	15.61	25,313	244	122	6,556	3,951
Johnston Atoll	0.785	0.536	24.40	15.60	217,777	1,710	1,167	53,138	33,973
EEZ sub-total	—	—	—	—	349,788	2,585	1,631	80,510	58,253
Excluded from initial offer									
D	0.380	0.224	16.05	14.96	34,215	130	77	5,492	5,119
E	0.888	0.368	21.46	16.83	52,076	462	192	11,176	8,764
F	0.652	0.365	19.56	14.17	156,499	1,020	571	30,611	22,176
Johnston Atoll	0.785	0.536	24.40	15.60	14,016	110	75	3,420	2,186
EEZ sub-total	—	—	—	—	256,806	1,722	915	50,699	38,245

Source: Department of Planning and Economic Development, State of Hawaii, and Minerals Management Service, DOI, *Mining Development Scenario for Cobalt-Rich Manganese Crusts in the Exclusive Economic Zones of the Hawaiian Archipelago and Johnston Island*, Ocean Resources Branch Contribution no. 38, (DPED and MMS, Honolulu, HA, Jan. 1987), p. 7.
Note: a. Assumptions: mean crustal coverage in permissive area is 40 per cent in A, B, C, H, I, and Johnston Atoll; and 25 per cent in D, E, F, and G; crustal thickness: Areas A, B, C, H, I, and Johnston Atoll – 2.5cm; Area D – 0.5cm; Area E – 1.0cm; Area F – 1.5cm; Area G – 2.0cm.

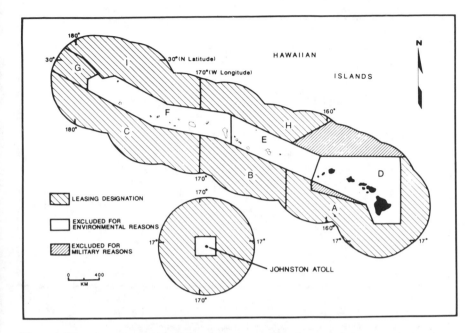

Figure 5.7 Hawaiian Island and Johnston Atoll EEZ study area

Source: Department of Planning and Economic Development, State of Hawaii, and Minerals Management Service, DOI, *Mining Development Scenario for Cobalt-Rich Manganese Crusts in the Exclusive Economic Zones of the Hawaiian Archipelago and Johnston Island*, Ocean Resources Branch Contribution no. 38 (DPED and MMS, Honolulu, HA, Jan. 1987), p. 6.

2.6 million tonnes of cobalt.[29] If only 10 per cent of the HA's crusts were exploitable, they would supply the US with enough cobalt and manganese to last for about twenty-five years, based on that country's estimated demands in the year 2000. Although these data appear highly favourable, decision-makers in both industry and government should be wary. Indeed, the authors of the resource assessment portion of the HA-JA study pointedly caution that their data are not complete enough to make predictions about the exploitability of the HA's ferromanganese crusts, and furthermore, their 'study is not intended to define specific mine sites'. They do suggest, however, that if exploitation does occur, it will be in off-axis areas and will likely require 'a number of individual crust deposits . . . to sustain a commercial mining operation' during the fifteen- to twenty-year period needed to recover a major mining investment.[30] The mining area required to support one twenty-year operation is likely to be about 400km^2 to 600km^2.[31]

The authors of the HA-JA report also emphasise that 'gross in-place metal cannot be equated with the possible recovered metal value or profit'.[32] For example, Figure 5.8 does not take into account roughness and size of the seamounts. Seamount mine sites with less than about

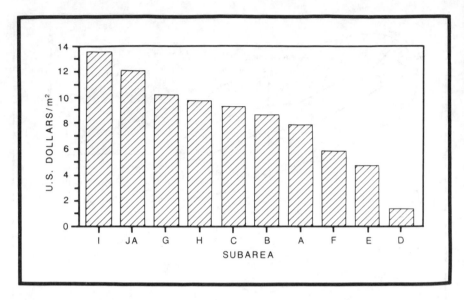

Figure 5.8 Estimated gross in-place value of cobalt-rich ferromanganese crusts

Source: Department of Planning and Economic Development, State of Hawaii, and Minerals
Management Service, DOI, *Mining Development Scenario for Cobalt-Rich Managanese Crusts in the
Exclusive Economic Zones of the Hawaiian Archipelago and Johnston Island*, Ocean Resources Branch
Contribution no. 38 (DPED and MMS, Honolulu, HA, Jan. 1987), p. 71.
Note: JA denotes Johnston Atoll's EEZ. Dollar values are based on the following assumptions: Co, Ni,
and Mn equal $22,000, $8,000, and $600/tonne (respectively) in 1985 US$.

200km[2] may be too small to encourage extraction and those with
especially rugged surfaces will be much more difficult and expensive to
mine.[33]

Once again, to assume the ferromanganese crusts will soon be added to
the world's metal supply would be naive. On the oher hand, we should
not dismiss their potential but rather push forward in developing mining
and processing techniques that can economically provide the world community with cobalt, and perhaps other minerals as well. This effort is
especially appropriate for countries having deposits within their 200-nmi
EEZs.

Crust mining and processing

Because crusts are attached to underlying rocks, they will be difficult to
extract. One conceptualised mining vehicle is self-propelled at a speed of
approximately 1km/hour. Articulated cutting devices would break up the
crust and avoid, when possible, removal of the substrate. Hydraulic
suction devices, rakes, or mechanical scrapers would recover the loosened
crust, to be fed into a gravity separator and then lifted to the surface

(Figure 5.9). After this rudimentary beneficiation and lifting to the mining vessel, the ore would be transferred (by ships) as a slurry to an onshore processing site. A CLB system has also been suggested for possible use in mining cobalt-rich crusts.[34]

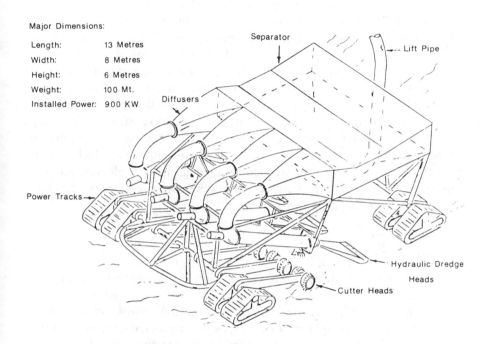

Major Dimensions:

Length:	13 Metres
Width:	8 Metres
Height:	6 Metres
Weight:	100 Mt.
Installed Power:	900 KW

Figure 5.9 Ferromanganese crust miner

Source: Minerals Management Service, DOI, *Draft Environmental Impact Statement: Proposed Marine Mineral Lease Sale in the Hawaiian Archipelago and Johnston Island Exclusive Economic Zones*, Ocean Resources Branch Contribution no. 40 (Department of Planning and Economic Development, Honolulu, HA, Jan. 1987), p. A26.

Crusts are thought of as primarily a cobalt ore, whereas nodules are considered a nickel ore, given the relative amounts of the metals in the two ore types. Crust processing techniques will probably be similar to those used for nodules. One recent study found that a sulphuric acid leach process extracted more than 90 per cent of the cobalt, nickel, and copper while rejecting most of the manganese and iron. A smelting process had an even higher recovery rate – 98 per cent for cobalt and nickel and 96 per cent for copper – while rejecting 18 per cent of the iron and 99 per cent of the manganese. The same analysis found that the slags and tailings produced were not toxic under the constraints of US government toxicity tests.[35]

Environmental and socio-economic impacts

Like all mining industries, deep seabed mineral production will have impacts on its immediate environment. Mining's impacts include alteration of the seabed topography, destruction of biota and their habitat, and damage to physical processes necessary to ecosystem maintenance. If processing occurs onshore – which is probable – local communities will be affected environmentally, economically, and socially.

From 1975 to 1981 a major US pilot-scale research effort – Project DOMES (Deep Ocean Mining Environmental Study) – focused on obtaining pre-mining data for the CCZ. Researchers looked at deep seabed nodule mining's potential effects on the oceanic environment. Much of this work, done under the auspices of NOAA, was directed to the photic zone where discharges associated with mining and ore processing may occur. Beginning in 1983 NOAA, in a co-operative effort with Scripps Institution of Oceanography and with OMA's service contractor (Deepsea Ventures), began a broad study of mining's effects on the benthic zone. In an effort to identify analogs from ecological recovery processes in deep-water benthic environments, investigators also have been examining shallow-water areas.[36]

Benthic disturbance

Seabed mineral extraction will badly damage the seabed's upper layer along the collector's path, especially in the first cm where most benthic life forms live. Whether mining occurs on abyssal plains, on seamounts, or in spreading zones, it may destroy as yet unstudied and undiscovered organisms. Nodule mining, for example, the main focus of past deep seabed environmental research, may destroy a major part of tiny nematode and polychaete worm populations, as well as test-forming protozoans (foraminifera).[37] Many scientists think that if miners leave undisturbed strips interspersed with mined areas, benthic organisms should successfully recolonise the disturbed habitat.[38] Studies begun by NOAA in 1983 are designed to determine the types and speed of organisms' recolonisation. OMA's 1978 collector-testing track was located and studied, using acoustically navigated box corers for sampling macrofauna and meiofauna.[39] If mining for polymetallic sulphides occurs near active volcanic vents in spreading zones, entire biotic communities could be damaged or destroyed. These systems are relatively unstudied, both for species present[40] and their interrelationships and for energy cycling which is dependent upon chemosynthesis.[41] It seems unlikely, however, that mining will occur near active vents, because enough inactive areas should be available and easier to exploit.

Biotic organisms will also be affected by the collector's disturbance of seabed sediments. Some may become entirely buried or have their food supplies covered. The sediments may not be fatal to mobile animals, but sessile organisms adapted to low particulate concentrations may not survive. Some species, such as the benthic clam, are very slow to mature sexually, which might jeopardise their survival. Most sediments should settle within a few hundred metres of the collector's track, but depending upon bottom currents and the type of sediments, some depositon will occur far from the mine site. Using pilot-scale mining test data, researchers in one study of a hypothetical seabed-collector plume suggest that sediments could be advected (assuming a velocity of 4cm/s) nearly 160km.[42]

Mortality of the benthic (and pelagic) biota raises the question of whether there will be an increased oxygen demand for degrading organic matter in pelagic and benthic waters and seabed sediments. One study indicates that the increased load of dead organic matter should not make a detectable demand on seabed oxygen supplies, but would be measurable in the uppermost 50m of the water column and within about 15km of the mining ship.[43]

Pelagic disturbance

If a CLB system is used to lift nodules or other seabed minerals directly through the water column, a large volume of sediments will be released into it. Hydraulic or pneumatic systems will discharge sediments at the surface. One study analysed sediment (turbidity) plume data obtained during prototype mining tests done by OMI in 1978 in the North Pacific. From the data, researchers extrapolated that discharged wastes in a full-scale mining operation could create a sediment plume of up to 100km with a width of 10 to 20km, but would cause substantial reductions in light levels only for 80 to 100 hours. The authors, J. Lavelle and E. Ozturgut, stress that their findings are tentative and do not establish that this reduction in light would or would not affect phytoplankton.[44] In an earlier study, however, Ozturgut *et al.* note that a

> reduction of light caused by an increase in suspended particulate matter ... in the upper layer will result in a decrease in primary productivity. This reduction may be about 80% at 5km distance from the mining ship along the plumè axis.

Although sediments in the photic zone may decrease photosynthesis, these same nutrient-rich materials could contribute to an increase in productivity (assuming adequate light is available), especially if 'the introduced nutrients remain in small permanent gyres (50 to 100km in diameter) which may be present in the area'.[45]

If nodule processing occurs onboard ship and the tailings are dumped overboard, even more sediments would be introduced into the photic zone, although these could be pumped into deeper waters. Studies done in the 1970s indicated that the sediment plume should not be a major problem to photosynthesis processes. More recent investigations, however, indicate that these tailings may be a hazard because of the toxic trace elements released. When mining begins, these trace elements (lead, copper, nickel, cadmium, and aluminium, among others) should be monitored to determine if they are building up in nearby biota – zooplankton, small fishes, and larger necton such as sharks and tunas.[46]

Onshore waste disposal

Processing manganese nodules and crusts will generate considerable waste, both in nodule and crust residues and in processing chemicals. The waste could total from 70 to 97 per cent of the material processed,[47] depending upon whether manganese is extracted. A two-million-tonne, three-metal nodule processing plant each year could produce enough waste to cover nearly 250ha of land to a compacted depth of one metre.[48] Most manganese nodule and crust specialists in industry and government expect processing to occur onshore, although wastes could be pumped as a slurry or hauled by barge to deep offshore areas, where they cannot affect ocean surface waters and onshore aquifers.[49]

Three main types of waste will occur, depending on the processing techniques used. These include

(1) leached tailings from three-metal hydrometallurgical processes and lime boil solids, if produced; (2) slag from smelting and from silico- or ferromanganese produced when manganese is recovered from tailings; and (3) leached tailings from four-metal hydrometallurgical processes and residues from electrolytic manganese reduction steps.[50]

Several onshore disposal methods could be used – backfilling of mines, injecting of liquid tailings into deep wells and constructing specially designed impoundments and excavated pits; the latter technique is the most likely method to be used. Slag from smelting should be inert (chemically bound) and cause little problem and could be used as a construction aggregate.[51] The main concerns have been possible aquifer contamination and wildlife exposure to toxic metals and chemicals in the waste water and sediments.[52] A 1985 USBM study, however, shows that nodule tailings will not exceed the allowable limit of the eight regulated heavy metals – arsenic, barium, cadmium, chromium, mercury, lead, selenium, and silver. Researchers reached this conclusion after using the Environmental Protection Agency's (EPA) EP (extraction procedure) test mandated under the US Resource Conservation and Recovery Act.[53] One study has demonstrated that usable maize can be grown on these tailings, if adequate phosphorus is added to them.[54]

114

Socio-economic impacts

Because the State of Hawaii (especially its Hawaii County) could be the site for a ferromanganese nodule and/or crust plant(s), studies have been done to evaluate the consequences of an offshore mining enterprise and onshore processing facility on the socio-economic conditions in surrounding communities and the region. Estimates put direct new job employment for a crust mining and processing enterprise (1 million tonnes annual capacity) at 600 for the processing facility, 105 for materials handling and miscellaneous functions, and 280 for ship crews. Indirect employment of perhaps 2,180 workers would put the total jobs generated at about 3,165. Annually workers would receive an estimated $87 million in household income, the state's businesses $550 million in retail and wholesale purchases, and the state tens of millions in taxes.[55]

The construction phase could require 2,000 workers. When construction is completed there may be some problems of unemployment, especially so because construction of the processing plant will likely be in a rural area. The result could be a 'boom and bust' economic cycle often associated with large development projects.[56]

When a major industrial enterprise is introduced into a rural area, severe strains often occur in the area's life-style, with some disruption of traditional values and pace of life. In Hawaii many residents still practise subsistence farming and fishing, have extended families, and put less emphasis on material goods. A large influx of workers from outside the area could increase ethnic tensions, create demands for new homes and other buildings, and add to infrastructural needs and social services. These changes could have a negative impact on the tourist business, if the charm of the rural countryside is compromised.[57] Changes described here could apply to onshore activities of various types of seabed mining enterprises in nearly all areas proposed as processing sites where rural conditions prevail.[58]

Polymetallic sulphides in the Red Sea

Many marine minerals specialists believe that some polymetallic sulphides will be extracted before the ferromanganese crusts and nodules. Where polymetallic sulphides occur as loosely consolidated sediments, they should be easier to exploit than the massive deposits associated with venting systems. One enterprise has done full-scale testing of mining, processing, and tailings disposal equipment in a spreading centre – the Red Sea Rift – where it expects to mine. If production begins here, it will be the world's first truly deep seabed mining establishment.

In the 1950s scientists discovered temperature and salinity anomalies in the Red Sea; this finding led to further exploration in the 1960s and to the discovery of metalliferous muds within numerous deeps.[59] Of the seventeen deeps identified, ten contain metalliferous sediments;[60] the most

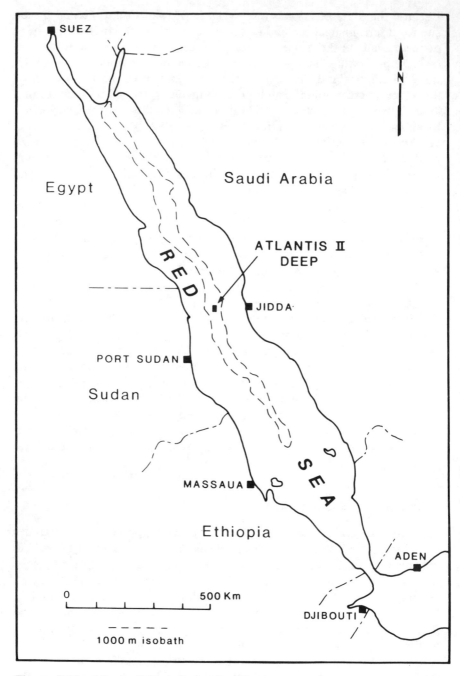

Figure 5.10 Atlantis II Deep in the Red Sea

Source: After H. Amann, 'Development of ocean mining in the Red Sea', *Marine Mining*, vol. 5, no. 2 (1985), p. 104. With permission, Taylor & Francis.

potentially productive is the Atlantis II Deep (Figure 5.10). It is also the world's largest hydrothermal mineral deposit known to occur along the world's sea-floor spreading-centre system. The sediments measure 10–20m deep witin a 5-km-wide and 13-km-long area of about 56 million m^2.[61] The sediments contain an estimated 32 million tonnes of metal and are about 35m thick at the main mine site,[62] (see Figure 4.10). Of this total (in dry weight) iron measures 29 per cent, zinc 1.5 per cent, copper 0.8 per cent, and lead 0.1 per cent. The deposit also contains about 54ppm of silver and 0.5ppm of gold.[63] One estimate made in 1986 (using 1981 world market prices) put the in-place value of the Atlantis II Deep sediments at from \$85 to \$348/tonne.[64]

Recognising the potential importance of the Red Sea's mineral resources, Saudi Arabia and the Sudan in 1974 negotiated a treaty that established a joint research and development programme for waters deeper than 1,000m, to be administered through the Saudi-Sudanese Red Sea Commission (see Figure 5.10). The two states previously had been disputing ownership and development rights. In 1976 the Commission awarded Preussag AG (a FRG firm that had already been active here for several years),[65] a contract to develop environmentally sound methods for mining and beneficiating the Atlantis II Deep's polymetallic sulphide ores.[66]

Using magnetometrical, seismical, and acoustical methods, along with extensive sampling, researchers developed detailed maps of the deposits. Although the Atlantis II sediments are very complex, lie at great depths, and are highly saline, Preussag seems to have overcome these problems. A cutterhead/suction system will probably be used to extract the ores, which have a consistency of shoe polish (Figure 5.11).[67] The sediment's fine-grained structure and high salt content requires a series of beneficiation steps, including flotation followed by chlorine and/or metal chloride treatment techniques.[68]

Preussag's detailed analyses, funded in part by the German Federal Ministry for Research and Technology, have demonstrated that miners can extract Atlantis II Deep deposits without major environmental damage; 50 per cent of the Red Sea Commission's development budget went to programmes designed to determine likely environmental impacts. Wastes should be amenable to redeposition via containment and/or controlled dilution in offshore sites. The graben itself will act as the containment structure, with the gangue being pumped down to about the 800m level in an area located away from the ore deposit. During experimental testing no surfacing of the sediment plume was detected.[69]

Just as the graben will be useful for waste disposal, proximity to land will be a decided economic advantage to a mining firm operating in the Red Sea. Logistical support from and easy accessibility to Europe's metal markets should enhance its competitive position. It appears, however, that commerical exploitation of the Atlantis II polymetallic sulphides may be delayed until petroleum prices rise to levels that allow the Saudi Arabian Government to invest in pilot operations.[70]

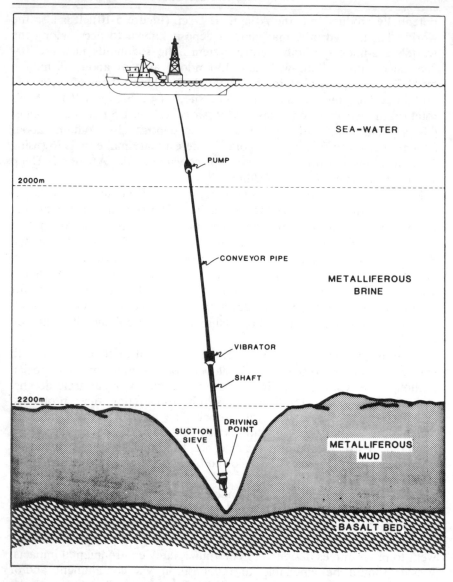

Figure 5.11 Proposed Red Sea polymetallic-sulphide mining system
Source: M. Cruickshank, J. P. Flanagan, B. Holt, and J. W. Padan, *Marine Mining on the Outer Continental Shelf*, Environmental Effects Overview, OCS Report 87–0035 (Minerals Management Service, DOI, Washington, DC, 1987), p. 26.

Conclusions

To locate, to mine, and to process ferromanganese nodules and crusts or the massive and unconsolidated polymetallic sulphides will take much capital and technology. If the analyses of Flipse and of Johnson and Otto offer a gloomy picture, a 1984 report of the Australian delegation to the

Preparatory Commission for the ISA presents an even more pessimistic view. The investigators show that (as of 1984) a 3-million-tonne/yr nodule producer extracting nickel, copper, and cobalt, as well as ferromanganese (with the latter obtained from manganese-rich wastes generated during the base metal processing) would have been unprofitable. To illustrate, of the Big Four metals, from 1974 to 1984 only cobalt showed an increase in 'real' price and since 1979 even cobalt has declined. The 'real' price, based on a discounted cash flow rate of return (DCFROR) on ferro-manganese went from $877/tonne in 1975 to $333/tonne in 1984. To have a minimum DCFROR of 18 per cent (a rate considered necessary to attract the needed investment capital), the per-tonne revenue from re-covered nodules (as of 1984) would have to be pushed up from $192/tonne to $366/tonne. A $192/tonne value would give a DCFROR of only 1 per cent. In sum, for the consortia or the Enterprise to produce nodules under these economic conditions is out of the question.[71] If nodules are exploited, the Clarion-Clipperton zone will be the main focus of activity.

Similarly ferromanganese crusts of the Pacific, mainly on the slopes of seamounts and islands, represent the best deposits known. Because crusts lie in shallower water and contain relatively high cobalt values (as well as the other Big Four metals), the marine mining industry is as interested in them as it is in nodules. The main difficulty in crust extraction is a relatively undeveloped technology applicable to it. Much of the nodule mining technology will not be usable. Processing crusts, however, lends itself to methods similar to those developed for nodules, depending on the metals desired; processing techniques combining hydro- and pyro-metallurgy are likely.

Considerable damage will occur to both pelagic and benthic eco-systems, although investigators seem not to see the degradation as irre-parable, because they expect a natural dissipation of surface and bottom sediment plumes and a biotic recolonisation of mined-out areas to re-establish normal conditions. Tailings waste disposal will be expensive, because onshore sites (adjacent to the processing plants) will probably be used. Locating marine mineral processing plants demands careful atten-tion to costs of waste disposal and energy. The business climate and political stability of an area may be the deciding factors in selecting the optimal process-plant location.

At present the best hope for mining the deep seabed is the polymetallic sulphide sediments of the Red Sea's Atlantis II Deep. Its future, how-ever, is somewhat problematic, depending especially on Saudi Arabia's economic recovery from a relatively depressed petroleum market.

Notes

1 R. Fellerer, 'Prospecting and evaluation: remote sensing techniques', Report no. 8, Seminar on the Exploitation of the Deep Seabed, presented under the auspices of the EEC for the benefit of the ACP experts to the United Nations

Conference on the Law of the Sea, Brussels, Belgium, 22–25 Feb. 1977, p. 121.

2 R. D. Ballard, 'The exploits of *Alvin* and *Angus*: exploring the East Pacific Rise', *Oceanus*, vol. 27, no. 3 (1984), p. 10.

3 R. W. Rowland, M. R. Goud, and B. A. McGregor, *The U.S. Exclusive Economic Zone – A Summary of its Geology, Exploration, and Resource Potential*, USGS Circular 912 (DOI, Alexandria, VA, 1983), p. 16.

4 Fellerer, 'Prospecting and evaluation: remote sensing techniques', p. 121.

5 See D. Felix, 'Some problems in making nodule abundance estimates from sea floor photographs', *Marine Mining*, vol. 2, no. 3 (1980), pp. 293–302.

6 Fellerer, 'Prospecting and evaluation: remote sensing techniques', pp. 123–4.

7 ibid., p. 125.

8 ibid., pp. 125–6.

9 R. Peters and M. Williamson, 'Design for a deep-ocean rock core drill', *Marine Mining*, vol. 5, no. 3 (1986), p. 321.

10 C. G. Welling, 'Polymetallic sulfides: an industry viewpoint', *Marine Technology Society Journal*, vol. 16, no. 3 (1982), p. 6.

11 P. J. C. Ryall, 'Remote drilling techology', *Marine Mining*, vol. 6, no. 2 (1987), p. 149.

12 D. W. Pasho, *Continuous Line Bucket Consortia: Appendix*, Internal Report RMB 1976–8 (Resource Management and Conservation Branch, Energy, Mines and Resources Canada, EMR, Ottowa, n.d.), n.p.

13 For an excellent discussion of problems associated with hydraulic lifting systems, see P. B. Grote and J. Q. Burns, 'System design considerations in deep ocean mining lift systems', *Marine Mining*, vol. 2, no. 4 (1981), pp. 357–83.

14 J. S. Chung, 'Advances in manganese nodule mining technology', *Marine Technology Society Journal*, vol. 19, no. 4 (1985), pp. 40–1.

15 For a useful examination of both pyrometallurgy and hydrometallurgy as applied to nodules, see J. C. Argarwal *et al.*, 'Comparative economics of recovery of metals from ocean nodules', *Marine Mining*, vol. 2, nos 1/2 (1979), pp. 119–30.

16 D. W. Pasho and D. E. C. King, *Processing Sector Cost Estimates for the Ocean Management Inc. Manganese Nodule Processing Facility*, Mineral Policy Sector Internal Report MRI 82/7 (Department of Energy, Mines and Resources Canada, Ottawa, Aug. 1980), p. 3.

17 A. M. Post, *Deepsea Mining and the Law of the Sea* (Martinus Nijhoff, The Hague, 1983), p. 30.

18 See C. T. Hillman and B. B. Gosling, *Mining Deep Ocean Manganese Nodules: Description and Economic Analysis of a Potential Venture*, USBM Information Circular 9015 (USGPO, Washington, DC, 1985); C. T. Hillman, *Manganese Nodule Resources of Three Areas in the Northeast Pacific Ocean: With Proposed Mining-Beneficiating Systems and Costs*, USBM Information Circular 8933 (USGPO, Washington, DC, 1983).

19 C. J. Johnson and J. M. Otto, 'Manganese nodule project economics', *Resources Policy*, vol. 12, no. 1 (1986), pp. 20, 22.

20 J. E. Flipse, *An Economic Analysis of a Pioneer Deep Ocean Mining Venture*, TAMU–SG–82–201, COE Report no. 262 (Sea Grant College Program, Texas A&M University, College Station, TX, Aug. 1982), pp. 1–4.

21 ibid., p. 42.

22 ibid., pp. iii, 32–4.

23 B. V. Andrews, J. E. Flipse, and F. C. Brown, *The Economic Viability of a Four-Metal Pioneer Deep Ocean Mining Venture*, TAMU–SG–84–201 (Sea Grant College Program, Texas A&M University, College Station, TX, Oct. 1983), p. iii.

24 See C. J. Johnson, *Economic and Business Investment Climates for Manganese Nodule Processing in Six Pacific Countries*, Part II of *Analysis of Laws Applicable to Manganese Nodule Processing and Investment Climates in Representative Pacific Areas*, prepared for DOC Institutional Grant no. NO81AA–D–00070 (University of Hawaii Sea Grant Program, Honolulu, HA, 1985).

25 See Q. D. Stephen-Hassard *et al.*, *The Feasibility and Potential Impact of Manganese Nodule Processing in Hawaii* (Department of Planning and Economic Development, State of Hawaii, Honolulu, Feb. 1978), pp. vi, and section 2, pp. 6–7.

26 J. M. Broadus, 'Seabed materials', *Science*, vol. 235, no. 4,791 (1987), p. 855.

27 F. T. Manheim, 'Marine cobalt resources', *Science*, vol. 232, no. 4,750 (1986), p. 607.

28 R. G. Paul, 'Development of metalliferous oxides from cobalt-rich manganese crusts', *Marine Technology Society Journal*, vol. 19, no. 4 (1985), p. 46.

29 Department of Planning and Economic Development, State of Hawaii, and Minerals Management Service, DOI, *Mining Development Scenario for Cobalt-Rich Manganese Crusts in the Exclusive Economic Zones of the Hawaiian Archipelago and Johnston Island*, Ocean Resources Branch Contribution no. 38 (DPED, and MMS, Honolulu, HA, Jan. 1987), pp. 5, 23.

30 C. J. Johnson *et al.*, *Resource Assessment of Cobalt-Rich Ferromanganese Crusts in the Hawaiian Archipelago*, Report of the East-West Center and the Research Corporation of the University of Hawaii for the Minerals Management Service, DOI, Co-operative Agreement no. 14–12–0001–30177, 'Environmental studies in support of the proposed leasing for cobalt-rich crusts' (Resource Systems Institute, East-West Center, Honolulu, HA, May 1985), pp. 101, 104, 106, 110–11.

31 Department of Planning and Economic Development, *Mining Development Scenario*, p. 72.

32 ibid., p. 70.

33 Johnson *et al.*, *Resource Assessment of Cobalt-Rich Ferromanganese Crusts*, p. 108.

34 Paul, 'Development of metalliferous oxides from cobalt-rich manganese crusts', p. 47.

35 B. W. Haynes, M. J. Magyar, and E. G. Godoy, 'Extractive metallurgy of ferromanganese crusts from the Necker Ridge Area, Hawaiian Exclusive Economic Zone', *Marine Mining*, vol. 6, no. 1 (1987), p. 23.

36 J. Padan, 'Development of metalliferous oxides from manganese nodules', *Marine Technology Society Journal*, vol. 19, no. 4 (1985), pp. 35–6.

37 P. A. Jumars, 'Limits in predicting and detecting benthic community responses to manganese nodule mining', *Marine Mining*, vol. 3, nos 1/2 (1981), pp. 215–17.

38 Padan, 'Development of metalliferous oxides from manganese nodules', pp. 35–7.

39 Office of Ocean and Coastal Resource Management, NOAA, DOC, *Deep Seabed Mining: Draft Environmental Impact Statement* (DOC, Washington, DC, May 1984), p. 77; 'meiofauna' normally refers to animals that will pass through a 0.3 to 1.0mm mesh sieve and be retained on a 0.05mm mesh sieve.

40 As of 1985, at least sixteen previously unknown families of animals had been discovered in vent communities. See J. F. Grassle, 'Hydrothermal vent animals: distribution and biology', *Science*, vol. 229, no. 4,715 (1985), pp. 713–17.

41 See H. W. Jannasch and M. J. Mottl, 'Geomicrobiology of deep-sea hydrothermal vents', *Science*, vol. 229, no. 4,715 (1985), pp. 717–25. See also H. W. Jannasch, 'Chemosynthesis: the nutritional basis for life at deep-sea

vents', *Oceanus*, vol. 27, no. 3 (1984), pp. 73–8; G. N. Somero, 'Physiology and biochemistry of the hydrothermal vent animals', *Oceanus*, vol. 27, no. 3 (1984), pp. 67–72.

42 J. W. Lavelle, E. Ozturgut, S. A. Swift, and B. H. Erickson, 'Dispersal and resedimentation of the benthic plume from deep-sea mining operations: a model with calibration', *Marine Mining*, vol. 3, nos 1/2 (1981), pp. 60, 90.

43 R. J. Ozretich, 'Increased oxygen demand and microbial biomass', *Marine Mining*, vol. 3, nos 1/2 (1981), p. 109.

44 J. W. Lavelle and E. Ozturgut, 'Dispersion of deep-sea mining particlates and their effect on light in ocean surface layers', *Marine Mining*, vol. 3, nos 1/2 (1981), p. 187.

45 E. Ozturgut et al., *Deep Ocean Mining of Manganese Nodules in the North Pacific: Pre-Mining Environmental Conditions and Anticipated Mining Effects*, NOAA Technical Memorandum, ERL MESA–33 (DOC, Washington, DC, 1978), p. 17.

46 J. Hirota, 'Potential effects of deep-sea minerals mining on macrozooplankton in the North Equatorial Pacific', *Marine Mining*, vol. 3, nos 1/2 (1981), pp. 19–57.

47 *The Ocean and the Future*, 'Hearings', before the Subcommittee on Oceanography of the Committee on Merchant Marine and Fisheries, House of Representatives, 99th Cong., 1st sess., 24 Oct. 1985 (USGPO, Wasington, DC, 1986), p. 104.

48 Stephen-Hassard et al., *Feasibility and Potential Impact of Manganese Nodule Processing in Hawaii*, section 4, p. 20.

49 J. C. Wiltshire, 'Environmental impacts of proposed manganese crust mining', in C. Johnson and A. Clark (eds) *Pacific Mineral Resources: Physical, Economic and Legal Issues* (East-West Center, Honolulu, HA, 1986), p. 477.

50 Office of Ocean and Coastal Resource Management, *Deep Seabed Mining; Draft Environmental Impact Statement*, p. 85.

51 Minerals Management Service, DOI, *Draft Environmental Impact Statement: Proposed Marine Mineral Lease Sale in the Hawaiian Archipelago and Johnston Island Exclusive Economic Zones*, Ocean Resources Branch Contribution no. 40 (Department of Planning and Economic Development, Honolulu, HA, Jan. 1987), p. A–62.

52 Office of Ocean and Coastal Resource Management, *Deep Seabed Mining*, p. 85.

53 B. W. Haynes et al., *Laboratory Processing and Characterization of Waste Materials from Manganese Nodules*, USBM Report of Investigations 8938 (USBM, Washington DC, 1985), p. 15.

54 See S. A. El-Swaify and F. W. Chromec, 'The agricultural potential of manganese nodule waste material', in P. B. Humphrey (ed.), *Marine Mining: A New Beginning, Conference Proceedings, 18–21 July 1982, Hilo, Hawaii* (Department of Planning and Economic Development, State of Hawaii, Honolulu, 1985), pp. 208–27.

55 Wiltshire, 'Environmental impacts of proposed manganese crust mining', p. 478.

56 ibid.

57 ibid.

58 Stephen-Hassard et al., *Feasibility and Potential Impact of Manganese Nodule Processing in Hawaii*, pp. 1–2 and 5–11; F. C. F. Earney, *Petroleum and Hard Minerals from the Sea* (Edward Arnold, London, 1980), pp. 221–4.

59 R. D. Bignell, 'Genesis of the Red Sea: metalliferous sediments', *Marine Mining*, vol. 1, no. 3 (1978), pp. 209–10.

60 G. A. Gross, 'Mineral deposits on the deep seabed', *Marine Mining*, vol. 6, no. 2 (1987), p. 110.

61 P. A. Rona, 'Hydrothermal mineralization at slow-spreading centers: Red Sea, Atlantic Ocean and Indian Ocean', *Marine Mining*, vol. 5, no. 2 (1985), pp. 130–1.
62 M. Cruickshank, J. P. Flanagan, B. Holt, and J. W. Padan, *Marine Mining on the Outer Continental Shelf*, Environmental Effects Overview, OCS Report 87–0035 (Minerals Management Service, DOI, Washington, DC, 1987), p. 24.
63 Rona, 'Hydrothermal mineralization at slow-spreading centers', pp. 130–1.
64 V. E. McKelvey, *Subsea Mineral Resources*, Chapter A of USGS Bulletin 1689 (USGPO, Washington, DC, 1986), p. 85.
65 Earney, *Petroleum and Hard Minerals from the Sea*, p. 171.
66 H. Amann, 'Development of ocean mining in the Red Sea', *Marine Mining*, vol. 5, no. 2 (1985), p. 103.
67 Cruickshank, Flanagan, Holt, and Padan, *Marine Mining on the Outer Continental Shelf*, p. 24.
68 Amann, 'Development of ocean mining in the Red Sea', pp. 108, 113.
69 ibid., pp. 103–6, 113.
70 Letter: H. Bäcker, Pruessag AG, Hannover, FRG, 7 July 1987.
71 Delegation of Australia, Special Commission 2, Preparatory Commission for the International Sea-Bed Authority and for the International Tribunal for the Law of the Sea, *The Enterprise: Economic Viability of Deep Sea-Bed Mining of Polymetallic Nodules*, LOS/PCN/SCN.2/WP.10 (UN, NY, 14 Jan. 1986), pp. 2, 38.

The continental margins

Introduction

Although controversies and uncertainties associated with deep-seabed mining's future have captured much of the general public and world statesmen's interest, presently the most important oceanic mineral-producing areas are the continental margins, especially their shelf and slope areas. This situation is unlikely to change for several decades.

A significant part of the world's petroleum (natural gas and crude oil) production comes from continental shelf regions. As petroleum exploration teams push into the new oceanic frontiers of the outer continental shelf and continental slope, as well as the Arctic and Antarctic Oceans, the offshore will become even more important as a supplier of energy.

We often think of petroleum as *the* continental margins' premier mineral resource, but hard minerals also have long been important there and will become more so in the future. Some hard mineral occurrences on the continental margins are a function of onshore erosional processes and offshore deposition, such as the metallic and non-metallic placers, the industrial sands, and the construction sands and gravels. Other continental margin hard minerals come from the death of biotic organisms such as molluscs (shells) and algae or form as precipitates (aragonite), all of which accumulate on the seabed. Although these minerals and the sands and gravels (as unconsolidated accumulations) could be classified as placers, they are discussed here as industrial or construction materials. Still other continental margin minerals occur as seaward extensions of bedded materials (phosphorites and coal) and veined metallic deposits (scheelite and tin).

Chapter six
Placers and subseabed metallics

Mining on the continental shelf is not new. In the past, miners used land-based tunnels to obtain lead, zinc, and tin from nearshore continental shelves. Today miners produce only one subseabed metallic mineral, scheelite. Much more important are the placers, either now produced or that may be produced both from the offshore seabed and from beaches.

Subseabed metallics

The Greeks, via land-based tunnels, mined lead and zinc from beneath the sea and for centuries miners of the Cornwall Peninsula, in England, mined seaward extensions of tin ores. The long inoperative Levant Mine was reopened in 1970, but closed again after only a few years of operation. In mid-1986 the Geevor Mine (at Pendeen, near Land's End) suspended mining.[1] Several difficulties account for this closure. During 1985 efforts to extend the Geevor's undersea workings were slowed by extremely hard rock that intersected the extension area. Also in 1985, the national government terminated its Mineral Exploration Grants programme.[2] The most important factor, however, was a collapse of the international tin market.[3] In the late 1970s approximately 60 per cent of the mine's production came from beneath the sea, and it employed some 300 workers.[4] To keep the mine operational, engineers now maintain continuous pumping, an expensive endeavour. To pay for the pumping, the mine has been opened to tourists. The Geevor's future looks rather gloomy, because the National Department of Trade and Industry has denied three company requests for financial assistance.[5]

For seven decades, miners extracted iron ore from under the sea in Conception Bay, at Wabana, Bell Island, just north-west of St John's, Newfoundland. The mine closed in 1966 because its ores could not be beneficiated to produce a product that met required standards at a competitive world-market price. In the Gulf of Finland 80km southwest of Helsinki, Finland, near Jussaro Island, a mining establishment produced magnetite ores from beneath the seabed. This mine opened in 1961, but closed after only six years, because of severe water-leakage problems. Japan, too, once had an undersea iron mine in the offshore of western Kyūshū, and at one time, dredging for iron sands in Kyūshū's Ariake Bay and in Tokyo Bay off southern Honshu was important to Japan's mining industry, but these producers no longer operate.[6]

A major operational underground offshore metallic mining establishment is located on King Island in Tasmania, Australia. King Island Scheelite Proprietary Ltd owns the mine, employing about 60 workers. An onshore decline shaft extends under the Bass Strait. Scheelite, a tungsten ore, functions well as a steel alloy for wear-resistant cutting tools. The mine produces about 25,000 tonnes of ore a year.[7]

Placers

When pebbles, sands, and silts become sorted through the action of moving water, minerals with higher specific gravities and resistance to weathering may also become concentrated, especially in beaches and drowned river mouths (Figure 6.1). It is here in the onshore-offshore transitional environment that geologists seek most of the marine placers (Figure 6.2).[8] Locating potential marine placer resources is often less difficult than is determining their exploitability, because varying littoral drift rates and directions, altering wave-energy distributions and changing water levels contribute to regional and local variations in placer distributions.[9]

A frequent lack of uniformity within placers requires appropriate sampling strategies. They may hold high mineral concentrations in one place and low concentrations in another, even nearby. Mining problems may occur because of the size of the placer materials and the presence of boulders within it, the amount of cementation and the relief of the underlying bedrock, as well as the occurrence of an overburden.[10]

Despite these problems, many marine geologists feel that placers hold much promise. J. Robert Moore, of the University of Texas, has long been an advocate of their potential.[11] Peter Rothe, of the University of Mannheim in the FRG, believes that placers are among the least speculative of the underwater minerals.[12] David Cronan, of the Imperial College of Science and Technology in London, contends that engineering difficulties of deep seabed mining and political issues associated with the 'Area' and the Law of the Sea Convention will encourage miners' interest in placers.[13]

Many mineral-bearing offshore and beach placers (diamond, gold, platinum, tin, chromite, iron sand, zircon, ilmenite, rutile, and monazite) are now mined or have been mined in the past. For example, during the Second World War beach sands in uplifted terraces of southern Oregon were mined for chromite.[14] According to a study done by a US DOI Outer Continental Shelf Mining Policy Task Force, nearly 30 million tonnes of chromite are potentially available in Oregon's offshore. Like this chromite deposit, many other placers remain unexploited, and often, inadequately explored. Marine geologists, however, are at work throughout

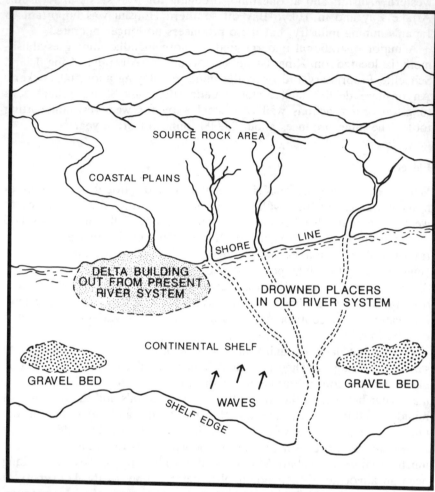

Figure 6.1 Schematic representation of placer locations

Source: After E. H. Madonald, *Alluvial Mining: The Geology, Technology and Economics of Placers* (Chapman & Hall, London, 1983), p. 19. With permission.

the world, both in frontier regions (as in the Philippines[15] and the PRC[16]) and in known regions, as in the Mediterranean Sea (Figure 6.3). Passive continental margins are the most favourable sites,[17] but active margins also hold many placers. Because of their wide distribution and great variety, only the most important placers are examined here.

Dredging techniques and environmental constraints

Different operating conditions, seabed characteristics and economic demands have contributed to the development of a variety of dredges, none of which is necessarily superior to another. Shallow-water dredges

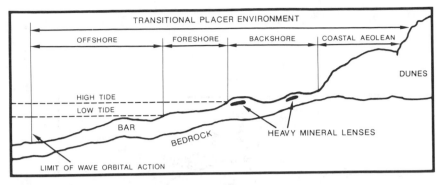

Figure 6.2 The transitional placer environment
Source: After E. H. Macdonald, *Alluvial Mining: The Geology, Technology and Economics of Placers* (Chapman & Hall, London, 1983), p. 18. With permission.
Note: Wave and tidal actions influence the distribution of placer mineral deposition.

may be broadly classified into hydraulic and mechanical types. Hydraulic units siphon (by pipe) the desired material onboard as an ore and water slurry, whereas mechanical units cut into the ore and lift the fragments onboard by a dragline, a dipper, a clam shell or an endless-chain bucket system. Mechanical dredges can operate in waters of unlimited depth, but ordinary hydraulic dredges are limited to about 60m, with most operating at depths of no more than 30m. When a water or an air-jet is introduced at the extraction surface to create a density difference in the siphon pipe, a powerful suction is created that lifts the placer materials to the surface. These units can operate at depths of 600m.[18]

Dredges operating in sea-water are subject to severe corrosion and when working in open waters are exposed to wind, current, and wave actions that create motions which put severe stress on mining equipment and the dredge. Surface wind and waves can be avoided by using remotely controlled underwater dredges, a technique proposed by engineers throughout the past five decades; one proposed system would use seabed currents and gravity to segregate gold or other heavy metals from the coarsest and heaviest materials, as well as the lightest sand particles (Figure 6.4).[19]

Seabed dredging creates sediment plumes that, after settling out, may affect some sessile and mobile benthic biotic communities. On the other hand, many biota in nearshore areas have developed considerable tolerance to sediment fallout from wave- and current-induced turbidity. Reef communities are the most sensitive to turbidity, but clams, crabs, and lobsters are also affected. If dredges make only long and deep furrows or holes, the seabed's irregularity and surface area increases, which could improve biotic habitat, but if the entire seabed surface is removed, mobile and sessile benthic organisms can suffer major damage. [20]

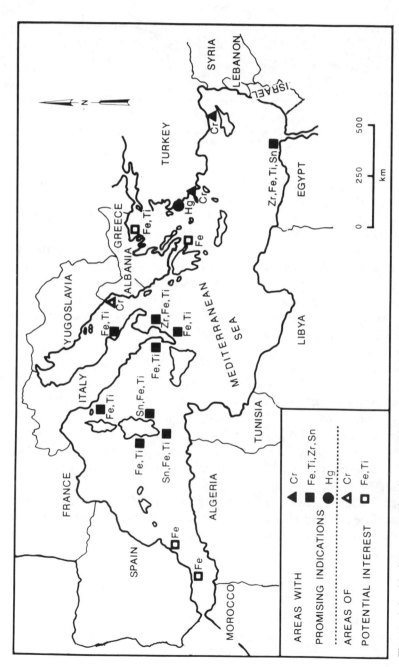

Figure 6.3 Mediterranean basin – promising potential for placers

Source: After S. P. Varanavas, 'An Fe-Ti-Cr placer deposit in a Cyprus beach associated with the Troodos Ophiolite Complex: implications for offshore mineral exploration', *Marine Mining*, vol. 5, no. 4 (1986), p. 407. With permission, Taylor & Francis.

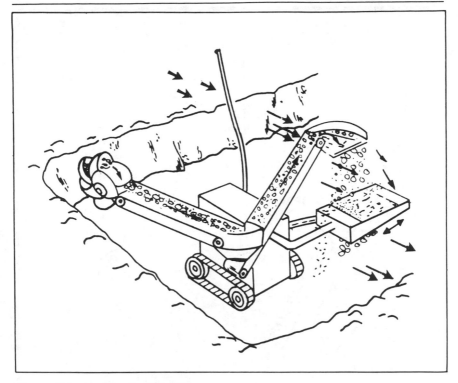

Figure 6.4 Underwater dredge

Source: After L.A. Lindelof, inventor, Figure 1 of United States Patent no. 3,731,975, patented 8 May 1973.

Note: Dredged materials are discharged at a specifically selected height and direction so that the coarsest particles fall upstream from the collector, with finer sediments being carried downstream toward the collecting device. The lightest sediments are carried downstream beyond the collector.

Diamonds

For nearly seven decades miners have won diamonds from sand and gravel beaches in Namaqualand in the north-west of the Republic of South Africa and in what is today South West Africa/Namibia. Outer-surf-zone exploitation commenced some twenty-five years ago. In Namibia the mining area presently lies in a 100km coastal strip between Oranjemund and Chameis Bay. Longshore currents carry diamond-bearing gravels and sands northward along the coast from the mouth of the Orange River (Figure 6.5).[21] In places, the diamonds become trapped within gullies and potholes in the bedrock.[22] The diamonds may also have originated from kimberlite pipes offshore, although this theory is not widely supported.[23]

From 1962 to 1971 Marine Diamond Corporation (a subsidiary of the De Beers Group since 1965) operated outer-surf-zone dredges and numerous support vessels and aircraft in the Hottentot Bay and Chameis Bay areas. These operations were dangerous and expensive. For offshore

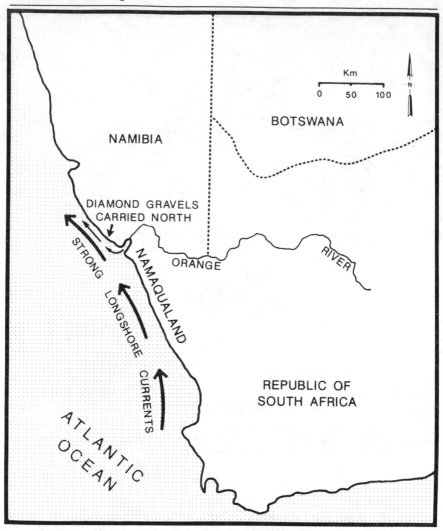

Figure 6.5 Location of diamond placers associated with longshore currents off the coasts of Namibia and the Republic of South Africa

Source: After F. C. F. Earney, *Ocean Mining: Geographic Perspectives*, Meddelelser fra Geografisk Institutt ved Norges Handelshøyskole og Universitet i Bergen, no. 70 (Geografisk Institutt, Bergen, Norway, 1982), p. 12. With permission.

surf-zone diamond extraction to be competitive, the deposits must be several times richer than onshore deposits.[24] Although no diamonds have been produced here in this way for several years, geologists are continuing to map and to analyse the area's offshore diamond-bearing placers, and the South African government is considering new exploration ventures, with modern equipment, to locate new deposits.

For more than two decades, CDM (Consolidated Diamond Mines), another De Beers subsidiary, has exploited beach sands and gravels in this region. Earliest efforts used earth-moving equipment to obtain the diamonds in the foreshore (surf) zone. At low-tide, equipment operators hurriedly extracted as much gravel as possible before the next high-tide reclaimed the work area. Later, steel sheets 'were pile-driven into the sand and anchored to a concrete buttress, beyond which sand fill and rubble had been tipped into the sea as an advance barrier'. This technique allowed sand and gravel removal below the high-tide mark but not to the low-tide level. Water seepage was a major problem, because the sheet piles did not fit snugly on the irregular bedrock below the sand. In the late 1960s engineers built an experimental skew prism wall of concrete blocks placed on a built-up area of overburden, with rubble piled on both sides as reinforcement. The low-tide mark region could now be mined, but the sea frequently reached the wall. CDM abandoned this technique in 1971. Then in 1972 engineers 'used deep overburden, removed to expose ore on land adjacent to the foreshore, to construct a [new] sea

Figure 6.6 Overburden removal from bedrock, with the sea wall in the background
Source: Courtesy CDM (Proprietary) Ltd. With permission.
Note: After removing the overburden, workers use giant vacuum cleaners and manual methods to clean the potholes and gullies where the diamonds tend to concentrate.

wall'; this wall was raised higher and extended farther into the surf than the skew prism wall. Wellpoints connected to vacuum pumps were built into the wall to dewater and stabilise it. Earth-moving machines worked twenty-four hours a day to repair erosion damage on the outer slope. By 1977 miners had pushed some 200m seaward from the high-tide mark. In 1985, after detailed studies of sediment transport and wave action along the shore, CDM moved the sea wall out to 350m beyond the high-tide mark, 'with mining taking place at about 12m below mean sea level' (Figure 6.6). The wall extends along the beach for 600m and is 60m thick at its base (Figure 6.7).[25]

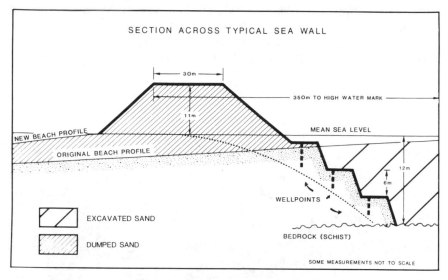

Figure 6.7 Section across a typical sea wall

Source: After D. L. Hodgson, 'Mining the beach for diamonds at CDM', *Engineering and Mining Journal*, vol. 178, no. 6 (1977), p. 151. With permission. Updated information supplied by Letters: C. Cowley, Public Relations Manager, CDM (Proprietary) Ltd. Windhoek, Southwest Africa/Namibia, 5 May and 12 Feb. 1987.
Note: The wall must be maintained continuously, because of wave action.

Titanium

The strategic element titanium, derived mainly from ilmenite and rutile ores, is an important placer material. About 5 per cent of the world's annual production is consumed as metallic titanium, mainly in metal alloys needing special strength and corrosion resistance, as in aircraft engines. The other 95 per cent (mostly titanium dioxide) provides a white pigment for paper, rubber, and plastics, among many other products.[26]

Titanium placers occur in West Africa, along the coast of Mauritania, and recently, on the continent's opposite coast, near the south-eastern shore of Madagascar, geologists have identified extensive titanium beach sands that also extend offshore. A large-scale pilot-plant sampling programme was to begin in mid-1987. The titanium is high enough in grade that it can be sold to some markets (such as pigment producers) without processing. If feasibility studies prove favourable, commercial production should begin in the second half of 1990.[27] On the continental shelf off the Zambezi River in Mozambique, scientists from the FRG have done detailed geophysical and vibrocoring studies of the distribution and economic potential of ilmenite, rutile, and zircon sands. Their findings show that the area sampled has an estimated 50 million tonnes of ilmenite, 0.9 million tonnes of rutile, and 4 million tonnes of zircon, all situated in waters of 30m to 60m depth. Several liabilities may preclude for some time the exploitation of these deposits, including the water depth, the long distances to ports (Beira 200km and Quelimane 170km) and the open-sea environment in an area subject to severe sea and weather conditions.[28]

In North America, the US has several offshore areas important for titanium – the southern shore of the Seward Peninsula in Alaska, southern California, the Gulf Coast and the Atlantic seaboard from north-eastern Florida to Cape Cod.[29] The Alaskan and Atlantic coastal zone areas combined have an estimated titanium placer potential worth $35,000 million (1986$), although these deposits are not necessarily exploitable presently.[30] Canada has some titanium placers along the shore of north-eastern Newfoundland.[31]

Several other areas have significant offshore or beach titanium placers. For example, on Brazil's south-eastern coast, intermittent deposits extend from the Vitoria region in the State of Espirito Santo to the Salvador region in the State of Bahia. Panama's south-eastern shore, just north of where it joins South America, contains titanium deposits. The mouth of India's Gulf of Khambhat, the western shore of Sri Lanka, the south-eastern littoral zone of Burma, and the shore of the west-central portion of New Zealand's North Island also contain titanium.[32] In Malaysia dredgers produce titanium-bearing ilmenite as a co-product during tin-mining operations.[33]

The titanium mining industry, in the past, experienced 'boom and bust' periods. Its fortunes are tied especially to the amount of activity in military and commercial aircraft industries. For example, between 1979 and 1981, consumption reached a new peak, declined precipitously in 1982, then rebounded in 1983, all in conjunction with changes in engineering industries. If airlines should decide, in the near future, to replace ageing fleets, demand for titanium could increase significantly. The US Bureau of Mines projects that in the year 2000 the probable world demand for primary titanium for both metal and non-metal uses will be about 2.8 million tonnes. From 1983 to 2000 annual growth in demand should be approximately 6.2 per cent.[34]

Although the world has very large supplies of titanium ores, a recent statement from the US Bureau of Mines indicates that low-cost rutile resources are rapidly declining, and 'to maintain future ... high-grade titanium resources, alternate sources will need to be developed'.[35] Perhaps oceanic sources can help fill this gap.

Platinum

One of the precious metals, platinum, is a critical metal to industry. Important users include the automotive, electronic, petroleum refining, chemical, and medical services industries. Although world platinum reserves are large, two states – the USSR and the Republic of South Africa – produce more than 90 per cent of the world's annual output of platinum-group metals, including (in addition to platinum) palladium, rhodium, ruthenium, iridium, and osmium. Each is vital to industry. South Africa supplies the major share of the western world's annual output of primary platinum (Table 6.1).[36] Considering the current political instability of South Africa and the likely unreliability of the USSR in supplying the market-economy states with platinum if other supplies were unavailable, it behoves consumers to consider carefully every potential source of platinum, including offshore and beach placers. Most platinum now produced is a co-product or by-product of copper, nickel, and gold lode mining.

Table 6.1 Platinum supply and demand in the western world, 1982–6 (kilograms)[a]

Supply[b]	1982	1983	1984	1985	1986
South Africa	60,963	64,384	70,916	72,782	73,093
Canada	3,732	2,488	4,666	4,666	4,666
Others	933	1,244	1,244	1,244	1,244
USSR sales	11,819	9,020	7,776	7,154	9,020
Totals	77,448	77,136	84,602	85,846	88,023
Demand [c]					
Western Europe	10,264	10,264	12,752	12,441	14,930
Japan	32,659	29,548	35,458	38,879	31,415
North America	22,083	22,395	28,304	31,415	37,013
Rest of western world	7,154	5,599	5,288	5,288	5,288
Western sales to					
Comecon/PRC	933	622	933	933	1,244
Movement in stocks[d]	4,354	8,709	1,866	(3,110)	(1,866)
Totals[e]	77,448	77,136	84,602	85,846	88,023

Source: G. G. Robson, *Platinum: 1987* (Johnson Matthey, London, May 1987), p. 6. With permission. Notes: a. Converted from troy oz, with 1kg equal to 32.15 troy oz; b. supply figures are estimates of sales by the mines of primary platinum; c. demand estimates are net figures, with demand in each sector being the total purchases by consumrs less any sales back to the market; d. movements in stocks in a given year reflect changes in stocks held by other than primary refiners and final consumers, such as metal in the hands of fabricators, dealers, banks and individuals; totals may not add because of rounding. A positive figure indicates an increase in stocks, including some platinum bought for investment, and a negative figure indicates a rundown in stocks; e. annual totals represent the amount of newly mined metal acquired by consumers in a given year.

Presently no platinum placers are mined in the world's oceans, but at one time, dredgers worked them in waters near the villages of Platinum and Goodnews Bay, Alaska. Field studies by two marine placer specialists, R. M. Owen and J. Robert Moore, have also demonstrated a good potential for offshore platinum at Chagvan Bay, approximately 24km south of Goodnews Bay.[37] To date, no efforts seem to have been made to pursue further the work of Owen and Moore. Such neglect may be unwise for the US's future platinum supply. To illustrate, in 1983 the US's total demand for platinum was 24,790kg (797,000 troy oz.). US mine output was 31kg (1,000 troy oz.), a by-product of copper production![38] The western world's estimated demand (including sales to Comecon and the PRC) for platinum in 1986 was 88,829kg (2,855,915 troy oz); this demand was greater than the supply. Analysts expect that the shortfall for 1986 will have been about 1,866kg (60,000 troy oz). This shortfall, however, is not statistically significant, given the large total demand.[39]

What would happen if platinum imports were suddenly cut off to the US? Canada might help fill the deficit, but that country's production in 1983 was only 2,240kg (72,000 troy oz). Why not tap the US strategic stockpile? Its reserves would last only about seven months, and President Reagan, during his second term in office, stopped further stockpiling of platinum. Consumers must have inventories. Usually these are maintained at about a four-months' supply. Perhaps substitutes could be used. Unfortunately to use substitutes will necessitate plant process modifications and will reduce industrial production efficiencies in many sectors (as in petrol refining).[40] In sum, under normal conditions of demand, the country might manage without platinum imports for nearly a year. What then?

The US consumes about one-third of the world's total annual platinum mine production, and its probable consumption is projected to increase by 63 per cent from 1983 to the year 2000, for a total demand of 40,435kg (1.3 million troy oz). Yet the US has done little to develop its known onshore deposits in Alaska, Minnesota, or Montana, although one platinum mine did begin operating in Montana's Beartooth Mountains in 1987. And, neither industry nor government has adequately explored the potential of the offshore and beach placers of Alaska. Because of the long lead times needed to bring mines into production, exploitation programmes begun now would require several years before even 1kg of platinum could be produced.

Tin

The best known and most important marine metallic placer is tin found primarily in the mineral cassiterite, the only commercially important source. Metalsmiths, as early as 3500 BC, added tin to copper to produce bronze for tools and weapons. Today tin goes into the fabrication of containers, solders, engine bearings, and air-cleaner and oil-filter cartridges.[41] Offshore or beach tin placers occur in the UK, USSR, US (Alaska), Burma, the PRC, the Philippines, Thailand, Malaysia, and Indonesia.

South-east Asia

The world's most important area for offshore tin production is in waters surrounding the Malay Peninsula and those lying between Sumatra and Kalimantan. For more than a 100 years, primitive tin mining operations have worked the offshore of Thailand[42] and Indonesia. Large-scale offshore placer mining, however, is a relatively recent development.

Indonesia's modern offshore operations began about twenty-five years ago, especially in waters near the islands of Bangka, Belitung, Singkep, Karimum, and Kundar. The Indonesian Government in 1986 expected to produce some 13,000 tonnes of tin in the offshore, nearly half the country's annual total output of 27,000 tonnes. As of early 1987 five large, government-owned dredges were mining tin in Indonesian waters.[43] The most recent addition to Indonesia's dredging fleet was in May 1985. It was a 0.6m³ (22ft³) capacity bucket dredge valued at $28 million which is working in the offshore of the Karimum and Kundar Islands in the Riau Archipelago. Most Indonesian dredgers are working in waters of 9m to 45m.[44]

Thailand's large-scale offshore production began only a decade ago (1977), with producers active in both the Andaman Sea and Gulf of Siam. The offshore now accounts for 50 per cent of Thailand's total annual tin production. Until recently some 1,000 small, privately owned suction boats, employing perhaps 60,000 workers, plied the shallow offshore of Thailand.[45] These craft (now numbering about 600)[46] – many unlicensed and without concession areas – work only seasonally, when the monsoon is not strong. Only a dozen registered for the 1986–7 season.[47] A typical suction boat may recover only 15kg of cassiterite concentrate per day,[48] and may do so in concession areas belonging to Thailand's six large, heavily capitalised dredging firms, one of which is state owned. Tin-ore poaching by the suction boats has jeopardised the economic viability of some of the large firms, especially when the poachers mine the pockets of high-grade ore.[49] Although the large dredging firms have begged the national government to make a greater effort to stop the poaching, the problem persists.

Suction-boat crews use a diver-directed suction hose to dredge usable ore. Onboard, workers employ gravity methods to beneficiate the tin ore, using a series of screens to remove large rocks and other debris. High-pressure water hoses break up the remaining clay balls and loosely aggregated material. After further screening, the small materials go to a series of jigs. The final yield is a concentrate of 20 to 30 per cent tin which is then ready for further processing onshore. Suction-boat producers have only a 95 per cent tin recovery efficiency, thus some of the best offshore tin ores are being wasted. Collectively these small producers extract a significant portion of Thailand's total tin output, even though they work only about nine months during the year.[50]

Malaysia, the world's most important tin producer, in 1984 had 31 per cent (12,805 tonnes) of its total tin output (metal content) come from the offshore.[51] Marine dredges in 1987 were operating off the western shore of the Malay Peninsula, near Kuala Lumpur.[52] Malaysian dredges have

Table 6.2 Estimated mining and beneficiating costs for producing tin mines by mining method

Mining method and country	Cost in US$/tonne of ore[a]	
	Mining	Beneficiating
Underground		
Bolivia	28.60	10.40
Republic of South Africa	15.80	12.70
United Kingdom	32.60	14.90
South-east Asia: Burma, Indonesia,		
Malaysia, Thailand	38.00	12.60
Others: Argentina, Australia, Japan,		
Peru, Zimbabwe	19.80	12.40
Open pit		
Australia	11.60	6.30
Thailand	8.80	3.20
Others: Brazil, Malaysia, Namibia	4.40	3.10
Dredge		
Indonesia	1.00	b
Malaysia	0.50	b
Thailand	0.80	b
Others: Australia, Bolivia, Brazil,		
Nigeria	1.10	b
Gravel pump		
Brazil	2.30	b
Indonesia	1.70	b
Malaysia	0.90	b
Thailand	1.70	b
Others: Australia, Bolivia, Burma,		
Zaïre	4.70	b

Source: D. I. Bleiwas, A. E. Sabin, and G. R. Peterson, *Tin Availability – Market Economy Countries: A Minerals Availbility Appraisal*, USBM Information Circular 9086 (DOI, Washington, DC, 1986), p. 33.
Notes: a. Based on 1982 data, updated to January 1984 US$; b. beneficiating and mining are grouped together, because most dredge and gravel pump operations were vertically integrated through the beneficiation stage.

the greatest production efficiency of all tin producers worldwide, and Thailand and Indonesia are close behind. As of 1984, on the average, South-east Asia's offshore dredges could mine and bencficiate a tonne of tin ore some sixty-two times more cheaply than could the UK's underground mines; Malaysia's cost was ninety-five times cheaper (Table 6.2).

World tin market

The foregoing critique might seem to indicate that the world tin industry is experiencing a vigorous demand for its product, and that producers in South-east Asia are especially prospering. On the contrary, the industry has been in a deep depression, epitomised by events on 24 October 1985, when the tin industry experienced a major crisis and the London Metal Exchange (LME), metal traders and the banking industry absorbed a severe shock, one from which the industry has yet to recover.

The origin of the problem dates from the early 1970s when a general 'commodity boom' occurred. In two years' time (1973 and 1974), the price of tin doubled from about £1,960 per tonne (where it had been for nearly a decade) to £3,493 per tonne. By 1979 the price had escalated to an average of £7,276 per tonne.[53] The price rise stimulated production and encouraged new producers to enter the market. Simultaneously, however, consumers developed techniques to conserve tin (such as thinner layers) and they sought and found substitutes (such as aluminium). Markets became oversupplied, prices began to fall and the International Tin Council (ITC) used a buffer-stock programme to purchase tin at a fixed price. Although world mine production from 1981 to 1984 declined from 236,052 tonnes to an estimated 207,842 tonnes, prices continued to fall.[54] In September 1985 ITC members met in an attempt to agree on their monetary share for continuing the support of the buffer-stock programme, but the effort failed.[55] Finally, on 24 October 1985 the buffer-stock manager had depleted funds available for tin purchases, forcing him to request that the LME suspend tin trading.

Suspension of tin trading represented a default by the ITC's member states.[56] Banks were left holding large tonnages of relatively worthless buffer-stock tin as collateral for loans to the ITC. Many metal traders were, also, caught in the squeeze. A London metal-trading firm, MacLaine Watson & Company (among several others) brought suit in the mid-1980s to obtain repayment of £6 million (US $9.6 million). The Maclaine Watson suit differed from other suits, because it was the 'only one to single out the British Government'. As in other suits, a High Court judge ruled that Maclaine Watson gave credit to the ITC itself and not to individual members. One estimate puts the ITC's total debt at £900 million (about US$1.4 thousand million in 1987 dollars).[57] One estimate puts the banks and metal traders' total losses at £420 million, with the latter losing the most (perhaps £260 million). John Spooner, a senior editor for *Mining Magazine*, contends that the debacle was, in part, a failure of the ITC to (1) persuade countries newly entering the world tin market to join the ITC and (2) lower the buffer-stock price to a point that the world's inefficient producers would have been forced out of the market.[58]

It might seem that the ITC have done nothing beyond its buffer-stock programme to prevent the downward economic slide of the tin market. Not so. In 1982 the ITC – under a Sixth International Tin Agreement

(ITA) – imposed export quotas on its tin-producing member states, Australia, Malaysia, Indonesia, Nigeria, Thailand, and Zaïre. By imposing an additional 39.6 per cent quota cut-back during the third quarter of 1983 and maintaining this limitation in 1984, the ITC reduced the world tin surplus by 10,700 tonnes. Industry sources estimate that in early 1984 there was a world market surplus of 70,000 tonnes, 39 per cent of the total market demand of 180,000 tonnes.[59] The ITC's tin quota caused much hardship for many tin dredgers. In 1984, for example, at least one dredging operation in Thailand suspended mining for several weeks, because it was not profitable to operate well below production capacity. Thailand's cut-back quota for 1984 was 40 per cent.[60]

Although the ITC managed to reduce output through its quota system, its efforts were badly diluted by tin smugglers, who avoid payment of taxes and royalties by clandestinely moving tin out of producing states and into the world market. The problem is especially severe in South-east Asia, with Singapore being a primary destination for smuggled tin.[61] Smelter operators in Singapore accept tin concentrates without checking on their origin.[62]

One author, Norani Visetbhakdi, writing for the *Bangkok Bank Monthly Review*, contends that the ITC's imposed quotas 'increased the smuggling of ore out of Thailand to Singapore'. The offshore sector was the first to smuggle tin; onshore producers joined in when the tin price fell. Smugglers undercut the buffer-stock floor price, making uncompetitive that tin produced and exported under the quota system. The Thai Government seems impotent in its efforts to staunch the flow of illegal tin exports, reportedly because these 'activities are under the protection of "influential" people and corrupt government officials'. This predicament has led many critics to call upon the Thai Government to quit the ITC, mainly because of the revenues it is losing via the smuggled tin.[63] The Thai Government estimates that in 1984 some 5,000 tonnes of tin concentrates were illegally moved out of the country,[64] to avoid both the ITC export quota and government royalties. ITC export quotas were finally lifted in April 1986, but their removal has done little to help Thailand's tin industry.[65]

Malaysia and Indonesia have a problem similar to Thailand's. ITC estimates put the total amount of tin smuggled from these three states during 1983 and 1984 at 16,500 tonnes each year. According to Elizabeth Mayo, editor of *Tin International*, the total may have been as much as 19,000 tonnes. By 1985 tin smuggling may have been reduced to between 10,000 (ITC estimate) and 13,000 tonnes (*Tin International* estimate). Malaysia considers the smuggling problem so severe that in 1984 it increased criminal penalties for those convicted of possessing tin illegally. The fine went from M$5,000 to M$50,000 and the possible prison term from six months to two years. In 1986 Mayo predicted that, if depressed market conditions persisted, smuggling would decline significantly,[66] and according to a recent report by Sivavong Changkasiri, Director-General of Thailand's Department of Mineral Resources, smuggling has declined since the collapse of the tin market.[67]

Smuggling is not the only contributory to the tin surplus. The surplus problem is exacerbated because not all tin-producing states are members of the ITA. For example, Bolivia and Brazil, the world's fourth and fifth largest market-economy producers continue to produce without quotas, with the latter actually increasing production recently.[68] The US, too, has contributed to the glutted tin market, because its General Services Administration has been selling stockpile tin surpluses. In 1983, however, the US signed an agreement with Malaysia, Thailand, and Indonesia to limit its sales during 1983–4 to no more than 6,000 tonnes;[69] more recently, US sales have been about 3,000 tonnes per year.[70] Despite these efforts, the tin market remains in oversupply, and the outlook for offshore tin mining, in the near term, looks grim. World Bank officials believe the tin price will continue depressed until 1988, when it will bring an average of US$6.00 per kg, a price similar to more prosperous times. The price is expected to then increase to US$7.50 kg in 1989, US$8.00 in 1990, US$11.00 in 1995, and US$13.00 in 2000.[71]

The glutted tin market has neither deterred Thailand, Indonesia, or Malaysia from seeking additional tin reserves, nor discouraged the UN from giving exploration assistance. Through a matching grant programme, the UN recently contributed $2.6 million to an exploration effort in Thailand.[72] The Thais investigated tin-associated granites in both the Gulf of Thailand and the Andaman Sea.[73] Their efforts resulted in the discovery of significant reserves in relatively deep waters (61m) at about 35km off the Phangnga-Phuket coast.[74] Developments of deposits at this depth will demand new technologies and equipment and may be delayed well into the future, especially with the glutted tin market. The Thai Government's search for new reserves reflects its commitment to the tin industry in the long term, but it, also, recognises the need to defer development programmes for the short term. In mid-1985 with the country's state-operated Offshore Mining Organization producing at only 50 per cent of capacity, the government suspended plans to provide additional mining concessions off the Phangnga-Phuket coast.[75]

Indonesia, also, has made recent major exploration surveys. Its state-owned firm, PT Tambang Timah, did exploratory work from 1979 through 1983 (with help from the Netherlands and France) to determine the extent of its tin reserves. In 1984 reserves stood at 740 million tonnes, with 50 per cent occurring offshore in waters of less than 45m.[76]

United Kingdom

In April 1984 a US-backed effort, estimated to have cost £2.9 million ($4.35 million), began to build transport and processing facilities to recover tin-bearing sands along the north coast of Cornwall, especially 'off Cligga Head, west of St. Agnes Head, and along the eastern half of St. Ives Bay'. Tidal action has concentrated tailings sands from earlier tin mining in the district. Grab-sampling and vibrocoring studies indicate the presence of more than 20 million tonnes of reserves of well-sorted sand, with a thickness of about 1.0m to 1.5m. The economically exploitable zone lies slightly less than 1km offshore and varies from 0.4 to 0.8km in width. The placers assay 0.25 per cent tin. A suction dredge will extract the ore and then transport it about 32km to an unloading terminal where a pipeline will carry the slurried ore to Gwithian for processing into concentrate for sale to smelters. The dredge should be able to extract up to 7,000 tonnes of sand per day. When (if) production begins, weather problems may limit mining to about six months during each year, a difficult prospect considering present market conditions.[77]

Gold

One of the world's most recent entrants into placer mining – Inspiration Gold Incorporated (IGI) – has benefited from the recent trauma of the tin industry. Corporate officials decided in 1985 to purchase an idled, fourteen-storey-tall tin dredge, the *Bima*, for $4 million, about one-tenth its original price. After towing the *Bima*, from Singapore to Nome, Alaska (a distance of more than 10,000km), the new owners did test-dredging in 1986. With the test-dredging a success, IGI towed the *Bima* to Seattle, Washington, where it received a $15-million overhaul. Upon returning the dredge to Nome's offshore, it began commercial production during the short summer weather-window in 1987.[78]

The battle-ship-sized *Bima* is stabilised by 80 tonnes of anchors. A bucket-fitted conveyor system daily lifts about 11,000 tonnes of sand and gravel onboard. Forty-eight workers (in twelve-hour shifts) monitor the screening and washing operations, which the owners hope will annually net the company some 900kg of gold, valued at approximately $14 million. During the 1987 work season, the worst problems encountered by the dredgers were weather-related down-time and an occasional walrus that boarded the ship via the escalator-like conveyor system![79]

Conclusions

Because placers are available in marine beaches and nearshore environments, they are more accessible than the deep sea minerals. Overall, the physical and economic problems confronting placer producers should be less severe than those faced by deep seabed miners, and more importantly, most placers lie within the politically safe 200-nmi EEZs. In future, marine placer extraction should form an increasingly significant part of the offshore mining industry. If this development is to take place, however, both industry and government must provide more funding for placer exploration and evaluation programmes.

A continuing vulnerability of the US's platinum supply, and an anticipated growth in demand for the metal, make it imperative that a greater effort be made to locate and develop domestic sources, both onshore and offshore. Further pressure on platinum supplies will occur in coming years, because of a growing use in automobile catalysts in western Europe.[80] Significant increases in demand for platinum's sister metals – palladium, rhodium, ruthenium, iridium and osmium – will also occur.[81]

With future growth likely in the worldwide aerospace industry, as well as in plastic, rubber, and ceramic products production, titanium will continue to play an important role in world markets. According to the USBM, total titanium metal consumption (primary and secondary) is expected to increase annually in the US by 5.5 per cent between 1983 and 2000; the probable annual rate for the rest of the world will be 26.2 per cent. Demand for primary non-metallic titanium in the US could increase by 1.8 per cent each year between 1983 and 2000; the rest of the world should have a rate of 3.1 per cent per annum.[82]

If predictions of the World Bank prove correct, the tin industry should recover in the 1990s, although many of the less efficient producers (as in the offshore of South-east Asia) may have been eliminated. If, however, tin prices reach levels suggested by the World Bank, new producers will probably enter the industry, perhaps even in the offshore of western Cornwall in the UK. Complicating the future tin market, however, is the large stock of tin held by banks and metal dealers when its price collapsed. As tin prices start to rise, these holders will likely begin selling off some of their stocks. This action could help dampen the recovery of the tin market. In sum, caution is the word![83]

Demand for gemstone diamonds is likely to expand as personal incomes increase in the US and other industrialised states. Thus, Namibia's future beach placer operations would seem secure, especially so given that the producing firm is a subsidiary of the De Beers Group which controls 80 to 85 per cent of the gem and natural industrial diamond markets.[84]

Notes

1 P. Stokes, 'Decision soon on tin mine', *Daily Telegraph* (8 Oct. 1986), p. 10.
2 'Geevor tin mines expansion proceeds at a slower pace', *Engineering & Mining Journal*, vol. 186, no. 10 (1985), p. 22.
3 'UK's Cornwall tin mines, teetering on brink of closure, seek capital', *Engineering & Mining Journal*, vol. 187, no. 6 (1986), p. 16.
4 Letter: P. M. Harris, Institute of Geological Sciences, London, 7 Nov. 1978.
5 Stokes, 'Decision soon on tin mine', p. 10.
6 F. C. F. Earney, 'New ores for old furnaces: pelletized iron', *Association of American Geographers, Annals*, vol. 59, no. 3 (1969), p. 532; F. C. F. Earney, *Petroleum and Hard Minerals from the Sea* (Edward Arnold, London, 1980), pp. 12, 18.
7 Letter: C. E. C. Wearne, Manager of Operations, King Island Scheelite Pty, Grassy, Tasmania, Australia, 5 April 1979.
8 Placers may be classified into three groups; each group forms under different physical conditions. Heavy-heavy placers, including gold, tin, and platinum,

have specific gravities of 6.8 to 21.0 and usually are deposited within about 16km of the source rock. Light-heavy placers such as ilmenite, rutile, zircon, chromite, and monazite have specific gravities of 4.2 to 5.3, and are often concentrated in beach sands and may occur hundreds of miles from their parent material. Gemstones (diamonds, for example) have low specific gravities (2.5 to 4.1), are hard and resistant to weathering and occur among gravels and sands. See T. J. Rowland, 'Non-energy marine mineral resources of the world oceans', *Marine Technology Society Journal*, vol. 19, no. 4 (1985), p. 8.

9 See, for example, B. Dimok, *An Assessment of Alluvial Sampling Systems for Offshore Placer Exploration* (Oil and Gas Lands Administration, Energy, Mines and Resources, Canada, Ottawa, 1986), especially p. 2; E. H. Macdonald, *Alluvial Mining: The Geology, Technology and Economics of Placers* (Chapman & Hall, London, 1983), pp. 14-20, 171–2.

10 W. C. Peters, *Exploration and Mining Geology* (John Wiley, NY, 1978), pp. 169, 199.

11 Letter: J. R. Moore, Professor, Department of Marine Studies, University of Texas, Austin, TX, 12 March 1987.

12 P. Rothe, 'Marine geology: mineral resources of the sea', in J. G. Richardson (ed.), *Managing the Ocean: Resources, Research, Law* (Lomond Publications, Mt Airy, MD, 1985), p. 24.

13 D. S. Cronan, 'Marine mineral resources: reaping the mineral harvest of the deep', *Geology Today*, vol. 1, no. 1 (1985), pp. 15-19.

14 V. E. McKelvey, *Subsea Mineral Resources*, Chapter A of USGS Bulletin 1689 (USGPO, Washington, DC, 1986) p. 13; see also J. W. Good, 'Prospects for nearshore placer mining in the Pacific Northwest', unpublished paper, Marine Resource Management program, College of Oceanography, Oregon State University, Corvallis, OR (1987).

15 Letter: N. C. Halcon, Mineral Economics and Information Division, Bureau of Mines and Geo-Sciences, Ministry of Natural Resources, Manila, Philippines, 26 March 1987.

16 Economic and Social Commission for Asia and the Pacific, UN, *Committee for Co-ordination of Joint Prospecting for Mineral Resources in Asian Offshore Areas: Proceedings of the Twenty-First Session, Bandung, Indonesia, 26 November – 7 December 1984, Part I, Report of the Committee* (UN, Bangkok, Thailand, 1985), p. 49.

17 Rothe, 'Marine Geology', p. 26.

18 D. B. Duane, 'Sedimentation and ocean engineering: placer mineral resources', in D. J. Stanley and D. J. P. Swift (eds), *Marine Sediment Transport and Environmental Management* (John Wiley, NY, 1976), pp. 544–7.

19 M. C. Rockwell and K. A. MacDonald, 'Processing technology for the recovery of placer minerals', *Marine Mining*, vol. 6, no. 2 (1987), p. 173.

20 Duane, 'Sedimentation and ocean engineering', pp. 543–4.

21 Letter: C. Cowley, Public Relations Manager, CDM (Proprietary) Ltd, Windhoek, South West Africa/Namibia, 5 May 1987.

22 G. J. Coakley, *Namibia: Mineral Perspectives* (USBM, DOI, Washington, DC, 1983), p. 27.

23 Cowley, Letter, 5 May 1987.

24 B. Horsfield and P. B. Stone, *The Great Ocean Business* (Coward, McCann & Geoghegan, NY, 1972), pp. 264–5.

25 Letter: C. Cowley, Public Relations Manager, CDM (Proprietary) Ltd, Windhoek, South West Africa/Namibia, 12 Feb. 1987; for a helpful discussion of diamond mining in Namibia's coastal zone, see P. M. Bartlett, 'Republic of South Africa coastal and marine minerals potential', *Marine Mining*, vol. 6, no. 4 (1987), pp. 361-9.

26 L. E. Lynd, 'Titanium', in *Mineral Facts and Problems 1985*, USBM Bulletin 675 (USGPO, Washington, DC, 1985), p. 859.
27 'Beach sand bonanza in Madagascar', *Mining Journal*, vol. 308, no. 7,914 (1987), p. 309.
28 H. Beiersdorf, H-R. Kudrass, and U. von Stackelberg, *Placer Deposits of Ilmenite and Zircon on the Zambezi Shelf, Geologisches Jahrbuch*, Reihe D, Heft 36, Bundesanstalt für Geowissenschaften und Rohstoff und den Geologischen Landesämtern in der Bundesrepublik Deutschland (Alfred-Bentz-Haus, Hannover, FRG, 1980), pp. 5, 57.
29 'In search of: minerals', *Marine Resource Bulletin*, vol. 18, no. 3 (1986), p. 5.
30 M. Cruickshank, 'Marine mineral resources survey (interview)', *Sea Technology*, vol. 27, no. 8 (1986), p. 29.
31 A. Couper (ed.), *The Times Atlas of the Oceans* (published by Van Nostrand Reinhold for Times Books, NY, 1983), p. 112.
32 ibid., pp. 111–13.
33 Lynd, 'Titanium', p. 871.
34 ibid. pp. 870, 875–6.
35 J. R. Pederson (comp. & ed.), 'Minerals availability: titanium', in *Bureau of Mines Research 1985: A Summary of Significant Results in Mineral Technology and Economics* (USGPO, Washington, DC, 1985), p. 76.
36 J. R. Loebenstein, 'Platinum-group metals', in *Mineral Facts and Problems 1985*, USBM Bulletin 675 (USGPO, Washington, DC, 1985), pp. 589, 608.
37 R. M. Owen and J. R. Moore, 'Sediment dispersal patterns as clues to placer-like platinum accumulation in and near Chagvan Bay, Alaska', paper presented at the Eighth Annual Offshore Conference, Houston, TX, 3–6 May 1976.
38 Loebenstein, 'Platinum-group metals', p. 601.
39 G. G. Robson, *Platinum: 1986 Interim Review* (Johnson Matthey, London, 1986), p. 5.
40 Loebenstein, 'Platinum-group metals', p. 602.
41 J. F. Carlin, Jr, 'Tin', in *Mineral Facts and Problems, 1985*, USBM Bulletin 675 (USGPO, Washington, DC, 1985), pp. 847, 850–1.
42 L. Bernier, 'Ocean mining activity shifting to Exclusive Economic Zones', *Engineering & Mining Journal*, vol. 185, no. 7 (1984), p. 57.
43 Letter: J. C. Wu, Division of International Minerals, USBM, DOI, Washington, DC, 27 Jan. 1987.
44 Bernier, 'Ocean mining', pp. 57–8.
45 D. I. Bleiwas, A. E. Sabin, and G. R. Peterson, *Tin Availability – Market Economy Countries: A Minerals Availability Appraisal*, USBM Information Circular 9086 (DOI, Washington, DC, 1986), p. 24.
46 Letter: G. L. Kinney, Division of International Minerals, USBM, DOI, Washington, DC, 20 Feb. 1987.
47 'Tin mines in quandary over price crisis', *Bangkok Post*, Economic Review Supplement (30 Dec. 1986), p. 72.
48 Bleiwas *et al.*, *Tin Availability*, p. 24.
49 Kinney, Letter, 20 Feb. 1987.
50 ibid.
51 J. C. Wu, 'The mineral industry of Malaysia', in *Minerals Yearbook, 1984, Vol. III, Area Reports: International* (USGPO, Washington, DC, 1986), p. 539.
52 Wu, Letter, 27 Jan. 1987.
53 E. Mayo, 'Tin', *Mining Annual Review 1986* (Mining Journal Ltd, London, 1986), p. 39.
54 C. J. B. Green, 'London Metal Exchange', *Mining Annual Review 1986* (Mining Journal Ltd, London, 1986), p. 41.

55 V. Vacharatith, 'Thai tin–weathering the worst crisis', *Bangkok Bank Monthly Review*, vol. 27, no. 2 (1986), p. 68.
56 Green, 'London Metal Exchange', p. 41.
57 'U.K. isn't liable for debts of tin council, judge says', *Wall Street Journal* (30 July 1987), p. 35.
58 J. Spooner, 'Tin–trial of errors', *Mining Magazine*, vol. 154, no. 6 (1986), pp. 475-7.
59 J. F. Carlin, Jr, 'Tin', in *Minerals Yearbook, 1984, Vol. I: Metals and Minerals* (USGPO, Washington, DC, 1985), p. 907.
60 G. L. Kinney, 'The mineral industry of Thailand', in *Minerals Yearbook, 1984, Vol. III, Area Reports: International* (USGPO, Washington, DC, 1986), p. 804.
61 Wu, 'The mineral industry of Malaysia', p. 542.
62 Bleiwas *et al.*, *Tin Availability*, p. 12.
63 N. Visetbhakdi, 'Mining', *Bangkok Bank Monthly Review*, vol. 26, no. 8 (1985), pp. 362–7.
64 Kinney, 'The mineral industry of Thailand', p. 804.
65 'Tin mines in quandary over price crisis', p. 72.
66 Mayo, 'Tin', pp. 38–9.
67 'Tin mines in quandary over price crisis', p. 72.
68 Carlin, 'Tin', p. 908.
69 Wu, 'The mineral industry of Malaysia', p. 542.
70 Vacharatith, 'Thai tin', p. 68.
71 'Tin mines in quandary over price crisis', p. 72.
72 Bernier, 'Ocean mining', p. 58.
73 'Offshore drilling for Thai Sn', *Mining Magazine*, vol. 148, no. 1 (1983), p. 12; 'Thailand', *World Mining*, vol. 35, no. 10 (1982), p. 168.
74 Kinney, 'The mineral industry of Thailand', p. 804.
75 As of late 1986, of the 596 registered tin mines, 221 had stopped operating and an additional 166 were not working full time. In 1986, four large Thai tin mining firms closed. In addition, one large dredging firm (Fairmont State Ltd) suspended operation of two of its three dredges. Similarly Jootte Tin Dredging Company stopped operating one of its two dredges, causing unemployment for several hundred workers. See: 'Making the best of a bad job', *Tin International*, vol. 59, no. 8 (1986), p. 267; 'Thai dredges halt operations', *Metal Bulletin*, no. 7,083 (7 May 1986), p. 8; 'Tin toll continues', *Mining Magazine*, vol. 55, no. 2 (1986), p. 77.
76 Bernier, 'Ocean mining', p. 58.
77 L. White, 'Cornish tin mining: 1984', Part II, *Engineering & Mining Journal*, vol. 185, no. 5 (1984), pp. 84–5; see also D. G. Osborne and K. Atkinson, 'Tin off the north coast of Cornwall and offshore testing of beneficiation equipment', *Marine Mining*, vol. 2, nos 1/2 (1979), pp. 45–57; W. W-S. Yim, 'Geochemical exploration for tin placers in St. Ives Bay, Cornwall', *Marine Mining*, vol. 2, nos 1/2 (1979), pp. 59–78.
78 K. Wells, 'On ship off Alaska, all that glitters is gold from sea floor', *Wall Street Journal* (18 Sept. 1987), pp. 1, 7.
79 ibid.
80 M. Siconolfi and N. Behrmann, 'Platinum prices reach 2½-month high amid Japanese and European demand', *Wall Street Journal* (31 July 1987), p. 18.
81 Loebenstein, 'Platinum-group metals', pp. 610–11.
81 Lynd, 'Titanium', p. 876.
83 'Tin: a word of caution', *Mining Journal*, vol. 308, no. 7,899 (1987), p. 17.
84 J. W. Pressler, 'Gem stones', in *Mineral Facts and Problems 1985*, USBM Bulletin 675 (USGPO, Washington, DC 1985), p. 305.

Chapter seven

Construction aggregates and industrial sand

Sand and gravel aggregates and industrial silica sand are the most important of the hard minerals now extracted in the near offshore. Although industrial sand goes mainly into industrial manufactures, it is included here because of its close relationship to sand and gravel aggregates. In the mid-1980s annual worldwide onshore and offshore output of construction sand and gravel totalled an estimated 7,600 million tonnes; industrial sand output was about 180 million tonnes. Only a small portion of these totals came from the offshore, probably not more than 1.5 per cent.[1] In time, however, offshore aggregate production will become more important.

Coastal zone sand and gravel aggregate and industrial sand producers will be increasingly attracted to the oceans. There are several reasons for this trend:

1 along many of the world's seaboards, onshore sources of aggregates and industrial sands have become depleted or built over during urban expansion;
2 producers have been forced out of many urban areas because of increasing taxes and land values;
3 stringent onshore environmental restrictions and zoning ordinances. have been imposed;
4 water is a relatively cheap transport mode when shipping products with high weight and bulk ratios relative to value;[2]
5 they often have superior compressive strength and better shaped particles than onshore aggregates.[3]

Michael Cruickshank, a marine minerals consultant, has stressed that each offshore deposit has relatively unique production constraints and decisions associated with it, all of which are in flux. He notes the importance of careful analyses of a specific resource, including examinations of its location relative to processing site needs and markets, as well as local social and political situations. In addition, seasonal and daily weather conditions and oceanographic characteristics (water depth, waves, and bottom currents) must be considered in selecting the specific mining system needed.[4] Aggregate producers in many world areas such as

Japan, the UK, Canada, and the US are either facing these problems now or will be in the future.

Japan

Japan is the world's most important producer of offshore aggregates. Production began in the early 1960s. By 1985 annual output of sand aggregates totalled about 57 million tonnes.[5] Between 20 and 25 per cent of Japan's supplies of natural aggregate (10 per cent of its total aggregates) comes from marine sources.[6] In 1985 279 firms operated 566 dredges.[7] Most dredging firms are small, with each dredging site employing (on average) about seven workers. In 1981 the industry produced about 18,500 tonnes of aggregates annually per worker, a ratio somewhat lower than in most European offshore aggregate producing countries.[8] Japan's dredgers use clamshell and hydraulic (suction pump) dredges, with the latter the more common. The dredge type used, however, depends on local conditions and environmental regulations. The main production areas are the Setonaikai (Inland Sea) off northern Shikoku and south-western Honshu (60 per cent) and off north-western Kyūshū (35 per cent).[9] Most of the remainder occurs off southern Shikoku and off Hokkaido (Figure 7.1).

Production in the Kyūshū region extends between Karatsu on the south-west and Kitakyūshū on the north-east, with the most important area centred on Hakata Bay near Fukuoka. Deposits lie 2 to 3km offshore at depths of 20 to 30m, with a thickness of 15 to 20m. Reserves here, however, have not been fully explored and evaluated. Strong winds, high waves, and heavy swells pose problems to dredgers.[10] This region anually produces about 5.6 million tonnes of aggregates.[11] Coarse, granitic sands in the Fukuoka area are used in cement blocks and ready-mixed concrete, whereas those sands composed of fine materials, farther to the east, make good plaster. Dredges working off Kyūshū make about two trips each day for a monthly average of fifty cycles. Most dredges here use a hydraulic system to load and a clamshell to unload[12]

In the Inland Sea region, aggregate dredging is most important off Hiroshima, because the area is protected from winds and rough seas. Many narrow straits between small islands, however, help create swift tidal currents. Where tidal currents are strongest, cobbles predominate, and where currents are weaker, sands of various grain sizes occur. The sands are mostly granitic and occur in deposits of 15m to 20m thickness. In contrast to the western Kyūshū region, operators use clamshell dredges for much of the work in the Inland Sea, mainly because they create less turbidity than hydraulic dredges. Dredgers make one trip each work day, averaging twenty to twenty-three trips per month. A typical work cycle (travel to and from the site and loading and unloading) takes about 4.5 hours.[13]

Japan's onshore aggregate industry is experiencing increasingly stringent

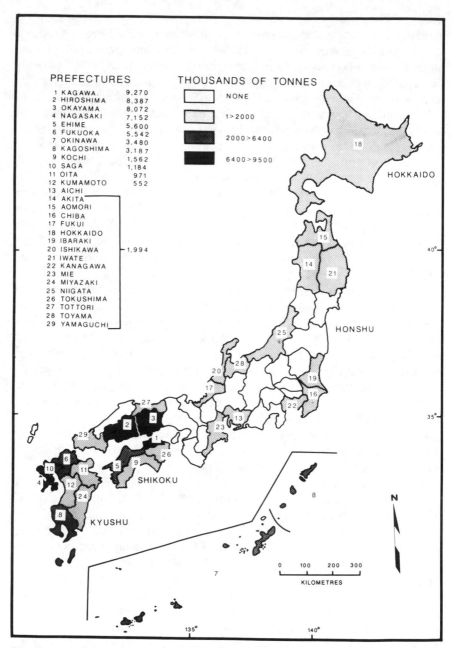

Figure 7.1 Offshore sand production in Japan

Source: Author; data provided by Letter: T. Takeshima, Director, Japan Sand and Gravel Association, Tokyo, 28 May 1987.

environmental regulations. Consequently offshore extraction of sand and gravel is becoming more important. Prefectures control offshore aggregate prospecting and production. They use national sea-water pollution, fishery, port and environmental conservation laws as guidelines for local regulations. Permission to extract aggregates may be given by a prefecture only after the dredger obtains approval from the prefecture's fishery committee. If the committee charges a high mining fee, as often occurs, the marine dredgers may be unable to operate at a profit.[14]

Turbidity plumes are the major problem created by both clamshell and hydraulic dredges, especially the latter. Some aquaculturists contend that previous harbour dredging has created poor conditions for vegetative growth and harvesting of seaweed. Similar difficulties may arise for commercial offshore sand and gravel dredging. Consequently Japanese dredgers have developed a system to de-aerate water discharges during loading, because air bubbles emitted entrain fine silt within them, which increases the size of the sediment plume.[15] Dredging firms cooperate with local fishermen by offering them mined-out sites for use in aquaculture.[16]

Conflicts with fishermen and depleting onshore aggregate supplies encouraged the Ministry of International Trade and Industry and the Geological Survey of Japan to explore for additional offshore aggregate reserves. During the early 1980s Inland Sea producers worked in waters of less than 30m. By the mid-1980s shallow areas were becoming depleted of usable aggregates and at least ten dredgers operated in waters of 45m to 50m.[17] A few operators are working at 70m and exploration geologists have surveyed deposits at depths of 80m. Dredgers expect (and probably will have the capability) to move into deeper areas, perhaps as much as 100m; these waters will require larger dredges and more powerful pumping systems.[18]

In some areas, the increasing use of deep waters occurs because local environmental regulations preclude mining closer than 4–5km from the coast line; shore erosion is the main concern. Japanese producers are constrained also by laws and taxes that make it comparatively expensive to use regular dredging ships. It is cheaper to use pusher barges fitted with detachable, stern-mounted propulsion systems. These barges are less efficient than craft fitted with their own internal propulsion unit.[19]

Besides possible environmental damage and legal restraints, the Japanese (like all marine aggregate producers) face another fundamental difficulty – chlorides. Most untreated marine sands in Japan contain 0.25–0.35 per cent chlorides. Japan's government allows a chloride content of only 0.04 per cent for fine aggregates used in general building and construction work and 0.1 per cent in public works projects. Japan's dredge operators hope to develop efficient methods for first removing from the sand not only salt but also calcareous and volcanogenic materials, and second, obtaining an optimum grain-size mix.[20]

The sand's salt content is primarily dependent on the amount of water it holds. Small-grained sand holds much more water than a coarse-grained sand. When unloaded, most of the water rapidly drains off, but some remains because of surface tension. Eventually this moisture evaporates, leaving a thin residue of salt. The salt can be removed from the sand with fresh water – by sprinkling, by field piling, or by mechanical washing. Onboard sprinkling done during the return trip saves time, but obtaining adequate fresh water is a problem. At-sea desalting is less efficient (especially within the load's lower portion) than is onshore desalting, because the sand often contains a large amount of very fine particles which retain the salt water. Onboard desalting usually must be supplemented with sprinkling onshore where the sand can be spread thinly enough for fresh water percolation to work effectively. Field piling involves storing the sand and allowing rainfall to remove the salt. Approximately 180mm of rainfall percolating through 80cm of sand will remove about 90 per cent of the salt. This method requires large storage areas, and, if rainfall is inadequate for proper leaching, may not always be able to supply the market. Mechanical washing provides the best washed sand, but requires large supplies of fresh water, normally 1.5 tonnes per 1.6 tonnes (1m^3) of sand.[21]

Canada

Recently researchers in Canada have studied that country's potential for offshore industrial sand and sand and gravel (aggregate) production. One investigator, Peter Hale, in the Ocean Mining Division of the Canada Oil and Gas Lands Administration, calculated projected minimum, maximum and most likely production-sales value scenarios for Canada and its individual coastal provinces for the year 2000. Projections were based on existing geological survey data for estimated reserves, their distribution, and their anticipated market values. Although gaps exist, Hale feels the date are good enough to make projections within wide ranges.[22] As of late 1984 Canada's offshore sand and gravel production was limited to the Arctic, where it was being dredged to build artificial islands for oil and gas drilling in the Beaufort Sea.[23] The Arctic region, however, is deleted from Hale's study because of its short and sporadic periods of demand. Government researchers expect offshore production of sand and gravel and industrial sand will soon begin on the east and west coasts of Canada.

Data on British Columbia's offshore aggregates are too fragmentary for investigators to outline its exact supply situation, but reserves should be adequate, because it does not take exceptionally large areas to provide significant quantities of aggregates. For example, sand and gravel deposits with an average thickness of 0.5m in an area of 5km^2 can provide 20 million tonnes of aggregates, given a 100 per cent recovery rate.[24]

On the east coast, Newfoundland-Labrador and Nova Scotia are projected to be the leading sand and gravel producers and sales-value recipients. Approximately 50 per cent of all aggregates produced in the east coast provinces are consumed there. Prince Edward Island (PEI) has a critical shortage of good aggregates. Consequently in recent years, aggregate producers have mined PEI's beaches; in 1982 beach sand and gravel provided a collective total of about 57,800 tonnes of product at twenty-one different sites. The province also imports aggregates from Nova Scotia and New Brunswick. In the distant future, PEI's beach production and imports may be reduced or avoided, because there are large reserves of aggregates in its offshore,[25] but for the near future, neither private industry nor government has plans to develop these offshore reserves. Beach sand removal continues (as of 1987), with about 100,000 tonnes being extracted annually.[26]

Beach mining has been practised in other east coast provinces. In Nova Scotia beach-protection legislation (enacted in 1976) has sterilised many onshore areas where aggregates were produced, causing some local shortages. Nova Scotia could benefit greatly from developing its offshore sand and gravel sources, but as of early 1987, no programmes were under way to do so.[27] Should Nova Scotia's continental shelf oil and gas development programmes become important, its offshore aggregates could be much in demand.[28] New Brunswick, too, has stopped (as of 1 April 1975) all commercial sand and gravel extraction in its coastal zone and islands (300m above and 300m below the high-water mark) and in all tidal waters of streams and estuaries. Before 1975 extensive quarrying of beaches occurred, especially on New Brunswick's eastern coast.[29]

Newfoundland-Labrador is self-sufficient in aggregate production, but local shortages occur. Consequently beach extraction has been resorted to in the past.[30] But beach aggregate removal is now strictly regulated under the province's 1976 Quarry Materials Act[31] and as amended in 1977.[32] The province's Department of Mines and Energy (DME) does not issue a permit for commercial beach removal unless no inland alternative exists. Small amounts (less than 1.6 tonnes) may be extracted for private use, where no alternative inland source is available. The DME maintains detailed inventory records for more than 1,000 beaches, including information on (1) the volume of aggregates removed; (2) recommendations and limitations for removal; (3) complaints from local residents concerning removals; and (4) follow-up inspections. DME geologists in 1987 were working jointly with the national government to identify and assess offshore aggregate reserves, but no offshore extraction is imminent.[33]

Collectively Canada's coastal provinces have good offshore aggregate potential. Canada's offshore sand and gravel production in the year 2000 may range between 1.8 million and 46 million tonnes, with a total value of between Canadian $3.1 million and $172 million (Table 7.1).

Known industrial sand deposits are less widely distributed than sand and gravel, but are of equal or greater value. By the year 2000, industrial sand production in the offshore of eastern Canada alone could range between 441,000 and 3.1 million tonnes, with a total value of Canadian $22 million to $152 million. Industrial sand is an important raw material

Table 7.1 Sand and gravel production scenarios for the Canadian offshore in the year 2000

Production scenarios	Production (Thousands of tonnes)[a]	Mininum value (thousands of 1984 Canadian $)
Minimum		
West Coast	321	400
East Coast		
Nova Scotia	507	948
Prince Edward Island	133	249
New Brunswick	177	330
Newfoundland-Labrador	253	1,200
Total	1,790	3,127
Most Likely		
West Coast	1,043	2,600
East Coast		
Nova Scotia	900	2,550
Prince Edward Island	332	934[b]
New Brunswick	308	865[b]
Newfoundland-Labrador	1,159	3,300
Total	3,743	10,248
Maximum		
West Coast	12,840	48,000
East Coast		
Nova Scotia	12,359	46,200
Prince Edward Island	594	2,220
New Brunswick	4,173	15,600
Newfoundland-Labrador	16,050	60,000
Total	46,015	172,020

Notes: a. Converted and rounded from m³ to tonnes, with 1m³ equal to 1.605 tonnes; b. Rounded to nearest thousand.
Source: P. B. Hale, *A Re-Appraisal of Offshore Non-Fuel Mineral Development Potential*, Ocean Mining Division Document no. 1984–2 (Energy, Mines and Resources Canada, Ottawa, 1984), p. A-16.

for industries with specific chemical and physical quality needs, as in glass and foundry enterprises. Other users include producers of solar cells, insulation, and fibre optical materials. As of 1981, most of Canada's national consumption (72 per cent) occurred in Ontario and Quebec, with much of the supply coming from regional sandstone deposits or from the US as imports (44 per cent).[34]

Known industrial sand reserves exist in the Bradell Bank region to the north-east of PEI; other reserves occur near the Hibernia oil and gas field on the Grand Banks south-east of St John's, Newfoundland. The best reserves lie near the Magdalen Islands in the Gulf of St Lawrence. A private firm, Magdalen Silica Inc., discovered and measured these reserves (estimated at 400 million tonnes). Magdalen Silica's economic analyses indicate that these industrial sands can be competitively delivered in Montreal, Quebec, in Hamilton, Ontario, and in the north-eastern US.[35]

David Pasho, with Canada's oil and Gas Lands Administration, has investigated the UK's experience in planning for offshore aggregate production. He sought to identify key management and regulatory problems that Canada should investigate before its offshore aggregate industry begins on a large scale. For example, dredges may become navigational hazards, cause coastal erosion and deplete sand supplies necessary to maintaining beaches,[36] as well as disturb cables and pipelines.[37] Pasho also noted that fishermen have complained about hazards created by dredges such as irregular seabed surfaces that contribute to difficulties in navigation and in using fishing gear. The navigation argument is rebuttable, because navigation routes can be changed. The dredging industry claims little damage occurs to fish stocks and habitat, if producers observe appropriate restrictions on timing and substrate removal. Some fishery biologists believe that in the North Sea the herring fisheries are the most potentially subject to damage by dredging. But the Ministry of Fisheries contends that past depletions of North Sea herring stocks occurred because of overfishing and had nothing to do with dredging.[38]

United Kingdom

Like Japan, the UK has a major offshore aggregate industry; it dates from the 1500s. When the industry began, empty sailing vessels took on aggregates as ballast to offset the weight of masts and rigging. By the 1600s sand production had become an important industry in south-eastern England.[39] In the 1800s miners towed small barges to sea during high tide, let them settle on the seabed at low tide, loaded them by hand and barrow and returned to port during the next high tide.[40] Later, barge-mounted grab cranes loaded the aggregates on to carrying barges, a method used until the 1920s. Only during calm weather and in relatively shallow areas could this method be used.[41]

In 1919 the annual aggregate output in the UK was 2.3 million tonnes;[42] by the mid-1970s production totalled 120 to 135 million tonnes, with the offshore accounting for 15.6 million tonnes – 12 to 13 per cent of the national total. By 1982 the offshore represented 15 to 16 per cent of the total (Figure 7.2).[43] In 1986 some 27 per cent of the total aggregates used in south-eastern England came from offshore; some counties' consumption of sea-dredged aggregates totalled more than 50 per cent of all aggregates used.[44] Because of diminishing onshore reserves, an increasing public concern for the environment and a likely growth in consumption, offshore dredgers probably will need to produce additional tonnage. Alan Miles, a UK marine geologist, projects production at more than 17 million tonnes by the year 2000.[45]

The main dredging areas are in the north-eastern Thames Estuary; off East Anglia; in the Humber Estuary; within the Southampton-Isle of Wight region; along the Bristol Channel; and to the west of Liverpool in the Liverpool Bay area of the south Irish Sea (Figure 7.3). Significant variations occur in regional demands for aggregate types. Because of geological differences, western regions' onshore aggregate producers must add sand to their products, whereas onshore producers in eastern

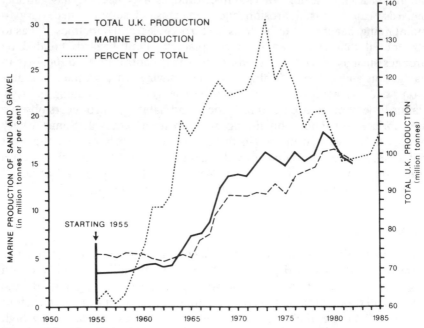

Figure 7.2 The United Kingdom's total and marine production of sand and gravel

Sources: D. W. Pasho, *The United Kingdom Offshore Aggregate Industry: A Review of Management Practices and Issues* (Ocean Mining Division, Canada Oil and Gas Lands Administration, Ottawa, 1986), p. 8; unconfirmed data for 1985 from: J. M. Uren, 'The marine sand and gravel dredging industry of the United Kingdom', paper presented at Offshore Sand and Gravel Mining Workshop, 18–20 March, 1986, Stony Brook, NY, n.p. With permission.

and southern regions require gravel as an additive.[46] The offshore dredging industry helps meet these special needs. The primary market for the UK's marine aggregates is Greater London; the area's dredgers account for more than 52 per cent of the total aggregates dredged from UK licensed areas.[47] During the early 1980s 20–27 per cent of all marine aggregates produced in the UK was exported to continental ports[48] – Dunkirk in France, Zeebrugge and Ostend in Belgium, and Flushing and Rotterdam in the Netherlands (Figure 7.4). The UK's export trade in aggregates has declined since 1980. Rising costs and changing currency values probably account for this trend, in part. Most important, however, is a decline in prices for aggregates on the continent. For example, in Belgium in 1977, prices were 10–15 per cent above those in the UK.[49] In 1982 the Belgian price was 25 per cent lower than in the UK.

Licensing occurs within a two-tiered system – exploration and production. Firms seeking an exploration licence must demonstrate competence (capital or experience), obtain all required permissions, and provide the national Crown Estates Commission (the licensing agency)[50] with a map/chart of the proposed exploration area. No licence may be granted in

158

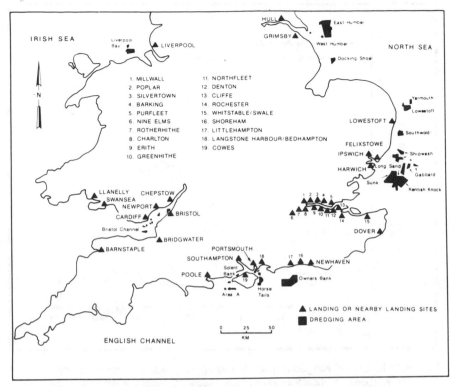

Figure 7.3 United Kingdom sand and gravel dredging and landing sites
Source: After W. J. Spreull and J. M. L. Uren, *Marine Aggregates: Offshore Dredging for Sand and Gravel* (St Albans Sand and Gravel Co. Ltd, UK, 1986), pp. 12,59. With permission.

existing licence areas, within 1.85km (1.0nmi) of a British Telecom/ Central Electricity Board cable or within 0.9km (0.5nmi) of an oil or gas pipeline. Applications for water depths of less than 18m are not usually granted because of the potential for sand depletion of nearby beaches. Most licences are granted for waters less than 37m deep. The dredging firms often hire companies specialising in aggregate prospecting to do their exploration surveys.[51] If economically exploitable deposits are discovered, the dredging firm may apply for a production licence. These licences cover much smaller areas than the exploration licences. The Departments of Trade, of Energy, and of the Environment, and the Ministry of Agriculture are among the various agencies that must review the mining application. Producers pay an annual rental fee and a royalty for each tonne of aggregate dredged from the licence area and landed in the UK. At one time, dredgers were required to remain anchored while dredging. Today, they may dredge while moving, using trailer dredge-pipe systems.[52]

In 1975 the UK had a total of thirty aggregate dredging companies with a fleet of about eighty dredges. By the mid-1980s the number of companies

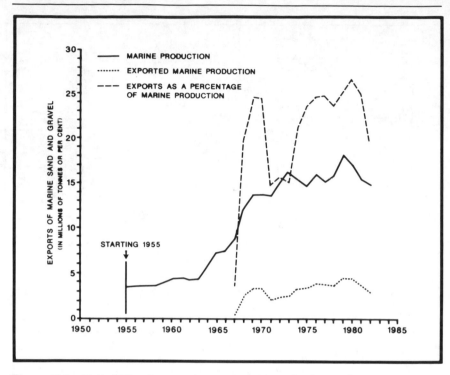

Figure 7.4 United Kingdom marine aggregate production and exports
Source: D. W. Pasho, *The United Kingdom Offshore Aggregate Industry: A Review of Management Practices and Issues*, (Ocean Mining Division, Canada Oil and Gas Lands Administration, Ottawa, 1986), p. 10. With permission.

had declined to about twenty with a total of sixty dredges.[53] Many dredgers, like their early-day counterparts, depend on the tides in their work. They cannot afford regular deep-water dockage (which is too costly in relation to the value of their product), so they use tidal wharves, wherever possible. This dependence on 12- and 24-hour tidal cycles can contribute to an inefficient use of capital, equipment, and labour,[54] although most North Sea dredging operations are on a 24- to 25-hour cycle which then fits the tides nicely, to give an economic cycle, both in time and cost. The future poses additional problems – diminishing shallow-water reserves and an increasingly antiquated dredging fleet. Significant reserves occur at depths below 50m, but centrifugal-pump dredges presently used are economically effective only in waters of 30–35m. It is expensive to replace dredges because construction costs have trebled since the early 1970s.[55] Of the thirty-nine dredges now owned by five major firms located on the east and south coasts, only three have been built since 1980.

Dredge size varies greatly; most have a capacity of 500 to 7,000 tonnes. The smallest are becoming increasingly uneconomic. Dredges with a 4,000 tonne capacity are considered optimal, because large dredges are unable to use many port facilities with shallow drafts. Self-unloading systems, including slurry-pumping and excavator bucket-wheel and conveyor-belt units, discharge the sand and gravel. Pumping systems may unload as much as 3,500 tonnes/hr but require onshore settling ponds and some method of water disposal.[56] If these large vessels cannot be unloaded rapidly, they are uneconomic because of the large capital investment per tonne of capacity. Dredges of 1,000 tonnes' capacity, for example, in 1962 were constructed for £250/tonne of pay-load. By the early 1980s after an increase in the size of the dredges (all dredges built in the UK since the mid-1970s have a capacity of 3,500 tonnes or more), the cost for a vessel with 4,000 tonnes' capacity was £2,000/tonne of pay-load. A dredge producing 1 million tonnes of aggregates annually must now earn approximately £1.00/tonne sold, merely to pay the interest on its capital cost.[57]

United States

US Bureau of Mines commodity specialists expect the demand for construction sand and gravel in the US to grow at an annual rate of about 2.9 per cent between 1982 and the year 2000. The US's industrial sand sector's demand curve should show a growth rate of 2.2 per cent annually during this same period.[58] When a presently imposed moratorium on dredging on the US's outer continental shelf (OCS) areas is lifted, offshore sand and aggregate production could become a major contributor in supplying the needs of construction industries, as well as those industries consuming industrial sand. For example, US consumers in 1986 were importing high-grade silica sands from as far away as Scotland.[59]

Industrial sand is one of the most widely used non-metallic commodities. Of the nearly 24 million tonnes of industrial sand produced in the US during 1983, glass manufacturers used 36 per cent, foundries 26 per cent, abrasives producers 8 per cent, and hydraulic fracturing sand consumers (as in the petroleum industry) 4 per cent. A variety of other industries consumed the other 26 per cent.[60]

Both the sand and gravel and industrial sand extraction industries are highly competitive, because of their relative ubiquity and the low price per unit of weight and volume of their product. Producers with access to relatively cheap and efficient transport systems will overshadow producers equally distant from markets but without competitive transport systems. Thus offshore sand and gravel and industrial sand producers (with their water transport systems) should compete effectively with onshore producers. Before aggregate producers can move into the offshore, however, they must know the extent and quality of resources available to them. To be usable, aggregates must have certain dimensions. Sand is defined as quartz grains of 0.06mm to 2.0mm in diameter with minor amounts of intermixed mica, feldspar, and iron oxides. Gravel includes materials of 2.0mm to 100mm in diameter.[61]

Unfortunately the US's continental shelf sand and gravel deposits, totalling an estimated 1.4 million million tonnes,[62] have not been adequately surveyed and analysed. The north-eastern US seaboard has large amounts of offshore aggregates deposited by fluvial processes of major rivers (as the Hudson) and tidal processes (as in the Georges Bank area), but their precise extent and quality are not known. Some 3,200 million tonnes of aggregates occur along the west coast of the US near Seattle, Washington, Portland, Oregon, and Los Angeles, California, and yet these areas are experiencing aggregate shortages, especially in California's coastal zone from San Diego to San Francisco.[63]

In anticipation of this region's need for aggregates, researchers sponsored by the US Geological Survey began in the early 1980s to develop engineering specifications for a special dredge capable of working in the offshore. One project has focused on the potential for dredging in waters of San Pedro, California, near Los Angeles. The proposed 85-m vessel would have onboard processing (crushing and washing for excess fines and salt) and storage facilities. The dredge (with a capacity of 3,700 tonnes) could handle several loads each day, totaling some 1.1 million tonnes annually, assuming a 300-day work year. Eighteen crew members would be required and the vessel's operating costs are estimated at $5.9 million a year. Investigators have used a computer model to determine the product that would optimise the dredge's economic efficiency. The study shows that (1) producing sand alone, rather than sand and gravel together, would be most profitable and (2) the dredge working out of San Pedro (up to a distance of more than 24km) could effectively compete with onshore producers.[64]

The southern California study's modelling effort did not, however, compare (1) relative efficiencies of other extraction technologies, (2) advantages of onshore processing facilities, or (3) possible variations in dredge size.[65] This study, however, points the way for similar investigations elsewhere, Such studies are needed, because demands for offshore aggregates will increase in future.

A fundamental concern for multiple use of coastal waters, including sand and gravel dredging, fishing, recreation, and navigation, exists in many potentially important offshore source areas for aggregates. The New York Harbor area exemplifies this problem. It is a good example of an aggregate-producing region of long standing and has a potentially important future. Joseph Dehais and William Wallace, who have an intimate knowledge of the aggregate industry in the Greater New York Metropolitan Area (GNYMA), have stressed that onshore reserves may be exhausted in 'ten to twenty years'.[66] Many major regional construction projects planned for the years ahead will demand readily accessible and economical aggregates, a demand that could confirm their prediction. They estimate that present demand in the GNYMA is 12.2 million tonnes/yr, an amount equivalent to extracting 2.6km^2 of usable material to a depth of 0.3048m (1ft).[67]

In the early 1900s sand dredging (mostly by clamshell) in the Lower

Bay of New York Harbor began to provide materials for roads and other public works. By the 1950s commercial producers were extracting an estimated 1.2 million tonnes/yr and were paying the state significant royalties. In 1950 public works extractive operations (which pay no royalties) removed nearly 4.2 million tonnes for use in building the Newark, New Jersey, Airport. Dredging in the harbour peaked during 1963 at about 23.9 million tonnes when 18 million tonnes were dredged to expand the airport. Production declined temporarily in the mid-1960s, then rose again in 1966. From 1966 to 1973 an average of 6.7 million tonnes of aggregates was removed each year.[68]

This sustained, large-scale removal began causing concern for the harbour's environment, especially for fishing and recreational beach use. Habitat alterations and beach sand replenishment were the main concerns. Consequently the New York Department of Environmental Conservation (DEC) and the New York State Office of General Services (OGS) began evaluations of the harbour's dredging activities. In 1966 dredging was confined to a major part of the West Bank of the Ambrose Channel, which passes between Staten Island on the west and Coney Island on the east. East Bank sites were made available in 1968. This action was taken to reduce potential sand loss from Staten Island beaches and to reduce possible damage to good fishing shoals farther south in the harbour. Dredging operations continued to expand in the 1970s; production in 1972 reached 8.5 million tonnes, with the state receiving more than $1 million in royalties from the commercial sector. Because of the dredging industry's increasing importance, the OGS hired a dredging specialist to manage its activities.[69]

Each year environmental groups had been becoming more concerned. In 1973 a major dredging violation was discovered, whereby one firm had dredged to a depth much greater than that allowed (13.7m). The company responsible was fined and made to refill the borrow pit, and all authorised West Bank dredging areas were closed to further mining and were limited on the East Bank until the DEC could make a thorough analysis of dredging's impact in the harbour. The DEC sought to answer six questions:

1 What is the effect of mining on water quality, circulation, and flushing?
2 What are the effects of borrow pits in the West Bank on the rates of shoreline erosion on Staten Island?
3 What are the biological effects of sand and gravel mining operations?
4 What is the quality and quantity of fill and aggregate resources in the Lower Bay?
5 What is the rate of replenishment of these resources?
6 What are the effects of mining isolated, deep pits on the water quality and aquatic life?[70]

While attempting to answer these questions, investigators found that the borrow areas had little impact on the speed of currents and only a small influence on tidal ranges and had somewhat mixed effects on wave-induced beach erosion. Biological impacts were limited to the immediate borrow pit areas. One of the most important findings came from core drilling which showed that good aggregate lies beneath several metres of sediments too find for use as aggregates, the possible reason for the dredging violation that initiated the overall investigation. Subsequently the OGS proposed a plan to (1) allow dredgers to work at a depth of 28m, to enable them to reach aggregate-quality sand and (2) have the borrow areas filled with fine sediments dredged from the harbour's navigation channels. Under this plan, the state would receive royalties on the aggregates dredged, as well as fees for the dredging/waste disposals. It would, also, avoid costs for distant disposal of dredged sediment taken from navigation channels.[71]

A demonstration project designed to prove the utility of sediment disposal in deep borrow pits has been halted by litigation initiated in 1981–2 by a Staten Island fishermen's association. The fishermen claimed the project would be using a productive fishery area. Subsequent fish-population studies have shown that the project's borrow pit does contain a significantly larger fish population than do nearby shoals, perhaps because of the pit's irregular bottom. Although the New York Harbor demonstration project remained stalled in 1986, similar studies (of mounded dredged muds capped with seals of sand) elsewhere have demonstrated the feasibility of safely using borrow pits for disposal of dredged silts.[72] Despite the prolonged litigation, public officials and dredgers want to use New York Harbor's aggregates. In time, when environmental questions are fully answered, the sand aggregate industry here will become active again. Dehais and Wallace have warned, however, that the salt content of the aggregates is near the limit allowed by the New York State Department of Transportation, and if washing the aggregates becomes necessary, further stress will be put on the area's already overtaxed fresh water supplies. They also feel that economies-of-scale requirements and near-shore environmental concerns (which will require large dredges capable of working in deep and distant waters) may result in only limited numbers of producers in the region. Indeed, an oligopoly or monopoly within the industry may develop. Still another problem may arise, that is royalty-payments disputes between various states in the region, as when aggregates are produced in New York waters but sold in New Jersey markets.[73]

Conclusions

In the years ahead, onshore aggregate producers throughout the world will experience increasingly rigorous environmental restrictions. If for reasons of coastal land- and water-use zoning restrictions, sand and gravel mining firms cannot shift to offshore production, local shortages may occur. Onshore depletion and sterilisation of deposits in urban areas will exacerbate the problem.

US Bureau of Mines personnel specialising in aggregates estimate that the cumulative world demand for construction-grade sand and gravel will be about 163,000 million tonnes by the year 2000.[74] Even with shifts to offshore production, good construction gravels may be in short supply, because deposits are usually limited to river mouths, rocky headlands, mountainous coasts and glacial outwash areas.[75] Shortages of gravel may require consumers to shift to crushed stone as a substitute, a more expensive material.[76]

Demand for industrial sand should follow a pattern similar to that of construction sand and gravel. From 1983 to the year 2000, cumulative world demand for industrial sand may be between 3,300 and 3,500 million tonnes. The US, for example, should experience an annual growth in demand of 2.2 per cent throughout the rest of this century. An overall increase in demand does not preclude the development of market problems for industrial sand producers. Substitutions for glass and cast metals may mean less demand for industrial sand in glassworks and foundries. On the other hand, use of fillers in plastic and glass fibres consumed in forced-plastic production could increase demand for silica flour. And if the price of petrolem rises, there will be additional needs for hydraulic fracturing sand used to improve petroleum recovery in oil wells.[77]

Notes

1 J. M. Broadus, 'Seabed materials', *Science*, vol. 235, no. 4,791 (1987), p. 855.
2 F. C. F. Earney, 'Mining, planning, and the urban environment', *CRC Critical Reviews in Environmental Control*, vol. 7, no. 1 (1977), p. 6; see also J. A. Dehais, P. L. Guyette, and W. A. Wallace, 'Onshore pressures make offshore mining viable', *Rock Products*, vol. 84, no. 6 (1981), pp. 72–6.
3 J. M. Uren, 'The marine sand and gravel dredging industry of the United Kingdom', p. 7, paper presented at Offshore Sand and Gravel Mining Workshop, 18–20 March 1986, Stony Brook, NY; this and other papers presented are available in *Marine Mining*, vol. 7, nos 1 and 2 (1988).
4 M. J. Cruickshank, 'Marine sand and gravel mining and processing technologies', pp. 2–4, paper presented at Offshore Sand and Gravel Mining Workshop, 18–20 March 1986, Stony Brook, NY.
5 Letter: T. Takeshima, Director, Japan Sand and Gravel Association, Tokyo, 28 May 1987.
6 T. Iwasaki, H. Okamura, A. Takata, and K. Tsurusaki, 'The status and prospects of sea bed mining in Japan', in J. S. Chung *et al.*, *Proceedings of the Fifth International Offshore Mechanics and Arctic Engineering (OMAE) Symposium, 13–18 April, 1986, New York, Vol. II* (American Society of Mechanical Engineers, New York, 1986), p. 524.
7 Takeshima, Letter, 28 May 1987.
8 M. Arita *et al.*, 'Exploration and exploitation of offshore sand in Japan', ESCAP, UN, *CCOP Technical Bulletin*, vol. 17 (Dec. 1985), p. 86.
9 Iwasaki *et all.*, 'Status and prospects of sea bed mining in Japan', pp. 524–5.
10 ibid., p. 525.
11 Takeshima, Letter, 28 May 1987.
12 Iwasaki *et al.*, 'Status and prospects of sea bed mining in Japan', p. 525.
13 ibid., p. 526.

14 Arita *et al.*, 'Exploration and exploitation of offshore sand in Japan', p. 86.
15 Iwasaki *et al.*, 'Status and prospects of sea bed mining in Japan', p. 527.
16 Arita, *et al.*, 'Exploration and exploitation of offshore sand in Japan', p. 98.
17 ibid., pp. 89, 96.
18 K. Tsurusaki, T. Iwasaki, and M. Arita, 'Seabed sand mining in Japan', pp. 4, 6–7, 19, paper presented at Offshore Sand and Gravel Mining Workshop, 18–20 March 1986, Stony Brook, NY.
19 D. G. Rogich, Internal Report on the Eleventh Joint Meeting of the Marine Mining Panel of the United States and Japan Program on National Resources, at Tokyo, Fukuoka, Nagasaki, and Tsukuba, 12–16 Oct. 1986, US Bureau of Mines, (Washington, DC, 1986, mimeographed), p. 6.
20 Arita *et al.*, 'Exploration and exploitation of offshore sand in Japan', p. 97.
21 Iwasaki *et al.*, 'Status and prospects of sea bed mining in Japan', pp. 527–8.
22 P. B. Hale, *A Re-Appraisal of Offshore Non-Fuel Mineral Development Potential*, Ocean Mining Division Document No. 1984–2 (Energy, Mines and Resources Canada, Ottawa, 1984), pp. 1–7; see also P. B. Hale, 'Canada's offshore non-fuel mineral resources – opportunities for development'. *Marine Mining*, vol. 6, no. 2 (1987), pp. 89–108.
23 D. W. Pasho, 'Canada and ocean mining', *Marine Technology Society Journal*, vol. 19, no. 4 (1985), pp. 26–30.
24 Hale, *Re-Appraisal of Offshore Non-Fuel Mineral Development Potential*, p. A6.
25 ibid., pp. A7–15.
26 Letter: W. MacQuarrie, Mineral Development Co-ordinator, Department of Energy and Forestry, Prince Edward Island, Charlottetown, PEI, Canada, 9 Feb. 1987.
27 Letter: G. Prime, Geologist, Nova Scotia Department of Mines and Energy, Halifax, NS, 7 Jan. 1987; for a useful discussion of the prospects for aggregate production in Nova Scotia, see C.K. Miller and J. H. Fowler, 'Development potential for offshore placer and aggregate resources of Nova Scotia, Canada', *Marine Mining*, vol. 6, no. 2 (1987), pp. 121–39.
28 J. H. Fowler, 'Aggregate resources in Nova Scotia', Open File Report 465 (Nova Scotia Department of Mines and Energy, Halifax, 1982), p. 10.
29 Letter: J. J. Thibault, Coastal Zone Geologist, Department of Natural Resources and Energy, Fredericton, New Brunswick, 8 Jan. 1987.
30 Hale, *Re-Appraisal of Offshore Non-Fuel Mineral Development Potential*, pp. A7–15.
31 *Quarry Materials Act, No. 45, An Act Respecting the Acquisition of Rights to Quarry Materials within the Province* (St John's, Newfoundland, 11 June 1976).
32 *Quarry Materials (Amendment) Act, Chapter 33, An Act to Amend the Quarry Materials Act, 1976* (St John's, Newfoundland, 5 April 1977).
33 Letter: D. G. Vanderveer, Senior Geologist, Mineral Development Division, Department of Mines and Energy, Government of Newfoundland and Labrador, St. John's, Newfoundland (Feb. 1987).
34 Hale, *Re-Appraisal of Offshore Non-Fuel Mineral Development Potential*, pp. A17, A22.
35 ibid., pp. A17–19.
36 D. W. Pasho, *The United Kingdom Offshore Aggregate Industry: A Review of Management Practices and Issues* (Ocean Mining Division, Canada Oil and Gas Lands Administration, Ottawa, 1986), pp. 1, 21–4, 26–7.
37 F. C. F. Earney, *Petroleum and Hard Minerals from the Sea* (Edward Arnold, London, 1980), p. 27.
38 Letter: J. M. Uren, Chairman and Managing Director, Civil and Marine Ltd, Greenhithe, UK, 16 Nov. 1987.
39 P. Webb, 'A look at changing methods of aggregate mining', *World Dredging and Marine Construction*, vol. 16, no. 23 (1982), p. 11.

40 J. L. Mero, 'Ocean mining: an historical perspective', *Marine Mining*, vol. 1, no. 3 (1978), p. 245.
41 Webb, 'Look at changing methods of aggregate mining', p. 11.
42 F. T. Manheim, *Mineral Resources off the Northeastern Coast of the United States*, USGS Circular 669 (USGS, Washington, DC, 1972), pp. 8, 18.
43 Pasho, *United Kingdom Offshore Aggregate Industry*, pp. 1, 8.
44 W. J. Spreull and J. M. L. Uren, *Marine Aggregates: Offshore Dredging for Sand and Gravel* (St Albans Sand & Gravel Co. Ltd, UK, 1986), p. iii.
45 A. J. Miles, 'The marine sand and gravel industry of the United Kingdom', *World Dredging and Marine Construction*, vol. 21, no. 8 (1985), p. 31.
46 ibid., p. 10.
47 Webb, 'Look at changing methods of aggregate mining', p. 13.
48 Pasho, *United Kingdom Offshore Aggregate Industry*, p. 10.
49 Webb, 'Look at changing methods of aggregate mining', p. 13.
50 Letter: R. Freer, Technical Executive, Sand and Gravel Association Ltd, London, UK, 1 April 1987.
51 The main methods employed are seismic reflection profiling (using sound pulses in the range of 12kHz to 20kHz); these profiles can reach 30m below the seabed. Side-scan sonar (where the pulsed beam is directed at an angle toward the seabed) also provides a useful exploration tool because it is highly sensitive to surface irregularities.
52 Miles, 'Marine sand and gravel industry of the United Kingdom', p. 32.
53 ibid., p. 31.
54 A. A. Archer, 'Sand and gravel demands on the North Sea – present and future', in E. D. Goldberg (ed.), *North Sea Science* (MIT Press, Cambridge, MA, 1973), p. 343.
55 Pasho, *United Kingdom Offshore Aggregate Industry*, p. 16.
56 Miles, 'Marine sand and gravel industry of the United Kingdom', p. 32.
57 Webb, 'Look at changing methods of aggregate mining', pp. 1, 14.
58 L. L. Davis and V. V. Tepordei, 'Sand and gravel', in *Mineral Facts and Problems 1985*, USBM Bulletin 675 (USGPO, Washington, DC 1985), pp. 694, 700).
59 Cruickshank, 'Marine sand and gravel mining and processing technologies', p. 11.
60 Davis and Tepordei, 'Sand and gravel', p. 201.
61 T. J. Rowland, 'Non-energy marine mineral resources of the world oceans', *Marine Technology Society Journal*, vol. 19, no. 4 (1985), p. 7.
62 M. Cruickshank, 'Marine mineral resources survey (interview)', *Sea Technology*, vol. 27, no. 8 (1986), p. 29.
63 D. H. Grover, 'Mining sand in the EEZ', *Sea Technology*, vol. 26, no. 2 (1985), p. 40.
64 ibid., p. 42.
65 ibid.; for a detailed analysis of the OCS off San Pedro, as well as San Diego, see J. R. Evans, G. S. Dabai, and C. R. Levine, 'Mining and marketing sand and gravel: outer continental shelf Southern California', *California Geology*, vol. 35, no. 12 (1982), pp. 259–76.
66 J. A. Dehais and W. A. Wallace, 'Economic aspects of offshore sand and gravel mining', p. 5, paper presented at Offshore Sand and Gravel Mining Workshop, 18–20 March 1986, Stony Brook, NY.
67 ibid., p. 8.
68 H. Bokuniewicz, 'Sand mining in New York Harbor: a chronology', pp. 4–6, paper presented at Offshore Sand and Gravel Mining Workshop, 18–20 March 1986, Stony Brook, NY.
69 ibid., p. 6.
70 ibid., pp. 6–7.

71 ibid., pp. 11–13.
72 ibid., pp. 19–21.
73 Dehais and Wallace, 'Economic aspects of offshore sand and gravel mining', pp. 9, 22.
74 Davis and Tepordei, 'Sand and gravel', p. 695.
75 V. E. McKelvey, *Subsea Mineral Resources*, Chapter A of USGS Bulletin 1689 (USGPO, Washington, DC 1986), p. 17.
76 Davis and Tepordei, 'Sand and gravel', p. 694.
77 ibid., pp. 699, 702–3.

Chapter eight

Industrial chemical materials and coal

Mining companies presently extract several industrial chemical materials in the offshore. These include a variety of calcium carbonates, sulphur, and coal, with the latter providing both industrial chemicals and energy. In future, offshore extraction of other industrial chemical materials may become important, especially if phosphorites can be economically exploited. Additional possibilities include potash and barite; both were at one time extracted in the offshore.

Potash and barite

Potash was once mined (1969–77) in the offshore of the People's Republic of the Congo. The mine closed after becoming flooded.[1] The offshore of the UK might also some day provide potash. Good potential exists in waters near the north-eastern shore of England.[2]

A barite dredge-mining establishment at Castle Island, about 23km south-west of Petersburg, Alaska, closed in the early 1980s, but geologists have continued to prospect for additional reserves nearby.[3] Because a major market for barite is the now-depressed petroleum industry – which uses it in drilling muds – demand has declined drastically in recent years. Since December 1981 drilling-related barite consumption declined by 60 per cent,[4] indicating that the Castle Island barite mine will likely remain closed for several years. Although offshore potash and barite mining have suffered recent setbacks, several other industrial chemical materials are continuing in production or have good prospects for exploitation.

Calcium carbonates

Marine calcium carbonate minerals include the shells of molluscs, the skeletons of corals, and cemented combinations of shells and corals that form a porous limestone called coquina. Calcium carbonates also occur as precipitates, as in the form of calcite or aragonite. These materials have a variety of uses, both in construction materials and in industrial manufactures. Exploitation occurs in numerous marine environments, from Iceland in the North Atlantic to the Bahamas in the Caribbean Sea.

Shells

Commercial dredgers extract calcareous shells in many areas of the world, and potentially usable shell and coral sands are commonly available in lagoons and surf zones along tropical coasts of Pacific and Indian Ocean islands and atolls, as well as the Caribbean. Deposits also occur in the Irish Sea and off Scotland and many other areas of the Atlantic Ocean, our main focus here. Shells have a variety of uses, including the production of agricultural lime, paper, soda ash, dietary supplement tablets, poultry feed, seedbeds for oyster propagation, road metal and cement.

Iceland

Because of Iceland's lack of sedimentary rock, the government has developed a state-owned cement-works (at Akranes) and an agricultural lime establishment (at Reykjavík) that use seashells as their raw material. The cement industry is the more important of the two. The cement-works, Sementsverksmiðja Ríkisins (SR), uses large deposits of shell sands that cover portions of Faxaflói, which lies between Reykjavík and Akranes (Figure 8.1). The main deposits lie at a depth of 30m to 35m; the shells vary in size from <0.5mm to >2.0mm, with most (69 per cent) measuring between 0.5 and 1.0mm.[5]

SR began producing cement in 1958; at first, it controlled all phases of operation – from dredging to distribution. In 1963 dredging operations were transferred to a private firm, Björgun Hf., headquartered in Reykjavík. Björgun uses a syphon dredge to produce an annual shell sand output of approximately 5,000m^3 for the agricultural lime plant and 115,000m^3 for the cement-works.[6]

When the shell sand is pumped on to the dredge, excess water flows overboard. Upon returning to port, the shells are re-slurried and pumped onshore for processing. Although the shells are 90 per cent calcium carbonate, some beneficiation is necessary at the cement-works. Small amounts of locally produced rhyolite and basaltic sand are added to the beneficiated shell sand. Grinding mills then pulverise the mixture which is sent to an inclined and rotating calcining-sintering kiln. The resulting clinkers have small and varying amounts of basaltic sand, gypsum, rhyolite, and ferro-silica dust added to them, whereupon they enter grinding mills. The resulting products provide different types of cement, depending on the additives used.[7]

Coastal vessels transport the cement products, bagged or in bulk, to Icelandic markets. Iceland is nearly self-sufficient in cement production, and to assure future supplies, SR personnel have for several years been locating and analysing other shell-sand deposits along Iceland's western coast. Present reserves in Faxaflói, however, are adequate for the country's needs until at least the year 2020. Although newly identified reserves are not as good as those now exploited, Iceland should have an economically viable seashell-based cement industry well into the future.[8]

Figure 8.1 Iceland's Faxaflói shell-sand reserves
Source: F. C. F. Earney, 'Seashells and Cement in Iceland', *Marine Mining*, vol. 5, no. 3 (1986), p. 309. With permission, Taylor & Francis.

Brazil

Like Iceland, Brazil began producing lime from seashells in the 1950s. Seashell (oyster) production occurs along the shore of the State of Bahia and in coastal lagoons near Rio de Janeiro, as well as at Laguna and Tubarão in the State of Santa Catarina.

Two establishments of Companhia de Cimento Aratu S/A extract fragmented calcareous shells, corals, and algae from Todos os Santos Bay, near Salvador, the capital of Bahia. Their product is used to manufacture cement. Annual shell production from 1984–86 totalled about 515,000 tonnes.[9] Four firms produce seashells in the Rio de Janeiro area. One, Companhia Nacional de Alcalis, a 31 per-cent government-owned enterprise, employs about 1,500 workers, mostly in its processing operations. The firm uses two dredges (each with a crew of ten) to

171

extract, and ten barges (each with a crew of three) to transport the shells to the processing plant. Before the shells are used, they must be screened to remove sand and other fines less than 3mm and then washed to remove the seasalt. The shells may then be processed into chemical compounds, including calcium carbide (used to manufacture acetylene) and sodium carbonate (used in glass, ceramics, petroleum refining, and metal processing). Collectively production by the four Rio de Janeiro firms from 1984 to 1986 annually averaged 82,000 tonnes. Output in the State of Santa Catarina averaged 38,000 tonnes between 1984 and 1986.[10]

Currently Brazil's entire seashell production is consumed within the country and overall output is climbing. Brazilian industrial planners hope to develop more seashell dredging and processing operations in other areas of the country where oyster shells occur.

United States

Several areas in the US produce, or have produced seashells, including California, Texas, Louisiana, Mississippi, Alabama, Florida, Virginia, and Maryland. In the Gulf Coast region road-builders surface roads with shells, where good construction sand, gravel, and stone are not available locally. Shells also provide raw material for many other industries.

Currently Louisiana has an important shell-dredging industry. The state has no major commercial limestone or other calcium carbonate resources. Shells help fill this gap in Louisiana's resource base, and the industry contributes significant revenues to the state's coffers.

Shells supply Louisiana with a truly superior construction material for roads,[11] levees and base pads for shallow-water oil well drilling barges, as well as for use in non-construction needs, including lime, pharmaceutical, poultry feed, and petrochemical production; in water purification; and in sulphur dioxide emission control. Many shells, also, are used as 'reef cultch' for development of live oyster beds. Shells compete well with limestone brought from other states (Mississippi, Alabama, Illinois, and Arkansas), selling (per m^3) for about half the price of limestone. Presently dredgers produce dead-reef oyster shells in the marine waters of the Cote Blanche-Atchafalaya-Four League Bays region, in south-central Louisiana (Figure 8.2). They also extract clam shells from the brackish waters of Lake Pontchartrain.

Although shells have been used in Louisiana's construction industries from the area's earliest history, leasing and commercial exploitation did not begin until 1914, when dredgers began working in Atchafalaya and East and West Cote Blanche Bays. Dredging began about 1924 in eastern and south-eastern Louisiana coastal waters (Lake Borgne, Barataria Bay, and Western Mississippi Sound) and in 1934 in Lake Pontchartrain. Operations were suspended in Lake Pontchartrain in 1944 but began again in 1959.[12] As early as 1940 Louisiana shell production totalled 1,147.000 m^3 annually; by 1950 it had doubled and by 1980 had reached nearly 6.6 million m^3. Production in 1985 dropped to 4,653,000m^3, a

Figure 8.2 The dredge *Mallard*, on oyster reefs in Atchafalaya Bay, Louisiana

Source: Photo by E. Londeree. Courtesy Dravo Basic Materials Company Inc. With permission.

decline of 29 per cent. Although shell output has declined recently, the industry continues to have a major economic impact locally and state-wide. The industry employs some 500 people directly, and indirectly it contributes to the support of another 9,000 jobs in shell-industry-related services and those industries using calcium carbonate as a raw material. Annual payrolls associated with these jobs total more than $50 million; present shell industry investments include more than $60 million in production equipment and dredges; and each year the state collects $2 to $3 million in severance taxes and royalties (Table 8.1).[13] Royalty rates are pegged to the national consumer price index.[14]

Much controversy has centred on the environmental impact of shell dredging in Louisiana waters. As early as 1939 a serious dispute had developed between shell dredgers and oystermen.[15] Shell dredges create considerable turbidity, because their hydraulic suction pipes take in a mixture of shell, mud, and water, with the mud and water being discharged after screening. But more than twenty studies (in Louisiana and elsewhere) have focused on the environmental effects of shell dredging and three complete environmental impact statements (EIS) have been done (Mobile Bay, Alabama;[16] Tampa Bay, Florida; and San Antonio Bay, Texas)[17] which show that shell dredging causes only minor and temporary damage. Overall, shell dredges seem not to have a significant environmental impact. They release no toxins or chemicals into the water, and benthic organisms destroyed by the dredge cut are soon replaced. Turbidity and sediment kills of live oysters seldom occur, because of a rigorous control of allowable dredging sites. Nearly all sediments settle out within about 120m of the dredging site. Furthermore, a reef-dredge travels only about 42m in a 24-hour period. Consequently a dredge disturbs a mere 0.54ha in a 24-hour period, whereas a shrimp boat (with a 12m trawl in tow) disturbs 70ha in only 12 hours. It would, therefore, require a reef-dredge more than four months to disturb the same number of hectares as will the shrimp boat in one work day.[18]

Nevertheless, in recent years, many disputes have developed between shell dredging companies and the State of Louisiana which has been attempting to alter dredging companies' coastal use permits. Environmental groups, such as the Sierra Club, recently joined the State Government in a lawsuit against the US Army Corps of Engineers (ACE), because it had not prepared an EIS before issuing shell dredging permits. The ACE has now prepared and released EISs for both Lake Pontchartrain and Atchafalaya waters.[19] Conflicts between the State of Louisiana and the federal government are long standing; in the late 1960s and throughout the 1970s, they disputed jurisdictional rights for control of coastal dredging. Associated environmental disputes continue, with the ACE being accused of indifference to environmental concerns.[20] Perhaps the recently released EISs will help reduce the steady flow of accusations from environmentalist groups.

In fairness to the ACE, a close examination of its record shows that it

Table 8.1 Louisiana clam and oyster reef shell industry, 1975–85[a]

| Year | Clam | | Oyster | | Total | | Total taxes |
	Production (m^3) (thousands)	Royalties (US$) (thousands)	Production (m^3) (thousands)	Royalties (US$) (thousands)	Production (m^3) (thousands)	Royalties (US$) (thousands)	collected (thousands) (US$)
1975	5,638	1,511	3,678	742	9,312	2,254	433
1980	3,873	1,089	2,722	584	6,595	1,673	286
1985	2,235	991	2,418	898	4,653	1,889	311

Source: G. Douglass, Jr (Comp.), *The Louisiana Shell Industry*, revised (Louisiana Shell Producers Association, n. 1., 1986), p. 4.
Notes: a. Production data converted from yd^3 to m^3, using $35.3147 ft^3$, equals $1 m^3$.

has not been insensitive to environmental problems associated with shell dredging. Because of shoreline erosion, the ACE has set limits for the distance offshore that dredgers can operate. In late 1977 the ACE set a minimum-distance limit of 457m for Atchafalaya Bay and for East and West Cote Blanche Bays. In 1983 the limit was moved outward to 900m. This action reduced by 24,282ha the total waters available for leasing. Lake Pontchartrain's 1977 limit was set by the state at 1.9km (5.6km in its southern portion in Orleans Parish) and remains unchanged. Although loss of land is attributed to erosion created by dredges, part of the problem is caused by a rise in the ocean's level and especially by land subsidence (now occurring in much of coastal Louisiana).[21] Loss of land is also related to the construction of oil field canals, storm erosion, and river changes in delta areas.

In addition to offshore distance limits, Louisiana dredgers may not operate within 254m of exposed sub-aerial reefs or active oil and gas drilling rigs or within 127m of an active oil- or gas-well platform. Shell dredgers have many other specific operating requirements both in Lake Pontchartrain and the Cote Blanche-Atchafalaya-Four League Bays area. In the latter region, for example, dredgers must

1 have location devices in operation at all times and maintain records for delivery to the Department of Natural Resources and Wildlife and Fisheries (the devices are in locked, tamper-proof boxes),
2 undertake offsite restoration, as recommended by the Secretary of the Department of Wildlife and Fisheries,
3 operate only two dredges at one time,
4 report archaeological and historical materials encountered during dredging,
5 discharge dredged materials back into the dredge cut,
6 co-operate in ecological impact studies on the effects of fossil shell dredging.[22]

The Louisiana Department of Wildlife and Fisheries in 1973 estimated that the state has between 70 million m^3 and 137 million m^3 of harvestable shell that should last from 18 to 35 years, assuming an annual harvest of 3.8 million m^3. There are, however, many other known dead shell reefs along coastal Louisiana not now open to exploitation.[23]

Until recently Texas had an active oyster shell dredging industry. The industry began in the 1890s. As dredging technology improved and with rapid development of the coastal region in the 1900s, shell production increased. By the late 1950s and early 1960s as many as twelve dredge operators were working in several coastal bays; they extracted up to 9 million m^3 of shell anually. Production first centred on the Galveston/Trinity Bays area, then shifted southward to Matagorda and San Antonio Bays. In the late 1960s and early 1970s production and the number of operators gradually declined,[24] perhaps because Texas Parks and Wildlife Department rules disallowed shell dredging from or near live or exposed reefs (that is those protruding above the bay bottom sediments), except in special cases.

From 1974 to 1978 Texas waters still supplied an annual average of 2,752,000m^3 of shell.[25] By 1980 only four firms retained dredging permits (all others had opted not to renew), and only one permittee was active. During 1981 and 1982 this last dredging company worked reefs in San Antonio Bay.[26] During 1982 it dredged just over 204,000m^3 of shell.[27] In November 1982 it ended its operations.[28].

Why should this once vigorous industry have declined so rapidly? Officials of the Texas Parks and Wildlife Department suggest that the overall decline came because of depleted shell supplies and competition from lower cost crushed limestone, materials produced by many of the shell dredging firms themselves at inland sites.[29] The *coup de grâce* may have been the recessionary economy of 1981–2. According to some observers, Texas assured the demise of its shell industry when in 1981 it increased royalties from $0.25 to $1.25 per yd^3 (0.77m^3). This increase, however, was related to a special permit granted to the one remaining firm. This permit, also, required the dredger to pay $50,000 annually for environmental monitoring, because the area to be used lay in an especially sensitive oyster reef area.[30]

Texas has made a special effort to maintain living reefs, and when dredging was occurring, carefully monitored siltation rates and associated problems of live reefs.[31] Dredgers were required to transplant living oysters before working at a given site, if there were five or more barrels of living oysters within an area of 232m^2 of the point of extraction.[32] Of the shell dredged in the early 1980s in San Antonio Bay, 3 per cent was used for oyster reef replacement in the immediate area and in other coastal zones of Texas.[33]

As early as 3,500 years ago, Indian peoples harvested oysters in the Mobile Bay and Mississippi Sound areas of Alabama and Mississippi. Commercial dredging for shells began in Mobile Bay in 1946 and continued until the spring of 1982.[34] A study published in 1971 put exploitable oyster shell reserves in Alabama at about 56.2 million tonnes. If the average annual production of 2.2 million tonnes occurring between 1947 and 1968[35] is used as a measure of resource depletion, reserves should have been adequate to support production until 1995. According to R. D. Palmore of Dravo Basic Materials Company, the firm which dredged shells in Mobile Bay, these reserves were not located in areas of the Bay within which the State of Alabama desired to make leases. When Dravo's major customer (a cement plant) began using limestone as its raw material, it ended its Mobile Bay dredging operations. The firm continues to operate dredges in Louisiana.[36]

Florida until recently also had a shell dredging industry in the Tampa Bay region. This activity ended when a cheaper source of lime was developed within about 80km of the shell dredging operation. Along the eastern seaboard's Chesapeake Bay region of Virginia, geologists are exploring for fossilised oyster shell beds. These shell beds are badly needed for seabeds to replace shell taken during each season's oyster harvest. The geologists use shallow penetration seismic equipment to locate the beds, a method they claim is an improvement over the old rod system used in the past. The seismic technique allows exploration in deeper waters and surveys over larger areas. Reef identification in Louisiana is still done mainly by the rod system.[37]

Coral

Extraction of coral (bottom-dwelling, calcium carbonate-secreting coelenterates) from beaches and the surf zone in Sri Lanka is an ancient practice, dating from perhaps 500 BC. In early times maharajas reserved the industry's products for themselves. Builders used the lime exclusively for construction of temples and palaces. Coral extraction continues, but today, the peasantry uses the lime in ordinary construction work, in agricultural liming and as a chew in combination with betel leaf, acrea nut, and tobacco.[38]

Coastal villagers mine fossil coral reefs buried under beaches and extract dead and living corals from fringing reefs (Figure 8.3). Although national legislation now forbids the taking of living coral, the practice continues. Consequently the reef ecology is disturbed and the surf can breach the reef and come ashore with full force, causing erosional damage.[39]

Where miners exploit fringing reefs, they may, with the aid of face masks, use crude tools (knives, crowbars, and rods,) to break off pieces of a reef's dead and living hard coral (*Diploria, Montastrea,* and *Scleractinia*); they then load it on coconut-wood rafts, tow the rafts ashore and then pile the coral to dry for a few days. After the coral dries, workers fire it for two days in small, crude kilns fuelled by wood or coconut shells. With firing completed, the lime is bagged for sale. Villagers along two main stretches of coast extract and process coral; the most important centres of production are Batticaloa, Matara, and Akurala.[40] The south-western coast, with some 300 kilns, is of greatest importance.[41]

At Akurala, relatively sophisticated operations mine buried reefs in the beach zone. A typical enterprise has fifteen to twenty workers who use sledges, crowbars, and baskets to extract and to transport the coral. Motor-driven pumps keep water out of the pits which have an average surface area of 12×12m and a depth of 10m.[42] Unfortunately such pits (especially whem pumping is used) alter ground water conditions. Pumping allows saline or brackish water to penetrate the onshore fresh-water lens. If the pumped water is not piped offshore, it can enter local paddies. Abandoned pits also create pools of stagnant water, making mosquito and fly control difficult.[43]

Coral extraction occurs (sometimes illegally) in many other world

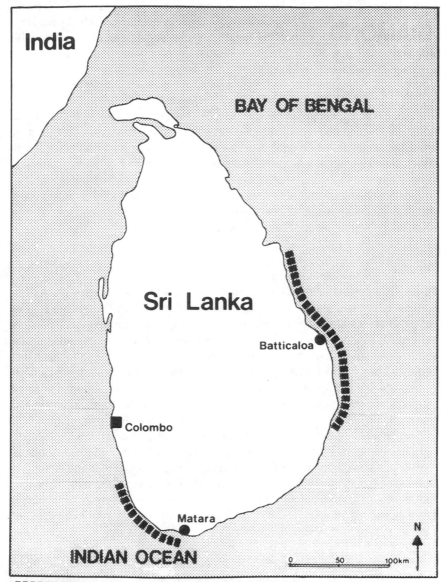

■■■■■■■■■■ CORAL REEFS (OFFSHORE POSITION EXAGGERATED)

Figure 8.3 Locations of beach and offshore coral reef extraction in Sri Lanka
Source: F. C. F. Earney, *Ocean Mining: Geographic Perspectives*, Meddelelser fra Geografisk Institutt ved Norges Handelshøyskole og Universitetet i Bergen, no. 70 (Geografisk Institutt, Bergen, Norway, 1982), p. 11. With permission.

areas. It is often used (as in Taiwan), in jewellery, religious artefacts, and figurines. Jamaican divers harvest black precious coral, and it is also produced in the Hawaiian Islands, off Mauii. Nearby Oahu (at Makapuu) has a delicate pink precious coral. Divers take pink and red coral

179

in waters off Morocco, Algeria, Tunisia, Italy (Sardinia and Naples), and France (Riviera and Corsica).[44] In Australia, at Moreton Bay, Queensland, cement producers legally extract coral from dead reefs. This industry has been active for many years.[45]

Coquina

To the north of Perth, at Shark Bay in the State of Western Australia, quarriers (since the early 1900s) have extracted coquina for use as a building stone. It occurs as consolidated shell concretions that form fossil-beach ridges and is composed almost exclusively of a salinity-stunted (1cm in diameter) bivalve *Fragum erugatum*. Quarrymen used chain saws to extract the relatively soft material. The state suspended production in the mid-1970s, because of beach erosion. Today only enough quarrying is allowed to repair existing buildings in the region.[46]

Aragonite

In the Bahamas the mineral aragonite or 'white gold' occurs in shallow waters lying off the islands of Andros, Bimini, and Eleuthera. The material forms as an oolitic precipitate during the interaction of calcium-saturated cold waters with the Gulf Stream and the warm, ubiquitous shallows surrounding the islands.[47] Spring tides carry the sea-water into shoals where the water mixes and rises in temperature, causing its calcium carbonate to precipitate around nucleic material into concentric laminates. The process contributes to a high and consistent calcium carbonate value (95 to 98 per cent) within the oolites.[48] Investigators estimate the collective tonnage of aragonite in the region at some 100,000 million.[49]

Marcona Ocean Industries Ltd, the Bahamas' only aragonite concessionaire, uses a suction dredge to extract the aragonite from the seabed at depths of 1m to 8m. Dredged material is transferred in slurry form to a barge where it is screened and sent by 1,500-tonne self-unloading barges to an onshore stockpile on a 96-ha combination of natural cays and interfilled area, collectively referred to as Ocean Cay. Here the final, stored product awaits shipment to US and Caribbean markets; about 1,250,000 tonnes of product are produced annually. Aragonite has many useful qualities. It is odourless, tasteless, dustless, and non-toxic. It is easy to handle, has a small grain structure, and a consistent size. Consumers include acid neutralising plants; steel, glass, and animal feed manufacturers; and agricultural and industrial chemical producers.[50]

Phosphorite

Phosphorite is a sedimentary material, containing various phosphate minerals such as francolite, chlorapatite, hydroxylapatite, and fluorapatite; the latter is the most commonly used. Phosphate minerals may occur

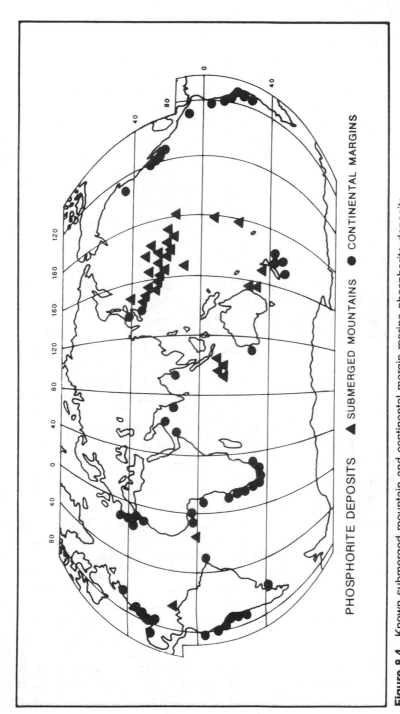

Figure 8.4 Known submerged mountain and continental margin marine phosphorite deposits

PHOSPHORITE DEPOSITS ▲ SUBMERGED MOUNTAINS ● CONTINENTAL MARGINS

Sources: After T. J. Rowland, 'Non-energy marine mineral resources of the world oceans', *Marine Technology Society Journal*, vol. 19, no. 4 (1985), p. 11. With permission.

as a cement in phosphatic-type rocks and form as oolites, granules, and nodules.

Deposits of phosphorite occur on several continental shelves, including the Atlantic waters of the US and those of Morocco, Gabon, the People's Republic of the Congo,[51] Namibia, and the Republic of South Africa. In the Pacific Basin, the US, Mexico, Ecuador, Peru, New Zealand (NZ), and many other islands have deposits near their shores, as do some areas of the Indian Ocean, extending from north-east Africa along the Asian coast to Australia (Figure 8.4).

Phosphorites occur mostly in shallow areas, usually at less than 1,000m (often only a few hundred).[52] They are most common in relatively tropical regions and seem linked to zones of coastal upwelling, divergence, and biological productivity,[53] but some deposits are relict and not necessarily related to contemporary oceanic environments.[54] Most deposits are of Miocene age, reflecting what were probably good environmental conditions for phosphorite deposition at that time.[5]

United States and Mexico

Florida and North Carolina have long been known for their onshore phosphate mining industries. Recently geologists have become interested in neighbouring deposits offshore that extend intermittently from the Onslow Bay area of North Carolina to the south-east of Miami. The Onslow Bay deposits (containing at least eight beds) lie approximately 95km offshore in a band 150km long and 25–30km wide. Five beds have a moderate to high phosphate content (28–30 per cent), totalling perhaps between 1,360 million and 4,000 million tonnes.[56] Farther south, on northern portions of the Blake Plateau – a terrace-like area located about 120km off the coasts of Georgia and Florida – investigators have demonstrated the presence of at least 2,000 million tonnes of phosphorites, valued at approximately $60,000 million in mid-1980s prices.[57] No adequate maximum quantity estimate has been made because sampling devices are unable to penetrate ferromanganese pavements covering much of the area. No primary phosphatisation is occurring there presently.[58] Along Florida's central Gulf of Mexico shore, phosphate-bearing formations extend seaward for 40km. These deposits, however, are not so well documented as are those on the Atlantic seaboard.

Offshore areas along California's central and southern coasts have phosphorites, but these have not been sampled other than superficially. The most significant deposits extend from Point Reyes, just north of San Francisco, to Mexico's Baja California Norte on the south, a distance of more than 2,000km.[59] Deposits occur as nodules, mainly in topographic highs and large offshore banks at depths of 30m to 370m and where sediments are largely absent. California's deposits alone contain an estimated 1,000 million tonnes, with perhaps 100 million tonnes phosphate-rich enough to be exploited. Mexico's deposits occur in a belt 80-km wide and cover an area of about 13,000km^2.[60] In the early 1980s Roca Fosfórica Mexicana SA hoped to develop a dredging operation for phosphoric beach sand. Output was to total 1.5 million tonnes annually. But the firm's engineers found that dredges imported from Singapore could not cut through hard limestone and coral deposits that had not been identified during exploratory drilling. No production has begun, to date.[61]

Other Pacific Ocean areas seem to have significant potential for phosphorites. Researchers studying manganese-cobalt crusts on seamounts and on the sides of volcanic islands belonging to the US (Johnston Island and the Line Islands) simultaneously discovered phosphorites.

New Zealand

Researchers in 1952 discovered an especially important phosphorite area in the South Pacific, the Chatham Rise (150km wide and 1,000km long) that lies at a depth of approximately 400m and extends between NZ's South Island and Chatham Island to the east. The deposits are composed of phosphorite gravel 'intermixed on the sea floor with glauconite foraminiferal muddy sand',[62] with the most promising area situated between 179°E and 180°E (Figure 8.5).

In 1967–8 Global Marine Inc., a California firm, surveyed and sampled the Chatham Rise area.[63] From 1975 to 1978, the New Zealand Oceanographic Institute (NZOI) made four research cruises to determine the phosphorites' distribution, age, geochemistry, and petrology. This research programme was followed by a joint effort that included NZOI and national resource and research agencies of the FRG. Their findings encouraged NZ's largest company (Fletcher-Challenge Ltd) to apply for a NZ prospecting licence; it received a three-year licence in May 1981. Although the NZ Government granted the exploration licence to Fletcher-Challenge, two FRG firms (Preussag AG and Salzgitter AG) were joint-venture members who contributed much of the needed marine technology expertise. Fletcher-Challenge was responsible for the economic and marketing analyses.[64] During 1981 NZOI, Preussag, and Salzgitter, along with the Federal Institute of Geosciences and Natural Resources (BRG) of the FRG, did further distributional mapping work and phosphorite analyses.[65] In 1984 Fletcher-Challenge let its licence lapse; management seemed to have lost interest despite extensive engineering design preparations for mining equipment and plans for pilot testing. Fletcher Challenge's waning interest may have come about because of a lacklustre situation in NZ's general and agricultural economies. According to Fletcher Challenge's management, their final analyses showed the Chatham Rise 'phosphorite resource would not be economic because the deposits (1) are too low a grade; (2) occur in a particularly hostile environment; (3) would have to be crushed before they could be used'. In addition, they expressed concerns for deep seabed mining's being 'a new untried technology'.[66]

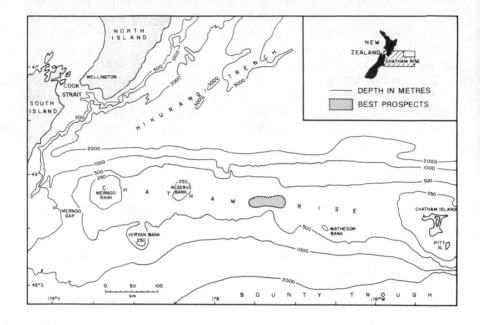

Figure 8.5 New Zealand's Chatham Rise phosphorite region

Source: After U. von Rad, 'Outline of SONNE Cruise SO-17 on the Chatham Rise phosphorite deposits east of New Zealand', in U. von Rad and H. R. Kudrass (comps), *Phosphorite Deposits on the Chatham Rise, New Zealand,* Geologisches Jahrbuch, Reihe D., Heft 65, Bundesanstalt für Geowissenschaften und Rohstoffe und den Geologischen Landesämtern in der Bundesrepublik Deütschland (Alfred-Bentz-Haus, Hannover, FRG, 1984), p. 8. With permission.

Reserve estimates vary from 35 million to 100 million exploitable tonnes, with the FRG participants offering the more conservative amount and NZOI subscribing to the more optimistic figure. Extraction cost estimates vary radically – US$3.50 to US$38.50 per tonne (in 1984 prices) – depending on the mining system used. Preussag and Salzgitter have determined that at depths of 400m a suction-dredge system would be most efficient.[67]

Several physical and economic problems may hinder exploitation of these phosphorites. They are underlain by about 1m of soft, sticky, chalk-like material that may create difficulties for suction dredges, and inter-mixed glacial materials may cause problems for crushers; but engineers seem to have solved both the overburden and glacial materials problems.[68] According to Ulrich von Rad, chief scientist for research cruises in 1981, the patchy distribution of the phosphorites could make exploitation diffi-cult.[69] Another problem includes a phosphate value of 21–22.5 per cent,[70] which is not as high as many onshore sources now exploited. However, because these materials can be applied directly to an agricultural soil without calcination[71] and because their contained phosphates break down slowly (which lets them enter the soil gradually), the low-phosphate-value problem is offset. Two economic difficulties stand out: (1) an excess capacity in the world's cargo vessel industry has contributed to relatively low shipping costs, making phosphate producers elsewhere more competitive in NZ and Australasian markets, and (2) supplier-consumer relationships in NZ's phosphate market are well established (a situation that might make it difficult for a new supplier to compete). On the other hand, the proximity of the Chatham Rise deposits to NZ and Australian markets could be a major advantage to producers.

The NZ Government hopes entrepreneurs will eventually exploit Chatham Rise phosphorites and make the country less dependent on superphosphate imports, but to date its efforts to interest other companies seem unsuccessful, although Lockheed Corporation did express some interest in 1984.[72] How significant the problems identified above were in discouraging Fletcher-Challenge is unclear, but it is likely that a glutted market was a major concern. Until the world market demand improves and unless the NZ Government becomes more receptive to offering private industry financial assistance, the Chatham Rise phosphorites (one of the world's most thoroughly researched marine mineral resources) will remain unexploited.

Future prospects

The world phosphate market is in oversupply. Large inventories in the mid-1980s caused production cut-backs and in 1985 two US firms filed for Chapter 11 protection under that country's bankruptcy law. Production capacity in 1985 was 189 million tonnes, whereas output was only 154 million tonnes. This situation continued in 1987. Difficulties in US agriculture have reduced demand for phosphates. Even in India and in the PRC, stockpiles precluded imports.[73]

Despite the presently saturated market for phosphates, the long-range view indicates an increase in their use.[74] World population growth with concomitant needs for more phosphates in agriculture and other industries (as by cleaning- and water-treatment-product producers) will contribute to an increase in demand. The US Bureau of Mines projects a probable annual world demand of 250 million tonnes in the year 2000.[75]

In future, a portion of world demand may be met by offshore phosphorite production. Why, with nearly a 23 per cent (as of 1985) overcapacity in production capability, would this be true? There are at least three reasons, all of which could encourage onshore producers to look to the offshore, especially in the US – the world's largest phosphate producer. First, onshore federal, state, and local environmental regulations are increasingly difficult to meet;[76] second, changing land-use ordinances associated with the poor public image of the phosphate mining industry are putting added stress on mining companies; and third, high-grade onshore deposits accessible by open-pit techniques are depleting,[77]

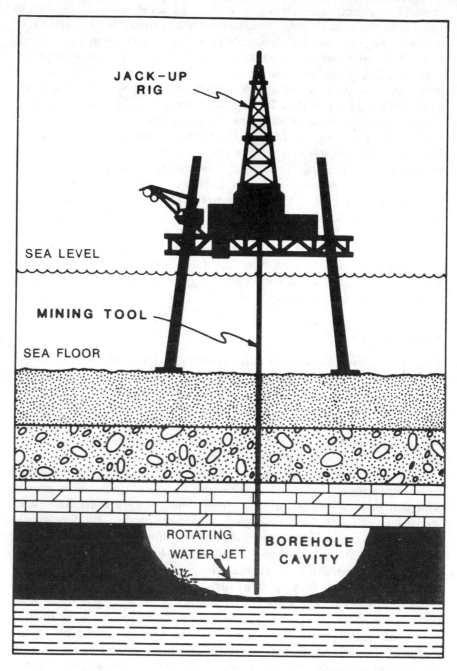

Figure 8.6 Proposed offshore borehole phosphorite mining system

Source: M. Cruickshank, J. P. Flanagan, B. Holt, and J. W. Padan, *Marine Mining on the Outer Continental Shelf*, Environmental Effects Overview, OCS Report 87–0035 (Minerals Management Service, DOI, (Washington DC, 1987), p. 31.

although applications of new technologies (such as borehole mining) may reduce this problem.

US researchers first experimented with borehole mining for phosphates in the 1960s, but no further work took place until 1980. In 1984–5 additional experiments gave encouraging results, indicating that the technique is potentially viable. Borehole technology extracts phosphates as a slurry, obtained by using a high-speed water jet to fragment the ore.[78] Borehole mining has several advantages whether used onshore or at sea[79] (Figure 8.6), although offshore testing has not yet been done. Advantages include

1 a minimal amount of pre-development work needed to begin production;
2 no need to remove overburden and the depth of the beds is not a major problem;
3 a reduction in environmental disturbances;
4 an ability to pump the wastes back into the cavity created, helping to reduce subsidence;
5 a fragmented-ore slurry that lends itself to pipeline transport for immediate delivery to processing sites;
6 a significant increase in usable phosphate reserves.[80]

Sulphur

Numerous manufacturing industries and agriculture depend on sulphur and sulphur products. Its primary uses are in phosphatic fertilisers and in manufacturing processes where sulphuric acid is needed such as chemicals, paper, pharmaceuticals, paints, and steel. Native sulphur is associated with salt domes in the offshore of the US Gulf Coast. Of the US's 329 offshore and onshore salt domes, 24 are capped by rock that contains sulphur. The sulphur results from complex bacterial and chemical reactions. Vincent McKelvey succinctly described the process:

> The cap rock forms when the top of a salt diapir penetrates an aquifer; the halite dissolves and leaves a residue of relatively insoluble anhydrite. Sulfate-reducing bacteria in the presence of hydrocarbons oxidize organic matter, reduce sulfate, and emit carbon dioxide and hydrogen sulfide as waste products. The carbon dioxide reacts with calcium ions from the anhydrite to form calcium carbonate, and the hydrogen sulfide oxidizes to elemental sulfur. The reaction is:
>
> $$CaSO_4 \rightarrow H_2S + CaCO_3 + H_2O$$
>
> $$2H_2S + O_2 \rightarrow 2S + 2H_2O.[81]$$

Most sulphur is produced onshore, but one large offshore mine – the Grand Isle – operates in the US within shallow (15m) waters, about 11km from the coast of central Louisiana. Freeport-McMoRan Inc. owns the facility and uses the Frasch method to extract the sulphur which occurs within sulphur-bearing limestones that cap the salt domes. Directional drilling at an angle of 60° from the vertical allows extraction of sulphur for a radius of about 500m. McMoRan produces offshore sulphur competitively, because it has developed a proprietary technology to use seawater (rather than fresh-water) in its extraction process, without scaling and corrosion.[82] Producing sulphur by the Frasch system requires the injection of superheated (163°C) water into the sulphur formations. The melted sulphur migrates to collection pipes where it is pushed to the surface by a compressed air pipe.[83]

Using the Frasch extraction process has several requisites. The sulphur deposits should have uniform permeability, yet be relatively separated from other permeable strata. The deposits also must lie at a depth adequate to prevent flashing while the injected water temperature is at 150°C to 170°C, but shallow enough that the well's cost is not prohibitive. Finally, because the Frasch process is thermally inefficient, deposits must be rich in sulphur.[84]

Freeport-McMoRan also owns another offshore Frasch sulphur mine (located 15km from the Grand Isle unit), the Caminada, which has been closed for nearly two decades. In 1969 after only one year of operation, the Caminada closed. The reason was glutted sulphur markets, stemming from by-products sulphur extraction[85] in environmental control systems, as in oil refining and gas processing. By-product sulphur production continues as a major factor in the sulphur market. In 1984 it accounted for 55 per cent (28.3 million tonnes) of the total world production. (51.9 million tonnes).[86] But by the mid-1980s sulphur markets were expanding enough that Freeport-McMoRan decided to reactivate the Caminada (at a cost of $35.5 million). It was due to begin production in 1987.[87]

The company from 1981 to 1985 annually averaged $231 million in elemental sulphur sales and also received an average of $197 million from sales of phosphoric acid and other products. About 10 per cent of its phosphoric acid sales came from raw materials purchased from oil and gas by-product sulphur producers, mainly from oil refineries in Texas, in northern Florida, and in southern Mississippi and Alabama. Florida's phosphate industry consumes a major part of McMoRan's sulphur sales.[88]

Proved developed reserves of sulphur reported by Freeport-McMoRan (as of late 1985) for its three mine sites (one of which – the Garden Island Bay – is onshore) totalled nearly 14.4 million tonnes.[89] Grand Isle's proved developed reserves were more than 3.9 million tonnes[90] and the Caminada had slightly less than 6.2 million tonnes. Production from the Grand Isle and Garden Island Bay facilities totalled almost 2.1 million tonnes, meaning that the company's reported reserves are rapidly depleting. Consequently it recently expanded exploration programmes for additional reserves, although not all its efforts are focused in offshore areas.[91]

Coal

Miners extract coal from beneath the sea (using land-based tunnels) in the UK, Turkey, Taiwan, Canada, Australia, NZ, Chile, and Japan. Recently producers in Japan have developed offshore shafts built on artificial islands.

Chile

Two state-owned undersea collieries of Empresa Nacional del Carbon S.A. produce coal in Chile – the Lota Mine at Lota and the Schwager Mine at Coronel; both locations are about 40km south of the city of Concepción.

Miners annually extract some 840,000 tonnes of usable coal, with the collieries operating three shifts each day. The Lota produces about 540,000 tonnes annually, 64 per cent of the two mines' total output.[92] Management hopes by 1988 to have expanded annual production of the Lota to 720,000 tonnes and the Schwager to 450,000 tonnes.[93] As these mines expand, officials hope eventually to capture methane gas[94] (for the generation of energy locally) as a by-product of the coal production process.[95] Coal production can easily expand, but the future could be somewhat clouded, if world oil prices should decline again. When the price of oil dropped to US$14.00/bbl these collieries could barely compete, because coal was selling at a little over US$50.00/tonne.[96] All coal produced is consumed within Chile; shipments go by rail, truck, and ship, with the latter carrying 60 per cent of the total traffic (Figure 8.7).

The Lota establishment employs about 3,000 men underground (offshore) and the Schwager about 2,100. Recently the two collieries' workforce has been cut nearly in half, primarily through increased mechanisation and changes in national labour laws. Still 55 per cent of production costs are for labour. Working conditions are difficult, especially in the Schwager, which has a moderately pitched (20°–22°) working-face where the coal seam's thickness varies from 1m to 60cm. Miners at the working-face (the best paid in the mines) earn about US$4.00/day, plus a piece-rate bonus. A miner's monthly wage, on average, may be about US$250. Each year approximately 1.5 per cent of the underground workers develops symptoms of pneumoconiosis, a rate common to many collieries. If a worker becomes 40 per cent disabled, he is eligible for a disability pension.[97]

Canada

Only one region in Canada has undersea collieries – Cape Breton Island in Nova Scotia. Undersea coal was once mined at Port Hood[98] on the western shore of the island, but the most important undersea coal mining area centres upon Sydney on the eastern shore. As early as 1673 settlers

Figure 8.7 Coal loading terminal for the Lota undersea colliery at Lota Alto, Chile
Source: Courtesy Empresa Nacional del Carbon S.A. With permission.

extracted coal in coastal outcrops near Mabou and Sydney Harbour, but not until 1720 did organised mining begin (at Morien Bay); this coal supplied fuel for workers constructing Louisbourg Fortress. Undersea mining commenced during the last quarter of the nineteenth century. In 1893 the Dominion Coal Company was formed, and under its administration – especially during the First and Second World Wars – coal output greatly expanded. With the post-Second World War decline in coal consumption, the Cape Breton mines experienced difficult times; in 1967 federal legislation established the Cape Breton Development Corporation (CBDC), which received a mandate to reorganise and rehabilitate the Cape Breton collieries. While CBDC was phasing down overall output, the petroleum crisis of 1973 developed. As the need for petroleum became more acute, the region's coal output expanded once again.[99]

Today, three collieries – the Lingan, the Prince, and the Phalen are operating; the latter began producing only in mid-1987. The Lingan started production in 1972 and employs nearly 1,000 workers. The Prince, opened in 1975, has 750 workers. Collectively the three collieries employ approximately 2,000 miners in their undersea operations and have about 3,700 employees, overall. A fourth undersea mine, the Number 26 Colliery, was closed by fire in April 1984 (Figure 8.8).[100]

The mines have considerable methane, both in the coal and within overlying sediments. Miners control the methane by drilling boreholes and connecting pipes and pumps to them to capture about 50 per cent of

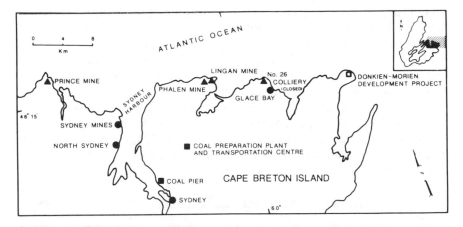

Figure 8.8 Cape Breton Island offshore coal mining region
Source: After Cape Breton Development Corporation, *Coal: The Energy Opportunity* (CBDC, Sydney, Nova Scotia, Canada, 1982), n.p. With permission.

the gas before it seeps into work areas. The collieries use both advance and retreat longwall mining techniques.[101] Past production has extended 8km offshore, but as of 1987 all operations were within about 5km of shore. In total, the Prince, Lingan, and Phalen collieries require about 60km of conveyor systems to move the coal to processing facilities onshore. Prior to the Phalen Colliery's opening, overall annual production of usable coal in these undersea mines varied from about 1.2 million tonnes in 1972 to 2.8 million in 1983–4 and 2.1 in 1986–7. When the Phalen is fully operative, the total annual tonnage for the three establishments will be nearly 4.0 million tonnes; the Phalen's production capacity can be expanded to 2.6 million tonnes per year.[102]

In the late 1970s the federal government did an offshore exploration programme, including core drilling, sonic and neutron and seismological-acoustical studies. These investigations identified large coal reserves (2,000 million tonnes) in Cape Breton Island's Donkin-Morien area. An estimated 50 per cent of these reserves is recoverable. Seam thickness varies, with a maximum of more than 3.5m. Prospects look good enough that studies have been done to determine the feasibility of opening a new mine here. When coal prices and the market merit it, this mine can be developed on a large or small scale.[103]

The Cape Breton collieries should continue producing well into the future. They have an excellent location relative to North Atlantic sea routes to coal-consuming regions of western Europe, South America, eastern Canada, and the north-eastern US, a region first supplied by Nova Scotian coal in 1724, when ships plied waters between Halifax and Boston.[104]

United Kingdom

Undersea collieries have operated in Scotland's Firth of Forth district at least since the 1600s,[105] and until the early 1980s miners extracted coal from beneath Solway Firth, a north-eastern arm of the Irish Sea in Scotland and England's borderland.[106] Today, seven major collieries operated by British Coal (formerly the National Coal Board) extract coal from under the sea. These collieries cluster around Newcastle, in England's north-eastern Durham and south-eastern Northumberland Counties. One colliery, the Seafield, operates on the north shore of the Firth of Forth, across from and due north of Edinburgh, Scotland.[107] These seven establishments employ about 12,500 miners, with a majority working in undersea areas (Table 8.2).

Table 8.2 Miners employed in British Coal's collieries with undersea mining areas

Colliery and county location	Number of workers
Scotland	
Seafield, Fife	850
England	
Ellington, Northumberland	2,000
Wearmouth, Durham	2,400
Easington, Durham	2,100
Westoe, Durham	2,000
Dawdon, Durham	1,800
Vane Tempest, Durham	1,400
Total	12,550

Source: Letter: P. Heap, Senior Press Officer, British Coal, London, 16 March 1987.

In 1980 employment in 14 undersea collieries totalled about 14,000.[108] According to Peter Heap, Senior Press Officer for British Coal, the reduced employment reflects a general decline in the coal industry brought on by the closure of worked-out and inefficient mines. A miners' strike in 1984–5 also forced the closure of some marginal establishments that, for a few more years, might have continued breaking even.[109]

A significant element in coal's decline is changing energy supply-demand relationships associated with North Sea petroleum production. Electrical generating facilities consume much (63 per cent in 1983) of the UK's undersea coal production. Other users include coking plants and aluminium smelters. Heap feels that North Sea crude oil production has had little impact on the UK's coal industry, because production there is mainly light crudes which do not compete directly with coal. North Sea gas is another matter. During the last fifteen years gas has captured more than 60 per cent of the UK's domestic central heating market and also much of the market for quality industrial-process heating. The demise of the UK's undersea coal industry, however, is not imminent. Two factors stand out that, in the long term, should sustain the industry: first, the

UK's gas and oil industry peaked in the mid-1980s and will now likely decline during the next forty to fifty years, whereas large reserves of coal will remain available; and second, shallow coal measures to the west, near the Pennines, have become depleted and the industry's centre of gravity has shifted to the east coast where the coal measures, although deep, have thicker seams and are known to extend far out under the sea.[110]

British Coal, between 1958 and 1965, financed major offshore exploration programmes to identify reserves and to assist in planning for future expansions. Using a special drilling rig, drillers put down eighteen bore-holes (covering 200km^2) in the offshore of Durham and Northumberland Counties. The bore holes' depth averaged 600m. This work identified large reserves of quality coal lying at 270m and 540m below the seabed.[111] The North Sea coal measures probably extend off the coast beyond economically feasible limits of undersea production which, through innovative engineering research, could reach 35km offshore. Deposits beyond this point may, in future, be exploitable by *in situ* gasification.[112] Although UK coal production will continue to face problems, newly developing technologies, along with uncertain petroleum supplies, should provide incentives to maintain, even expand, some undersea collieries.

Japan

Undersea coal mining in Japan began in 1920 in the Ariake Bay area of western Kyūshū. Mitsui Coal Mining Company continues work there today. To provide adequate ventilation for very hot working conditions and to afford easier access to the undersea workings, Mitsui has constructed three artificial islands that contain shafts and air circulation systems (Figure 8.9).[113] As of mid-1987 two other Japanese coal producers operated undersea collieries. Matsushima Coal Mining Company has a colliery in western Kyūshū (near Sasebo) and Taiheiyo Coal Company operates another in south-eastern Hokkaido, near Kushiro. In total, Japan's three operating undersea collieries employ approximately 6,700 miners and they annually produce about 7.9 million tonnes of coal (Table 8.3). Until December 1986 a fourth undersea colliery operated near Nagasaki on Kyūshū.[114]

Conclusions

Offshore production of industrial-chemical materials has a mixed future. Nearshore shell dredgers in the US face tough competition from onshore crushed limestone producers and are experiencing more restrictive environmental regulations. This trend will likely occur in other world areas where dredgers compete with oystermen and fishermen. Controversy will continue about the environmental damage done by shell dredgers, even though many studies in the US indicate that their damage is minimal, if properly regulated.

Figure 8.9 Hatsu-shima, an artificial island containing air ventilation systems for Mitsui's undersea collieries
Source: Courtesy Mitsui Coal Mining Company Ltd. With permission.

Table 8.3 Coal production and average number of employees in Japan's undersea collieries during 1 April 1986 to 31 March 1987

Colliery	Operator	Coal production (Tonnes of Clean Coal)	Average number of employees
Miike	Mitsui	4,142,080	3,347
Ikeshima	Matsushima	1,431,097	1,377
Kushiro	Taiheiyo	2,290,380	2,010
Total		7,863,557	6,734

Source: Letter: K. Endo, Manager, Energy and Mineral Resources Research Office, Mitsui Coal Mining Co. Ltd, Tokyo, Japan 8 Aug. 1987.

With increasing demands for energy and declining oil and gas reserves in many producing countries, undersea coal producers are doing well, with some areas actually expanding output. Coal's use as an industrial chemical may also become more important as petroleum reserves deplete.

If, in the short term, the petroleum industry expands, offshore barite production in Alaska may commence once again. On the other hand, an expanding petroleum industry (with its refining by-product, sulphur) may cause increased competition for the world's only offshore sulphur producer, in Louisiana. A likely expansion in demand for agricultural products, however, should increase consumption of phosphate fertilisers (a major consumer of sulphuric acid), a situation that should help sulphur producers with good access to markets, as from Louisiana to Florida.

Expanding demands for phosphates could mean declining onshore reserves, although new technologies such as borehole extraction may extend the life of these deposits. Eventually, however, onshore depletion should help make offshore production competitive. Industry analysts expect world demand for phosphates to increase by 3.6 per cent annually between now and the year 2000. Nevertheless, supply should exceed demand until at least 1995. Rising production costs in the US, however, may make that country a net importer by the turn of the century.[115] NZ's Chatham Rise phosphorites may yet be needed in world markets.

Notes

1 See P. A. C. de Ruiter, 'The Gabon and Congo Basin salt deposits', *Economic Geology*, vol. 74, no. 2 (1979), pp. 419–31.
2 See D. B. Smith and A. Crosby, 'The regional and stratigraphic context of Zechstein 3 and 4 potash deposits in the British sector of the southern North Sea and adjoining land areas, *Economic Geology*, vol. 74, no. 2 (1979), pp.397–408.
3 Letter: J. R. Moore, Professor, Department of Marine Studies, University of Texas, Austin, TX, 21 Dec. 1986.
4 'Barite and clay minerals', *Minerals and Materials: A Bimonthly Survey* (June/July 1986), p. 32.
5 F. C. F. Earney, 'Seashells and cement in Iceland', *Marine Mining*, vol. 5, no. 3 (1986), p. 308.
6 ibid., pp. 308–10.
7 ibid., pp. 310–315.
8 ibid., p. 316.
9 Letter: J. Tarcisio de Almeida, Chief Geologist, Economic Minerals Section, Departamento Nacional da Produção Mineral, Salvador, Bahia, Brazil, 11 May 1987.
10 Interviews: L. da Rocha Lima, Director, and A. C. Alcoforado do Couteo, Engineer, Companhia Nacional de Alcalis, Rio de Janeiro, Brazil, 18 Aug. 1986; Letter: J. Tarcisio de Almeida, Chief Geologist, Economic Minerals Section, Departamento Nacional da Produção Mineral, Salvador, Bahia, Brazil, 21 July 1987.
11 See J. L. Melancon and S. G. Bokun, *Evaluation of Reef Shell Embankment: Final Report*. Louisiana Highway Report no. FHWA/LA–81/129 (Louisiana Department of Transportation and Development, Baton Rouge, LA, 1980).
12 C. L. Juneau, Jr, *Shell Dredging in Louisiana, 1914–1984* (Louisiana Department of Wildlife and Fisheries, New Iberia, LA, 1984), pp. 5–6.
13 G. Douglass, Jr (comp.), *The Louisiana Shell Industry*, revised (Louisiana Shell Producers Association, n.l., 1986), p. 3.
14 Juneau, *Shell Dredging in Louisiana*, p. 22.
15 ibid., p. 5.

16 US Army Corps of Engineers, *Final Environmental Impact Statement: Permit Application by Radcliff Materials, Inc., Dredging of Dead-Reef Shells, Mobile Bay, Alabama* (US Army Engineer District, Mobile, AL, 1973).
17 Telephone Interview: R. D. Palmore, Dravo Basic Materials Co. Inc., Mobile, AL, 31 March 1987.
18 Juneau, *Shell Dredging in Louisiana*, pp. 38, 41.
19 US Army Corps of Engineers, *Clam Shell Dredging in Lakes Pontchartrain and Maurepas, Louisiana, Vol I, Final Environmental Impact Statement* (USACE, New Orleans, LA, Nov. 1987); US Army Corps of Engineers, *Oyster Shell Dredging in Atchafalaya Bay and Adjacent Waters, Louisiana, Vol. I, Final Environmental Impact Statement* (USACE, New Orleans, LA, Nov. 1987).
20 Juneau, *Shell Dredging in Louisiana*, pp. 7–10.
21 Douglass, *Louisiana Shell Industry*, pp. 19, 21.
22 ibid., pp. 21–2.
23 Douglass, *Louisiana Shell Industry*, p. 18; Juneau, *Shell Dredging in Louisiana*, p. 35.
24 Letter: J. R. MacRae, Wetland Resources Co-ordinator, Environmental Assessment Branch, Resource Protection Division, Texas Parks and Wildlife Department, Austin, TX, 19 May 1987.
25 Letter: J. Farris, Office of Public Information, Texas Department of Water Resources, Austin, TX, 25 Oct. 1978.
26 MacRae, Letter, 19 May 1987.
27 A. L. Crowe, *Shell Management Annual Report, September 1982-April 1983*, Management Data Series no. 70 (Coastal Fisheries Branch, Texas Parks and Wildlife Department, Austin, TX, 1984), p. ii.
28 Letter: C. E. Bryan, Director, Fisheries Resource Programs, Texas Parks and Wildlife Department, Austin, TX, 12 Feb. 1987.
29 ibid.; MacRae, Letter, 19 May 1987.
30 MacRae, ibid.
31 Crowe, *Shell Management Annual Report*, p. 1.
32 US Army Corps of Engineers, *Final Environmental Statement: Shell Dredging in San Antonio Bay, Texas* (USACE, Galveston, TX, 1974), p. 7.
33 Crowe, *Shell Management Annual Report*, p. 1.
34 Palmore, Letter, 6 March 1987.
35 E. B. May, 'A survey of the oyster and oyster shell resources of Alabama', *Alabama Marine Resources Bulletin*, no. 4 (Feb. 1971), p. 1.
36 Palmore, Telephone Interview, 31 March 1987.
37 'In search of: oyster shell', *Marine Resource Bulletin*, vol. 18, no. 3 (1986), p. 4. Some Louisiana producers continue to find the rod system more efficient. Letter: A. Jordan, Manager, Public Relations, Dravo Basic Materials Co. Inc., Kenner, LA, 14 Oct. 1987.
38 S. Neudecker, 'Coral mining in Sri Lanka', *Sea Frontiers*, vol. 22, no. 4 (1976), p. 215.
39 Letter: [name illegible], Director General, Geological Survey Department, Colombo, Sri Lanka, 8 April 1987.
40 Neudecker, 'Coral mining in Sri Lanka', pp. 215–23.
41 Letter: [name illegible], Director General, Geological Survey Department, Sri Lanka, 8 April 1987.
42 ibid.
43 Neudecker, 'Coral mining in Sri Lanka', pp. 218–20.
44 F. C. F. Earney, *Petroleum and Hard Minerals from the Sea* (Edward Arnold, London, 1980), p. 22.
45 Letter: E. Young, Senior Information Officer, Bureau of Mineral Resources, Geology and Geophysics, Department of Resources and Energy, Canberra, Australia, 11 June 1987.

46 Letters: P. E. Playford, Director, Geological Survey of Western Australia, Perth, WA, 11 May and 16 March 1987.
47 'Aragonite: white gold in the Bahamas', *Carib*, vol. 1 (1978), n.p.
48 T. J. Rowland, 'Non-energy marine mineral resources of the world oceans', *Marine Technology Society Journal*, vol. 19, no. 4 (1985), p. 8.
49 S. Balzer, *Survey of Foreign Offshore Development Activities for Minerals Other than Oil and Gas* (Oil and Gas Lands Administration, Energy, Mines and Resources Canada, Ottawa, May 1986), p. 10
50 'Aragonite: white gold in the Bahamas', brochure supplied by Marcona Ocean Industries Ltd, n.p.; Letter: D. Streicher, Executive Assistant, Marcona Ocean Industries Ltd, Apopka, FL, 30 July 1987.
51 See J. R. Woolsey and D. L. Bargeron, 'Exploration for phosphorite in the offshore territories of the People's Republic of the Congo, West Africa', *Marine Mining*, vol. 5, no. 3 (1986), pp. 217–37.
52 W. C. Burnett, 'Phosphorites in the U.S. Exclusive Economic Zone', in M. Lockwood and G. Hill, *Proceedings: Exclusive Economic Zone Symposium: Exploring the New Ocean Frontier, Washington, D.C., 2–3 October 1985* (DOC, Rockville, MD, 1986), p. 136.
53 Rowland, 'Non-energy marine mineral resources', p. 11.
54 For an excellent discussion of the many variables associated with phosphorite formation, see V. E. McKelvey, *Subsea Mineral Resources*, Chapter A of USGS Bulletin 1689 (USGPO, Washington, DC, 1986), pp. 21–7.
55 Burnett, 'Phosphorites in the U.S. Exclusive Economic Zone', p. 136.
56 S. R. Riggs, 'Geologic framework phosphate research in Onslow Bay, North Carolina Continental Shelf', *Economic Geology*, vol. 80, no. 3 (1985), p. 716; see also S. R. Riggs, 'Future frontier for phosphate in the Exclusive Economic Zone – Continental shelf of Southeastern United States', in M. Lockwood and G. Hill, *Proceedings: Exclusive Economic Zone Symposium: Exploring the New Ocean Frontier, Washington, D.C., 2–3 October 1985* (DOC, Rockville, MD, 1986), pp. 97–107.
57 M. Cruickshank, 'Marine mineral resources survey (interview)', *Sea Technology*, vol. 27, no. 8 (1986), p. 29.
58 W. P. Dillon and F. T. Manheim, 'Resource potential of the Western North Atlantic Basin', in P. R. Vogt and B. E. Tucholke (eds), *The Geology of North America, The Western North Atlantic Region: Vol. M*, (Geological Society of North America, Boulder, CO, 1986), p. 671.
59 Scc H. T. Mullins and R. F. Rasch, 'Sea-floor phosphorites along the Central California Margin', *Economic Geology*, vol. 80, no. 3 (1985), pp. 696–715; Rowland, 'Non-energy marine mineral resources', p. 12.
60 Rowland, ibid.
61 Letter: W. F. Stowasser, Division of Industrial Minerals, USBM, DOI, Washington, DC, 7 April 1987.
62 U. von Rad, 'Outline of SONNE Cruise SO-17 on the Chatham Rise phosphorite deposits east of New Zealand', in U. von Rad and H–R. Kudrass (comps), *Phosphorite Deposits on the Chatham Rise, New Zealand* Geologisches Jahrbuch, Reihe D, Heft 65, Bundesanstalt für Geowissenschaften und Rohstoffe und den Geologischen Landesämtern in der Bundesrepublik Deutschland (Alfred-Bentz-Haus, Hannover, FRG, 1984), p. 7.
63 L. Bernier, Ocean mining activity shifting to Exclusive Economic Zones', *Engineering and Mining Journal*, vol. 185, no. 7 (1984), p. 58.
64 Letter: R. K. H. Falconer, GeoResearch Associates, Wellington, NZ, 25 Feb. 1987.
65 von Rad, 'Outline of SONNE Cruise', p. 7.
66 Letter: D. H. Bryce, Project Co-Director, Fletcher Challenge Ltd, Auckland, NZ, 3 April 1987.

67 Bernier, 'Ocean mining activity shifting to Exclusive Economic Zones', p. 58.
68 ibid.
69 Interview: U. von Rad, Bundesanstalt für Geowissenschaften und Rohstoffe, Hannover, FRG, 26 Sept. 1986.
70 Letter: J. Arden, for the Secretary of Energy, New Zealand Ministry of Energy, Wellington, 30 April 1987.
71 Rowland, 'Non-energy Marine Mineral resources', p. 12.
72 Falconer, Letter, 25 Feb. 1987.
73 M. Mew, 'Phosphate rock', *Mining Annual Review 1986* (Mining Journal Ltd, London, 1986), pp. 101–2.
74 For a good review of current worldwide marine phosphorite research, see W. L. Stubblefield, 'Phosphate minerals on the sea floor: geologic, economic and social aspects of mining', *Sea Grant Research Advances*, Research Note 10 (NOAA, DOC, Aug. 1987).
75 Stowasser, Letter, 7 April 1987.
76 W. C. Burnett, 'Phosphorites in the U.S. Exclusive Economic Zone', p. 136.
77 Riggs, 'Future frontier for phosphates', p. 97.
78 V. E. McKelvey, 'The U.S. Phosphate industry: revised prospects and potential', *Marine Technology Society Journal*, vol. 19, no. 4 (1985), pp. 65–6.
79 Riggs, 'Future frontier for phosphate', p. 104.
80 McKelvey, 'The U.S. phosphate industry', p. 66.
81 McKelvey, *Subsea Mineral Resources*, p. 29.
82 *Freeport-McMoRan Resource Partners, Limited Partnership: Prospectus* (Dean Witter Reynolds Inc. *et al.*, 20 June 1986), pp. 3, 34.
83 Pumping stations send the melted sulphur by insulated pipes to a terminal onshore where it is loaded on to a 'thermos bottle' barge and then shipped (40km) to Port Sulphur on the Mississippi River where it is treated to remove impurities such as ash, carbon and hydrogen sulphide. Finally it may be stored or transferred to ocean cargo vessels for shipment throughout the world.
84 W. C. Peters, *Exploration and Mining Geology* (John Wiley, NY, 1978), p. 200.
85 *Freeport-McMoRan Inc., Form 10-K, Annual Report Pursuant to Section 13 on 15 (d) of the Securities Exchange Act of 1934* (New Orleans, LA, 14 April 1986), p. 5.
86 C. L. Kimbell and W. L. Zajac, 'Minerals in the world economy', preprint from *USBM Minerals Yearbook, Vol. I: Metals and Minerals* (USGPO, Washington, DC, 1986), p. 31.
87 *Freeport-McMoRan Inc., Form 10-K*, p. 5.
88 *Freeport-McMoRan Resource Partners*, p. 38.
89 *Freeport-McMoRan Inc., Annual Report 1985* (New Orleans, LA, 1985), p. 15.
90 *Freeport-McMoRan Resource Partners*, p. 36.
91 *Freeport-McMoRan Inc., Annual Report 1985*, pp. 5, 15–16.
92 Interview: P. H. Crorkan, Manager of Operations, Lota Mine, Empresa Nacional del Carbon SA, Lota Alto, Chile, 9 Aug. 1986.
93 Empresa Nacional del Carbon SA, *Proyecto Aumento Productividad Mina Lota, 1986–2000*, 3rd edn (ENACAR, Lota Alto, Chile, May 1986), p. 1.
94 Letter: P. H. Crorkan, Manager of Operations, Lota Mine, Empresa Nacional del Carbon SA, Lota Alto, Chile, 1 April 1987.
95 Crorkan, Interview, 9 Aug. 1986.
96 Empresa Nacional del Carbon SA, p. 2.
97 Interview: J. Cortez Latorre, Assistant Manager, Production Division, Lota Mine, Empresa Nacional del Carbon SA, Lota Alto, Chile, 9 Aug. 1986.
98 Letter: G. Prime, Geologist, Nova Scotia Department of Mines and Energy, Halifax, NS, 7 March 1987.

99 Cape Breton Development Corporation, *Coal: The Energy Opportunity* (CBDC, Sydney, Nova Scotia, 1982), n.p.
100 Letter: D. MacIssac, Information Officer, Cape Breton Development Corporation, Sydney, Nova Scotia, 9 March 1987.
101 In retreat longwall mining, miners first drive haulage roads and airways to the boundary of a coal seam then mine it in a single face (without pillars) back toward the shaft.
102 Letter: D. MacIssac, Information Officer, Cape Breton Development Corporation, Sydney, Nova Scotia, Canada, 9 April 1987.
103 ibid.
104 Cape Breton Development Corporation, *Coal*, n.p.
105 W. M. Holden, 'Miners under the sea – right now', *Oceans*, vol. 8, no. 1 (1975), pp. 55–7.
106 Letter: P. Heap, Senior Press Officer, British Coal, London, 16 March 1987.
107 British Coal, *Coal in Scotland* (BC, London, 1986), n.p.
108 Earney, *Petroleum and Hard Minerals from the Sea*, p. 13.
109 Letter: P. Heap, Senior Press Officer, British Coal, London, 30 March 1987.
110 ibid.
111 National Coal Board, *Coal in Northumberland and Durham* (NCB, London, 1983), n.p.
112 McKelvey, *Subsea Mineral Resources*, p. 28.
113 Letter: T. Ohuchi, Mining Engineer, Production Management Department, Mitsui Coal Mining Co. Ltd, Tokyo, Japan, 12 June 1987.
114 Letter: K. Endo, Manager, Energy and Mineral Resources Research Office, Mitsui Coal Mining Co. Ltd, Tokyo, Japan, 8 Aug. 1987; Coal Mining Research Centre, *Japan's Coal Mining Industry Today* (CMRC, Tokyo, Japan, 1986).
115 W. F. Stowasser, 'Phosphate rock', in *Mineral Facts and Problems 1985*, USBM Bulletin 675 (USGPO, Washington, DC, 1985), pp. 587, 592–3.

Chapter nine

Sea-water as an ore

The most accessible oceanic ore is sea-water, and a vast resource it is. Covering approximately 71 per cent of the earth's surface, its total volume is calculated at 1,370 million km^3 and its surface area at 361 million km^2. Of all waters on the planet, only a little more than 1.2 per cent occurs outside the oceans – 1.2 per cent as ice, 0.002 per cent as rivers and lakes, and 0.0008 per cent as vapour in the atmosphere.[1]

Contained within this vast amount of sea-water is a large number of minerals (Table 9.1). Most minerals we desire occur in combination with other substances, in this case as solutions within the sea-water. This situation is the major problem in our tapping sea-water for its mineral wealth. The minerals are so minutely disseminated in the water, that most do not meet the normally accepted definition of an ore – a mineral element or compound of elements of sufficient value in quantity and quality that it can be profitably extracted. Only a few minerals are presently extracted from sea-water for a profit – sodium chloride, magnesium (with gypsum[2] and potassium compounds as by-products and bromine as a co-product) and fresh water.

Fresh water

In the distant future, potable water may be the oceans' most valuable mineral resource. In extremely arid regions, it already is. Indeed, in some places, distilled sea-water is now delivered at a price below that of stored water.

Ocean-going ships have used on-board sea-water distillation plants for well over a century. In 1980 more than 2,200 land-based salt-water conversion plants (with a capacity greater than 114kl – 25,000 gal) were operating worldwide, although not all were using sea-water. These plants had a daily production capacity of about 9 million kl (nearly 2,000 million gal) of fresh water, and sea-water accounted for 75.7 per cent of all desalting plants' feed water.[3] The design of desalting plants varies greatly, but can be classified into four conversion process types – distillation, crystallisation, membrane, and chemical. Production of fresh water from sea-water is not a problem of feasibility but of economics. In recent years

Table 9.1 Concentration of various elements in sea-water

Element		Concentration (parts/1,000 million)	Element		Concentration (parts/1,000 million)
Oxygen	O	857,000,000	Nickel	Ni	2
Hydrogen	H	108,000,000	Vanadium	V	2
Chlorine	Cl	19,000,000	Manganese	Mn	2
Sodium	Na	10,500,000	Titanium	Ti	1
Magnesium	Mg	1,350,000	Tin	Sn	0.8
Sulphur	S	890,000	Cesium	Cs	0.5
Calcium	Ca	400,000	Antimony	Sb	0.5
Potassium	K	380,000	Selenium	Se	0.4
Bromine	Br	65,000	Yttrium	Y	0.3
Carbon	C	28,000	Cadmium	Cd	0.1
Strontium	Sr	8,000	Tungsten	W	0.1
Boron	B	4,600	Cobalt	Co	0.1
Silicon	Si	3,000	Germanium	Ge	0.06
Fluorine	F	1,300	Chromium	Cr	0.05
Argon	A	600	Thorium	Th	0.05
Nitrogen[a]	N	500	Silver	Ag	0.04
Lithium	Li	170	Scandium	Sc	0.04
Rubidium	Rb	120	Lead	Pb	0.03
Phosphorus	P	70	Mercury	Hg	0.03
Iodine	I	60	Gallium	Ga	0.03
Barium	Ba	30	Bismuth	Bi	0.02
Indium	In	20	Niobium	Nb	0.01
Zinc	Zn	10	Lanthanum	La	0.01
Iron	Fe	10	Thallium	Tl	0.01
Aluminum	Al	10	Gold	Au	0.004
Molybdenum	Mo	10	Cerium	Ce	0.005
Copper	Cu	3	Rare Earths		0.003–0.0005
Arsenic	As	3	Protoactinium	Pa	2×10^{-6}
Uranium	U	3	Radium	Ra	1×10^{-7}

Source: K. S. Stowe *Ocean Science* (John Wiley, NY, 1979), p. 184. With permission.
Note: a. Nutrient nitrogen only; dissolved gas is not included.

the cost of operating most desalting plants has risen significantly, mainly because of rising energy prices.[4] It is the cost of energy that will largely determine the future viability of sea-water desalination.[5]

Distillation

All sea-water distillation methods depend upon the volatility of water and the involatility (below 300°C) of salts dissolved in it. Distillation is a two-step process – evaporation, then condensation.

Solar still and evaporator condenser

The simplest technique is the solar still, designed so that the sun's energy evaporates sea-water, with subsequent condensation of the vapour and collection of the salt-free water. The energy in this system is free, but it has limitations in its operational efficiency, depending on the time of year, weather conditions, and latitudinal location of the still (Figure 9.1).

Another simple method is to create vapour by sending hot steam

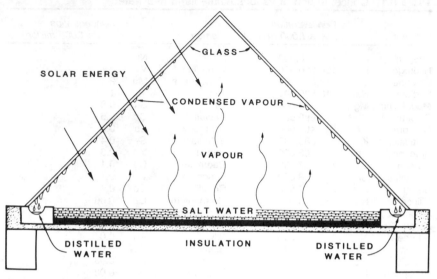

Figure 9.1 Solar still
Source: After Office of Water Research and Technology, DOI, *The A-B-C of Desalting* (USGPO, Washington, DC, 1977), p. 19.
Note: The glass or plastic top is not heated by the sun's rays as much as is the black surface layer below the salt water. As the water heats, some vapour rises, strikes the cooler glass or plastic surface, condenses, and then runs downward into the collecting trough.

through tubes which pass through sea-water in a partially filled evaporator. As the sea-water boils, the vapours leave the evaporator chamber and travel to a condenser, from which the now-fresh water is drained and stored. A major disadvantage of this distillation system is the precipitates left on evaporator and heat-transfer surfaces. This scaly residue reduces the efficiency of heat transfer from the steam to the entering sea-water. If either carbon dioxide or calcium is removed from the sea-water, scales will not form, thus improving the plant's efficiency.[6]

Flash systems

More complicated and expensive sea-water distillation methods include single- and multiple-flash systems. In the mid-1980s multiple-flash systems were the most commonly used distillation method.[7]

With flash-system techniques, pre-heated sea-water (kept under a pressure high enough to prevent boiling) enters a flash chamber that is maintained at a reduced pressure. When the sea-water encounters the reduced pressure, it partially vaporises, condenses upon contact with cold sea-water pipes and then collects in a distillate chamber. If a multiple-flash system is used, each flash chamber is maintained at a progressively lower pressure, resulting in additional vaporisation at each stage (Figure 9.2).[8]

A potentially useful although still uneconomic power source for operating sea-water flash-distillation systems is based upon ocean thermal energy

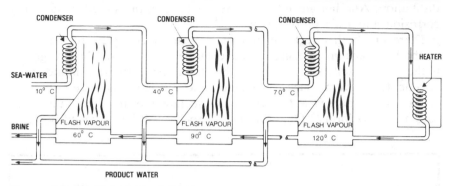

Figure 9.2 Multiple-flash distillation

Source: After Office of Water Research and Technology, DOI, *The A-B-C of Desalting* (USGPO, Washington, DC, 1977), p.15.

conversion (OTEC). Deep, cold ocean waters can be used to condense warm surface waters that have been fed into a flash chamber in the process of generating electricity. Researchers have calculated that with a surface-water to deep-water ratio of 3:1 (assuming a temperature difference of 20°C) a plant could produce 58.3 litres of fresh water for every m³ of deep sea-water pumped to the surface.[9] This technique could be especially useful in tropical areas where great temperature differences occur between deep and surface waters. For example, engineers in Hawaii hope to build a small commercial fresh-water OTEC plant in 1987. OTEC seems on the verge of commercial production, but several problems must be overcome before it can be a viable worldwide source of energy and fresh water. Among these problems are the high cost of plant construction and engineering difficulties associated with heat exchanger biofouling,[10] equipment corrosion, non-condensable gases emitted by the vaporising sea-water in the flash chamber,[11] and the hazards of the cold water pipe's weight.[12]

Crystallisation

Techniques using the freezing process have developed more recently than distillation methods. Like the vapour created during distillation, when ice crystals form they are nearly salt free. An advantage of the freezing or crystallisation principle is that during the melting of 1 kg of ice, it absorbs 335,000 joules (318 BTUs) of energy (the latent heat of fusion) but does not change temperature. This number of joules is exactly the same as that removed from the water in converting it into ice (the latent heat of crystallisation). The consequence of this balanced conversion is that the freeze process, theoretically, has the lowest energy cost of all desalination processes requiring a phase change, because it does not demand the use of costly heat-transfer surfaces to remove the heat from the feed water. It takes approximately 60 per cent less energy than does multiple-stage flash distillation. Another advantage of the low temperatures used is that fewer corrosion and scale problems occur.[13]

There are both direct and indirect freezing techniques. In the direct method or vacuum freezing-vapour compression process, the refrigerant is the water itself, whereas the indirect process employs a liquid more volatile than water, such as butane which, when evaporated while in contact with the sea-water, removes the latent heat of crystallisation, causing ice to form.[14] Only the direct process will be discussed here.

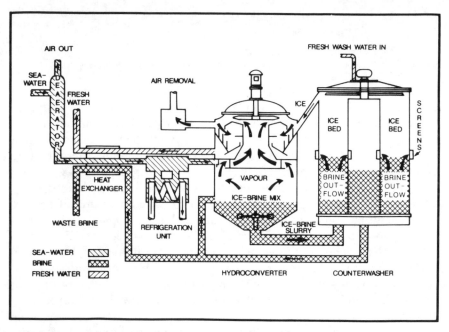

Figure 9.3 Vacuum freezing-vapour compression sea-water conversion
Source: After Office of Water Research and Tecnology, DOI, *The A-B-C of Desalting* (USGPO, Washington, DC, 1977), p. 22.

The direct freezing process feeds pre-cooled and de-aerated sea-water into a hydroconverter or crystalliser, which is under greatly reduced pressure. This action makes part of the sea-water vaporise, causing the rest to cool enough to become a slushy ice. The slush and brine travel to a counterwasher where the brine flows out through screens and the ice rises to the top where it is washed and sent to a melting tank above the hydroconverter. Melting occurs from heat generated by the condensation of vapour pulled out of the hydro-converter, a necessary step to maintain the evaporation and cooling process there (Figure 9.3).[15]

Membrane

Sea-water conversion may be accomplished by using membranes as selective separators, which (under certain conditions) act to separate one substance from another. Electrodialysis and reverse osmosis are two (among several) of the membrane desalination processes.

Electrodialysis

Sea-water can be converted to fresh water by using electrodialysis, which is achieved by constructing a series of concentrating chambers (up to several hundred) separated by alternating kinds of special membranes that are permeable to positively or negatively charged particles (ions). When electrodes are connected so as to provide a direct current passing through the chambers, the sea-water ions migrate in different directions, with positive ions (cations – as sodium, calcium, and magnesium) travelling one way and negative ions (anions – as chloride, sulphate, and bicarbonate) in the other. Both sets of ions travel through the membranes, but because each membrane is permeable to only one type of ion, the alternate compartments become fresh or brine (Figure 9.4).[16]

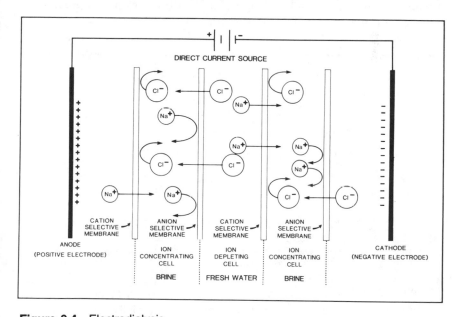

Figure 9.4 Electrodialysis

Source: After Office of Water Research and Technology, DOI, *Electrodialysis Technology* (USGPO, Washington, DC, 1979), p. 5.

Electrodialysis was first developed in the 1930s and 1940s, but was not widely used until the 1950s and 1960s. By 1977, worldwide, there were some 600 plants operating with a total capacity of 191,000kl (42 million gal). Electrodialysis systems have been used primarily in processing brackish rather than saline waters, because the electrical energy needed depends on how much dissolved material must be removed. The process is most efficient when dissolved solids range between 1,000 and 3,500ppm and when the water temperature is between 44°C and 71°C. This method thus lends itself to areas where sea-water temperatures are relatively high, which helps to avoid some of the feed-water heating costs. Overall, the larger the system, the more cost efficient it is.[17]

A major problem associated with electrodialysis is fouling and scaling of membrane surfaces. To avoid this problem, membranes may be exposed to chemicals such as acids, or if preferred, the electrical polarity may be reversed several times each hour with a simultaneous reversal of the direction of water flow. These techniques loosen the scale on the membrane walls.[18]

Reverse osmosis

When two liquids of different salt concentrations are separated by a semi-permeable membrane, part of the less salty liquid will diffuse through the membrane into the saltier liquid. The driving force is called osmotic pressure. By putting enough pressure on the saltier solution, the osmotic pressure is overcome and salt-free water diffuses from it (by reverse osmosis) through the membrane (Figure 9.5).

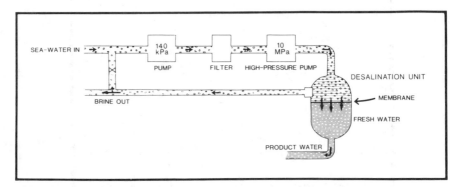

Figure 9.5 Reverse osmosis
Source: After Office of Saline Water, DOI, *The A-B-Seas of Desalting* (USGPO, Washington, DC, 1968), p. 29.

In the mid-1980s reverse osmosis had captured about 50 per cent of the total desalination market,[19] but it was not commercially applied to sea-water until 1981 when the city of Eilat, Israel, installed the world's first facility, which processes sea-water at 50 atmospheres (5.5 MPa) of pressure. Approximately 9kWh of electrical energy must be used for every m^3 (about 220 gal) of product water. This energy requirement, however, compares favourably with the city's multiple-flash distillation plant, the previous source for most of the city's potable water. Eilat's reverse osmosis plant has a daily output of approximately 100m^3.[20]

The advantages of reverse osmosis desalination are (1) initial plant investment costs are normally less; (2) energy consumption is relatively low, because no phase change is required; and (3) corrosion and scale problems are limited, given that the process works under relatively low temperatures. The major problem involves membrane fouling. Because the membranes must be replaced (at least every three years), reverse-osmosis systems can be more expensive to maintain than other types. Researchers, however, are making excellent progress in developing membranes resistant to particulate fouling.

Other sources

On a world scale, a large volume of fresh water is available on the sea-floor itself, if it can be captured before mixing with the sea-water. Numerous continental shelf areas have fresh-water springs and aquifers that can be tapped. The ancient city of Argolis, Greece, used submarine springs for fresh water. Springs there emit 863,000m^3 each day.[21]

Although continental shelf fresh-water springs and aquifers could be a prime resource for some areas, too heavy a withdrawal of these artesian sources might cause onshore fresh-water wells to become brackish, if the cone of depression becomes too extensive.[22]

Sodium chloride

Sea-water contains, as a dissolved solid, the world's largest reserve of sodium chloride or common salt. Dissolved solids account for 3.5 per cent of the weight of sea-water; most (71 per cent) of these dissolved solids is sodium chloride. US Bureau of Mines personnel estimate that the oceans hold approximately 40,000 million million tonnes of sodium chloride.[23]

Throughout history, common salt has been an important mineral. Cave dwellings in Belgium contain evidence indicating that salt was probably used in cooking wheat and barley some 5,000 years ago.[24] Salt has been traded among peoples for at least 4,000 years, probably longer. The Phoenicians, from 1200 BC to 300 BC, carried it in their maritime trade, and Roman soldiers received part of their pay in salt rations '*salarium argentum*', a Latin term from which the English word 'salary' evolved.[25]

Mankind's initiation to marine salt probably came from natural solar evaporation in shallow pools of sea-water. The first marine salt producers may have diked off small, tidal-filled coastal basins and then allowed time to do the rest. Because the evaporation stages were not separated, the salt was impure and bitter. Much later, salt-makers discovered that the product was improved if, before complete crystallisation occurred, the brine were removed from the first pond and put into a second for further evaporation.[26] Through time, producers learned to use several stages of fractional crystallisation. They have modern counterparts in many countries today. Presently India is by far the world's largest producer of marine salt. As of 1984, India produced 61 per cent of the identifiable total marine salt output (Table 9.2).

Where low precipitation and high evaporation occur along coastal zones, as on Ibiza, one of Spain's Balearic Islands, solar evaporation techniques work well, if slowly; the entire evaporation cycle can take approximately five years. The initial evaporation phase precipitates the iron, calcium, and magnesium compounds. The brine then goes to lime ponds where calcium sulphate is removed. Subsequently the salt workers transfer the brine into harvesting ponds where crystallisation begins. When about 85 per cent of the sodium chloride has precipitated out, the remaining brines go to yet another evaporating pond: here magnesium, potassium, and bromine can be removed. After washing the salt crystals with a dilute brine to remove remaining impurities, special tractors remove the salt from the crystallisation pond. With one more washing (fresh water this time), the salt is dried, screened, and shipped.[27]

Table 9.2 Marine salt producers, 1984

State[c]	Estimated Production[a] (Thousands of Tonnes)[d]
India	7,530
Mexico[b]	5,530
France	1,452
Spain	1,089[e]
Italy	1,000
Colombia	272
Ethiopia	110
Costa Rica	109
Portugal	98
Federal Republic of Germany	54
Yugoslavia	38[f]

Sources: a. D. E. Morse, 'Salt', in *Minerals Yearbook 1984, Vol. I, Metals and Minerals* (USGPO, Washington, DC, 1985), pp. 772–4; b. Production in Mexico totalled between an estimated 5 million and 6 million tonnes. Letter: O. Martino, Division of International Minerals, USBM, DOI, Washington, DC, 3 March 1987.
Notes: c. Not all marine salt producers are included here, because of production-data groupings; d. converted from short tons; e. data for Spain include marine and other evaporated salt; f. data for Yugoslavia are for 1982.

Solar salt production can create two major local environmental problems. The large land area required for evaporation ponds may take up space needed for other functions and cause damage to nearby ecosystems, if concentrated brine solutions are dumped into estuaries or bays. Salt also has an impact on the environment where it is used, because it facilitates corrosion, can kill vegetation and wildlife, and may damage potable water supplies.

Despite these problems, demand for salt will increase in future. Metal production, oil- and gas-well drilling, and water treatment facilities are major users of salt. The world's cumulative demand from 1983 to the year 2000 probably will be about 3,800 million tonnes, and annual production should be approximately 290 million tonnes.[28] To meet these needs, many producers must expand their production, build more solar plants, or open new mines; in locations where environmental or spatial problems do not preclude the use of coastal sites, sea-water can provide the resource base for some of the additional salt needed.

Bromine

This dark, reddish-brown substance is the only non-metallic element that, at an ordinary temperature and pressure remains liquified. it is the ninth most common element in sea-water. In total, the oceans are estimated to contain about 91 million million tonnes of bromine, yet it occurs at only 65,000 parts/1,000 million. Immense quantities of sea-water must be processed to obtain significant amounts of bromine. Sea-water bromine plants operated in the US for forty years, but unable to compete with subsurface land-based brines with a much richer bromine content, the last plant closed in 1969.[29]

Bromine compounds are used in anaesthetics, in anti-spasmodic medicines, in dye-making reagents, in photographic film emulsions, in paper, in fire retardants, in oil- and gas-well drilling, in fungicides and rodenticides, and (in combination with lead) in anti-knock fluids for petrol,[30] where the bromine (as ethylene dibromide) acts as a scavenger to prevent engine lead deposition.[31] Bromine is toxic, especially when put into compounds such as ethylene dibromide, methyl bromide, and vinyl bromide.[32] If not used properly, these products may create environmental problems during production, transport and consumption. Ethylene dibromide is so toxic that the US Environmental Protection Agency (EPA) in 1984 suspended its use as a fumigant for grain, citrus, and papaya and has instituted regulations phasing it out as a petrol additive.[33] In the case of methyl bromide, some scientists are now worried about its potential effects on stratospheric ozone.[34]

Consumption of bromine is declining in some industries, as in leaded petrol. An increasing need for fire-retardants and well-drilling fluids, however, will demand more bromine output in coming years.[35] Overall, there is no danger of an inadequate world supply of bromine, because it can be produced·as a co-product of salt-making (as noted above) and also in the production of magnesium.[36]

Magnesium

The ocean's fifth most common element is magnesium. Estimates indicate that each km^3 of sea-water contains about 1.3 million tonnes of magnesium. The oceanic sea-water magnesium industry produces both magnesium compounds (non-metallic) and metallic magnesium, the lightest (specific gravity 1.74) of the structural metals. To produce magnesium from sea-water, one needs a supply of (1) sea-water uncontaminated by algae or freshwater; (2) calcium carbonate or dolomite; and (3) cheap electricity. As of 1987, sea-water magnesium producers accounted for about 18 per cent of the world's total annual magnesium production; 74 per cent originated from mined magnesite and the remainder from lake and well brines.[37]

Magnesium compounds

As of the mid-1980s countries producing oceanic non-metallic magnesium compounds included Ireland, Italy, Japan, Mexico, Norway, the UK, the US, and the USSR. The world's total annual oceanic production capacity was estimated at slightly more than 1.6 million tonnes (Table 9.3). Beneficiating sea-water to obtain caustic-calcined or dead-burned magnesium oxide (magnesia) (both non-metallic forms) involves a series of steps whereby the calcium carbonate and sulphate materials are precipitated out, after which the resultant slurry of magnesium hydroxide is thickened, washed with fresh water, filtered, and calcined. Magnesium sulphate (epsom salts), another non-metallic magnesium substance, is produced by dissolving magnesia in sulphuric acid and allowing crystallisation to occur.[38]

In 1983, 85 per cent of the magnesium consumed in the US was used as non-metallic magnesium compounds. For example, Magnesium hydroxide is used in sugar refining, paper pulp production, and pharmaceuticals. Magnesium oxide is a raw material for manufacturers of refractories used in the iron and steel, copper, nickel, cement, and glass industries and for producers of cement, rayon, animal feed, construction materials, fertilisers, rubber, electrical insulators, and uranium. Magnesium carbonate goes into glass, ceramics, ink, pigments, and paint, and magnesium chloride serves as an additive in textiles, paper, and magnesium metal.[39]

There are many substitutes for magnesium as an additive, and the demand for magnesium worldwide declined considerably during the 1973–83 decade. This decline occurred, in large part, because of reduced consumption for refractories in the iron and steel industries. For example, in the Republic of Ireland's Meath County, the Premier Periclase magnesia plant at Drogheda began operating in 1980 but was forced to close temporarily during the depression in the iron and steel industry in late 1982 and early 1983. Ireland's other magnesia plant, the Quigley (70,000 tones capacity) at Dungavan, on the south coast, closed permanently in 1982.[40]

Refractory magnesia consumption in the US (the world's largest documented consumer) decreased from 828,000 tonnes in 1973 to 508,000 tonnes in 1983. World production capacity (6,532,000 tonnes) for all sources of magnesium compounds (sea-water, brines, and mined magnesia) in 1983 was well above output (5.0 million tonnes), a good indication that the market will remain saturated for the near future. US Bureau of Mines projections, however, show that the total probable

Table 9.3 Sea-water magnesium compound producers

Country and company	Location	Annual capacity (tonnes of MgO equivalent)[f]
Japan[a]		
Ube Chemical Industries Co. Ltd	Ube City, Yamaguchi Pref.	450,000
Shin Nihon Chemical Industry Co. Ltd	Minamata, Kumumoto Pref.	100,000
	Toyama, Toyama Pref.	80,000
	Onahama, Fukushima Pref.	20,000
Asahi Glass Co. Ltd	Iho, Niigata Pref.	15,000
United States[b]		
National Refractories & Minerals Corp.	Moss Landing, CA	136,000
Dow Chemical Co.	Freeport, TX	86,000
Basic Magnesia Inc.[c]	Port St Joe, Fl	45,000
Merck & Co. Inc.	South San Francisco, CA	13,600
Barcroft Co.	Lewes, DE	4,500
United Kingdom[d]		
Steetley Industries Ltd	Hartlepool, Durham Co.	250,000
Italy[e]		
Sardamag SpA.	Sant'Antioco, Sardinia	70,000
Cogema SpA-Cie Generale del Magnesia SpA.	Siracusa, Sicily	65,000
Republic of Ireland[d]		
Premier Periclase Ltd	Drogheda, Meath Co.	100,000
USSR[d]		
Sivash Works	Crimean Peninsula	100,000
Mexico[c]		
Industrias Penoles SA de CV.	Ciudad Madero, Tamaulipas	80,000
Norway[c]		
Norsk Hydro A/S	Porsgrunn, Telemark Co.	25,000
Total		1,640,100

Sources: a. G. M. Clarke, (ed.), *Industrial Minerals Directory* (Metal Bulletin Books, London, 1984), pp. 250, 258, 263; b. D. A. Kramer, 'Magnesium compounds', preprint from *Minerals Yearbook 1985, Vol. I, Metals and Minerals* (USGPO, Washington, DC, 1986), p. 2 and Letters: D. A. Kramer, Division of Nonferrous Metals, USBM, DOI, Washington, DC, 9 and 3 March 1987; c. G. M. Clarke, (ed.), *Industrial Minerals Directory – First Edition 1987* (Metal Bulletin Books, London, 1986), pp. 288, 518; d. H. M. Mikami, 'Refractory Magnesia', paper presented at the Conference for Raw Materials for Refractories, Tuscaloosa, AL, 8–9 Feb. 1982, pp. 28–34; e. Letter: J. Craynon, Division of International Minerals, USBM, DOI, Washington, DC, 27 Feb. 1987.
Notes: f. Converted from short tons and rounded, where required.

demand (6,423,000 tonnes) in the year 2000 will nearly equal 1983's production capacity.[41]

Metallic magnesium

One firm in the US, Dow Chemical Company, in Freeport, Texas, processes sea-water to produce magnesium metal.[42] The plant, constructed in 1940, uses an electrolytic process and is the world's largest

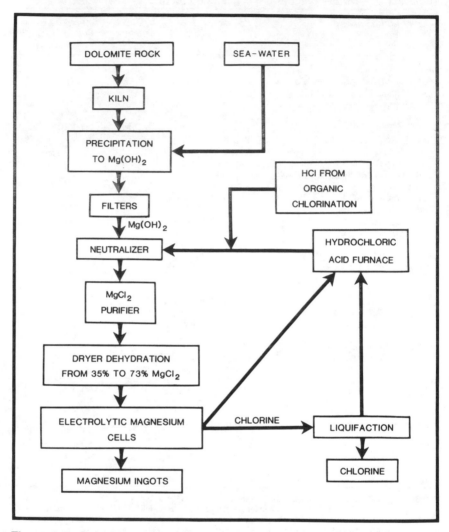

Figure 9.6 Dow Chemical sea-water magnesium metal extraction process
Source: Drafted from information provided by Letter: B. B. Clow, Executive Director, International
Magnesium Association, McLean, VA, 13 Jan. 1987. With permission.

producer of primary magnesium. Because of a worldwide overcapacity
relative to present market demand, in 1986 Dow's plant operated at only
81 per cent of its annual capacity (110,000 tonnes).[43] The only other plant
presently producing metallic magnesium from sea-water is in Porsgrunn,
Norway, an establishment with an annual capacity of 60,000 tonnes,
owned by Norsk Hydro. Its feed water is a combination of sea-water and
brine.

Magnesium metal is produced in two ways – electrolytically or sili-
cothermically. Only the electrolytic process will be discussed here,

Figure 9.7 Magnesium ingots.
Source: Courtesy Dow Chemical USA. With permission.
Note: After the electrolytic process, the molten magnesium is sent to casting where it is pumped into moulds, inspected for quality, stacked, strapped, and stored for shipment to markets.

because sea-water or a combination of brine and sea-water are the main feeds. Processors first produce either a hydrous or an anhydrous magnesium chloride. To make the hydrous material, dolomite is mixed with sea-water, which precipitates the dissolved magnesium as magnesium hydroxide. After being filtered, the magnesium hydroxide is neutralised with hydrochloric acid, resulting in a magnesium chloride solution that is then partially dehydrated. In preparing the anhydrous magnesium chloride, a concentrated chloride brine is treated with calcium chloride, causing sulphate impurities to precipitate out. More concentration occurs when the brine is put through a spray dryer, resulting in a magnesium chloride powder. Both the hydrous and anhydrous feeds are put into electrolytic cells that have a direct current anode (graphite rods) and cathode (steel rods) system. The direct current separates the magnesium chloride into chlorine gas and molten magnesium metal (Figure 9.6).[44] Sea-water magnesium extraction results in salt and bromine as co-products and gypsum and potassium compounds as by-products.

In 1983 15 per cent of the magnesium consumed in the US was as magnesium metal (Figure 9.7). Magnesium metal (and its alloys) is a highly versatile metal that is resistant to corrosion by alkalines. It is easy to machine at high speeds and can be soldered, brazed, riveted, or adhesively bonded. Because of its hardness and light weight, magnesium metal performs well as an aluminium alloy, as in beer and soft-drink cans and in military vehicle, aircraft, and automobile parts. To reduce weight, engineers use it in automobile headlamp assemblies and clutch housings, and power-tool producers make lawnmower and chainsaw housings from it.[45]

Magnesium metal consumption in the US during 1973 was 87,000 tonnes and in 1983 it was 129,000 tonnes, a 48 per cent increase for the decade. In 1985 the International Magnesium Association (IMA), a trade association for metallic magnesium, developed worldwide projections for primary production (all sources) and consumption of metallic magnesium. Production was projected for 1986 at 240,000 tonnes.[46] Actual production in 1986 was only 226,000 tonnes. The IMA expects that 88 per cent of the world's production capacity will be used in 1987 and only 71 per cent in 1991, mainly because of an anticipated increase in world production capacity (Table 9.4).

Table 9.4 Projected primary magnesium consumption and production capacity (thousands of tonnes)

Area	(End of 1987)		(End of 1991)	
	Consumption	Capacity	Consumption	Capacity
North America	108	158	121	234
Latin America	11	10	14	35
Western Europe	75	88	84	93
Africa/Middle East	3	—	4	—
Asia/Oceania	39	12	43	12
Total	236	268	266	374

Source: T. Aoyagi, 'Magnesium supply and demand report', in *Magnesium in the Auto Industry: Prospects for the Future, Proceedings 44th Annual World Magnesium Conference, Tokyo, Japan, 17–20 May 1987* (International Magnesium Association and Japan Light Metal Association, McLean, VA, 1987), pp. 32–3. With permission.

Uranium

Because many states are concerned for their future energy supplies, they have sea-water uranium extraction research programmes. A lack of energy sources, a fear of exorbitant energy costs in future, a failure of other energy technologies to move forward and a desire for a secure energy supply are some of the reasons that Japan, France, South Korea, Sweden, Taiwan, the UK, the US, and the FRG are working to obtain uranium from sea-water. Canada, India, and Italy have also had sea-water uranium research programmes.[47] Some researchers are now recovering significant amounts of sea-water uranium – enough that commercial firms, in the distant future, may begin production. But the discovery of large, high-grade terrestrial deposits in Canada and Australia and low spot-market prices (induced by overproduction) have deferred the day when oceanic uranium resources will be exploited.

As in the production of other oceanic minerals in solution, uranium must be concentrated before it is usable. There are only three parts of uranium per thousand million parts of sea-water, but it is enough to provide the oceans with an estimated total of nearly 4,000 million tonnes.[48] Several methods can be used to take minerals out of solution – precipitation, solvent extraction, and ion exchange. In precipitation processes, chemicals added to the water form insoluble solids with the desired elements. Solvent extraction involves transferring 'the desired elements into another liquid that does not mix with the water solution'. The ion-exchange technique passes a flow of water across resins which exchange ions for those desired in the water.[49]

Researchers in the US have been working with computer models to determine an optimum sea-water uranium recovery system. Their studies show that systems based on absorption of uranium by fibres or solid particles seem to have the best economic potential. The components of the modelled ideal-system include an absorption process using hydrous titanium oxide to extract the uranium from the sea-water and a moored oil-rig-type platform which supports a pump-diffuser that takes in water at a very low velocity (Figure 9.8). They note that

> The recovery system consists of a uranium loading period during which seawater passes through the bed, followed by a freshwater wash, followed by an ammonium carbonate elution to desorb the uranium, followed by a freshwater wash and return to a loading period. The uranium-rich ammonium carbonate eluate is stripped of ammonia and carbon dioxide, which are recycled to produce fresh ammonia carbonate solution. The uranium is ultimately recovered by an ion exchange process.[50]

Scientists in the UK are experimenting with sea-water uranium extraction, using a substance named 'poly(hydroxamic acid)', a resinous crystal material. These crystals are very stable, even after six months of immersion, and remove 76 per cent of the uranium from sea-water during only three minutes of exposure. The researchers' main problem is to reduce the energy costs for resin crystal production below the value of the energy obtained from the uranium produced. Investigators are attempting to design a 'poly(hydroxamic acid)' fibre which can be woven into a fabric endless-belt. A belt 200m long and 200m wide with a thickness of 1cm could provide up to 6 tonnes of uranium per year.[51]

Although the Japanese were late in beginning sea-water uranium-production research, they have made rapid progress and 'their efforts exceed that of all others combined'.[52] They are now out of the laboratory, and since April 1986 have been producing small amounts of uranium from the ocean near the village of Nio (in Kagawa Prefecture) on the shore of the Inland Sea, some 550km south-west of Tokyo. Construction began on the $19-million shore-based plant in 1981. Today, it employs 48 workers who monitor the daily processing of 4,000 tonnes of sea-water. The process involves putting sea-water through a titanic acid compound

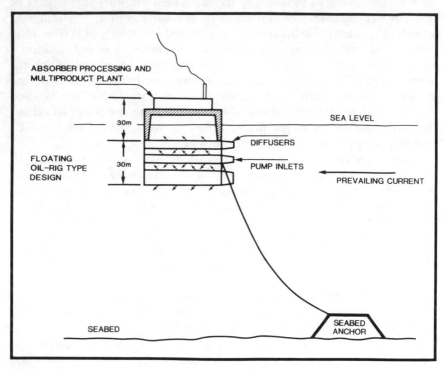

Figure 9.8 Sea-water uranium recovery plant module

Source: After F. R. Best and M. J. Driscoll, 'Prospects for the recovery of Uranium from seawater', *Nuclear Technology*, vol. 73, no. 1 (1986), p. 57. With permission.

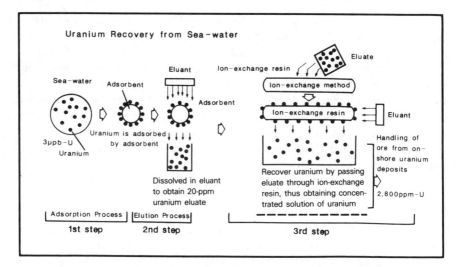

Figure 9.9 Schematic diagram of uranium recovery from sea-water

Source: After Metal Mining Agency of Japan, *Recovery of Uranium from Seawater* (MMAJ, Tokyo, Aug. 1986), n.p. With permission. This system is used at the Nio Institute for Uranium Recovery from Seawater, at Nio, Kagawa, Japan.

(which absorbs the uranium) and then dumping the water back into the sea. The material obtained is eluted with dilute hydrochloric acid to make a 20-ppm solution of uranium and sea-water. Ion exchangers then concentrate the yellow fluid into a 2,800-ppm solution of uranium and sea-water (Figure 9.9). The Metal Mining Agency of Japan (the plant's operator) had hoped to have produced (by March 1987) about 10kg of uranium, but only 5.3kg had been extracted. This extraction process is not yet economically competitive with conventional onshore producers. A decision on whether to continue with the project was scheduled for September 1987.[53] The Japanese hope sea-water will ultimately supply 1,000 tonnes of uranium a year, about 15 per cent of Japan's current annual consumption.[54]

The Japanese also have been experimenting with ion-binding resins such as acrylic amidoxime. The problem here is how to get adequate water passing through the system. Using the energy contained in currents and waves 'as the power to drive the influx of seawater into columns containing the amidoxime', they have managed to load this material with uranium at a level of 1,500 ppm. According to one researcher at Massachusetts Institute of Technology, to supply enough uranium to keep one reactor running steadily will require 'a water current something like that of the Nile River passing through [the] system'.[55]

Scientists are watching to see how technically and economically viable the Japanese experimental systems are, because uranium produced from sea-water could eliminate the need for breeder reactors, for nuclear fuel reprocessing, and for the use of plutonium.

Best and Driscoll, investigators in the US modelling project, suggest that development of economically viable techniques to remove uranium from the oceans might help make people more readily accept nuclear energy developments. Uranium produced from sea-water could also possibly provide co-products such as cobalt, vanadium, and chromium, all of which are absorbed by resins. These co-products could help offset sea-water uranium production costs.[56]

Deuterium

Deuterium is 'a hydrogen isotope, the nucleus of which contains one neutron and one proton', making it nearly twice as heavy as the nucleus of normal hydrogen, which has only one proton. Deuterium occurs in nature at 1.56 atoms/10,000 atoms of normal hydrogen. Yet, the oceans hold an estimated 46 million million tonnes of deuterium atoms.[57]

If the energy potential of deuterium in nuclear fusion processes could be made economically available to us, it could supply all our energy needs indefinitely. But that day seems to lie well into the future. Deuterium's main use presently is as a moderator/coolant/reflector in nuclear fission reactors.

Only one commercial effort – at Glace Bay, Nova Scotia, in Canada – has attempted to extract deuterium from sea-water. The programme was abandoned, because of severe corrosion of the processing equipment.[58]

Conclusions

By the late 1970s the US was withdrawing fresh water for industrial and municipal uses at about 1,600 million kl (350,000 million gal) per day – three times the volume withdrawn only three decades earlier. How long can this trend continue in the US and elsewhere before we must look to the oceans for additional supplies?

As desalination technology continues to improve, more coastal regions will come to depend on sea-water as a potable water source. Solar energy is now used to power both multiple-flash and freezing sea-water conversion systems. A company in the FRG is now producing a prototype system that has 75 multiple-flash stages which, when powered by solar energy, is 85 per cent less costly to operate than an ordinary multiple-flash plant.[59] The ocean itself can supply alternate sources of energy needed in sea-water conversion systems, such as OTEC. Tides, currents, waves, and salinity gradients are other potential energy sources for driving sea-water conversion plants.

When electrical power generating plants are constructed near coastal zones, engineers could design them to provide secondary benefits to local communities. The heat energy remaining in the steam after travelling through the turbines can supply heat for the distillation of sea-water. In very large plants it is possible, also, to harvest several usable minerals (salt, magnesia, and potash) from the brines produced during the desalination process. A desalting plant that produces 227,000kl (50 million gal) of fresh water/day (with a brine concentration of 3:1) could produce large volumes of salt, as well as about 136 tonnes of potash.[60]

Sea-water uranium technology has only begun to develop. To become truly economically viable, researchers must design much more efficient collector systems, although the Japanese are making significant progress in meeting this need. Investigators in the US have used biomass as an absorbent for sea-water uranium, but unfortunately those organisms tested so far do not have the absorptive capacity of synthetic inorganic materials.[61] The bromine industry will continue to depend on several past markets, but environmental bans and restrictions (as in the US) imposed on bromine compounds' use signal difficult years ahead. Bromine producers should, therefore, begin to vigorously develop alternative markets. The industry's economic situation, however, may not improve significantly until the world economy experiences another major upswing. Indicative of its precarious position is the closure of the US's only sea-water bromine plant, operated in conjunction with Dow Chemical Company in Freeport, Texas.

Magnesium producers, also, must be ready for market shifts. Currently they depend too much on aluminium alloy consumers. Greater emphasis should be placed on penetrating markets where a light-weight metal is needed.[62]

Future advancements in sea-water mining technology may make some minerals now produced more economically competitive, especially fresh water. And, with time, ocean uranium mining could emerge as an economically viable industry.

Notes

1 H. Brabyn 'Blue planet', *UNESCO Courier* (Feb. 1986), p. 6.
2 A small amount of gypsum is also produced as a primary product from sea-water at a site along the south-west coast of Taiwan. Letter: Y.F. Fan, Director, Department of Mines, Taipei, Taiwan, 31 March 1987.
3 *Desalting Plant Inventories*, Inventory no.7 (International Desalination Association, Topsfield, MA, 1980), p. 1.
4 S. A. Reed, *Desalting Seawater and Brackish Water: 1981 Cost Update*, Contract no. W-7405-eng-26, OWRT no. 14–34–0001–1440 (Oak Ridge National Laboratory, Oak Ridge, TN, 1982), pp. 32–41.
5 For a useful discussion of energy requirements in desalination processes, as well as basic extraction techniques, see K. S. Spiegler, *Salt-Water Purification*, 2nd edn (Plenum Press, NY, 1977), pp. 21–7.
6 Office of Water Research and Technology, DOI, *The A-B-C of Desalting* (USGPO, Washington, DC, 1977), p. 15.
7 J. Chowdhury and W.P. Stadig, 'Solar energy vies for a desalting niche', *Chemical Engineering* (US), vol. 91, no. 1 (1984) p. 42.
8 H. S. Hundal, R. B. Trattner, and P. N. Cheremisinoff, 'Freshwater from seawater, part 4: from screening to filtration – treating the water supply', *Water and Sewage Works*, vol. 126, no. 12 (1979), p. 25; Spiegler, *Salt-Water Purification*, p. 54.
9 O. A. Roels, 'From the deep sea: food, energy, and fresh water', *Mechanical Engineering*, vol. 102, no. 6 (1980), p. 40.
10 R. Mitchell and P. Benson, 'Control of marine biofouling in heat exchanger systems', *Marine Technology Society Journal*, vol. 15, no. 4 (1981), pp. 11–20.
11 F. J. Tanzosh, 'OTEC is charting two courses', *Chemical Engineering*, vol. 91, no. 22 (1984), p. 31.
12 A. T. Maris and J. R. Paulling, 'Analysis and design of the cold-water pipe (CWP) for the OTEC system with application to OTEC-1', *Marine Technology*, vol. 17, no. 3 (1980), pp. 281–9.
13 Office of Water Research and Technology, DOI, *The A-B-C of Desalting*, p. 21.
14 Office of Saline Water, DOI, *The A-B-Seas of Desalting* (USGPO, Washington, DC, 1968), p. 19.
15 Office of Water Research and Technology, DOI, *The A-B-C of Desalting*, p. 23.
16 Office of Water Research and Technology, DOI, *Electrodialysis Technology* (USGPO, Washington, DC, 1979), p. 5; K. S. Speigler, *Salt-Water Purification* (John Wiley, NY, 1962), pp. 84–5.
17 Office of Water Research and Technology, DOI, *Electrodialysis Technology*, pp. 4, 7, 11–12.
18 Chowdhury and Stadig, 'Solar energy', p. 42.
19 'Reverse osmosis desalination', *Mechanical Engineering*, vol. 103, no. 6 (1981), p. 72.

20 M. Sekino *et al.*, 'Reverse osmosis modules for water desalination', *Chemical Engineering Progress*, vol. 81, no. 12 (1985), p. 52.
21 R. H. Charlier, 'Other ocean resources', in E. M. Borgese and N. Ginsburg (eds), *Ocean Yearbook 1* (University of Chicago Press, Chicago, IL, 1978), p. 162.
22 W. P. Dillon, 'Mineral resources of the Atlantic Exclusive Economic Zone', in *Exclusive Economic Zone Papers: Exclusive Economic Zone Symposium '84* (Marine Technology Society and the Institute of Electrical and Electronics Engineers Council on Oceanic Engineering, n. 1, 1984), p. 77.
23 E. D. Morse, 'Salt', in *Mineral Facts and Problems 1985*, USBM Bulletin 675 (USGPO, Washington, DC, 1985), p. 679.
24 D. W. Kaufman (ed.), *Sodium Chloride: The Production and Properties of Salt* (Hafner Publishing, NY, 1971), p. 4.
25 Morse, 'Salt', p. 679.
26 S. D. See, 'Solar salt', in D. W. Kaufman, *Sodium Chloride: The Production and Properties of Salt* (Hafner Publishing, NY, 1971), p. 96.
27 Morse, 'Salt', pp. 681–2.
28 ibid., p. 687.
29 P. A. Lyday, 'Bromine', in *Mineral Facts and Problems 1985*, USBM Bulletin 675 (USGPO, Washington, DC, 1985), pp. 103–4.
30 ibid., p. 105.
31 F. Yaron, 'Bromine manufacture: technology and economic aspects', in Z.E. Jolles (ed.), *Bromine and its Compounds* (Ernest Benn, London, 1966), p. 39.
32 L. W. Margler, *Project Summary: Environmental Implications of Changes in the Brominated Chemical Industry* (Environmental Protection Agency, Cincinnati, OH, Sept. 1982), pp. 2–3.
33 P. A. Lyday, 'Bromine', *Mining Engineering*, vol. 37, no. 5 (1985), pp. 463–4.
34 Margler, *Project Summary*, P. 3.
35 ibid., p. 2.
36 Yaron, 'Bromine manufacture', p. 33.
37 Letter: H. M. Mikami, Consultant, Pleasanton, CA, 17 June 1987.
38 D. A. Kramer, 'Magnesium', in *Mineral Facts and Problems 1985*, USBM Bulletin 675 (USGPO, Washington, DC, 1985), p. 474.
39 ibid., p. 471–2.
40 R. H. Singleton, 'The mineral industry of Ireland', in *Minerals Yearbook 1984, Vol. III, Area Reports International* (USGPO, Washington, DC, 1986, pp. 443–4.
41 Kramer, 'Magnesium', pp. 471, 477, 479–80.
42 ibid., p. 472.
43 Letters: B. B. Clow, Executive Director, International Magnesium Association, McLean, VA, 23 June and 13 Jan. 1987.
44 Kramer, 'Magnesium', pp. 473–4.
45 ibid., pp. 471–2.
46 C. W. Nelson, *Magnesium Supply and Demand Report* (International Magnesium Association, McLean, VA, 1985), p. 5.
47 F. R. Best and S. Yamamoto, 'A review on uranium recovery processes and the international meeting on the recovery of uranium from seawater', *Scientific Bulletin*, Office of Naval Research, vol. 9, no. 1 (1984), p. 93.
48 D. Kennedy, 'Ocean uranium: limitless energy', *Technology Review*, vol. 87, no. 7 (1984), p. 74.
49 'Uranium ocean mining', *Engineering Digest*, vol. 31 (Jan. 1985), p. 33.
50 F. R. Best and M. J. Driscoll, 'Prospects for the recovery of uranium from seawater', *Nuclear Technology*, vol. 73, no. 1 (1986), pp. 55–6.
51 'Uranium ocean mining', p. 33.

52 Letter: M. J. Driscoll, Professor, Department of Nuclear Engineering, Massachusetts Institute of Technology, Cambridge, MA, 13 July 1987.
53 Letter: Y. Shoda, Director, Technical Development Department, Metal Mining Agency of Japan, Tokyo, Japan, 3 July 1987.
54 In 1987 Japan's annual uranium demand to fuel its reactors stood at 5,000 tones. By 1996 demand is expected to reach 8,400 tonnes. For the period 1987–96 the total cumulative demand (given a power plant load of 75 per cent) will be about 68,000 tonnes; see 'Japan's energy evolution', *Mining Journal*, vol. 308, no. 7,906 (1987), pp. 141–3.
55 Kennedy, 'Ocean uranium', p. 74; for a detailed discussion of Japan's sea-water extraction efforts, see H. Hotta, 'Recovery of uranium from seawater', *Oceanus*, vol. 30, no. 1 (1987), pp. 44–7.
56 Best and Driscoll, 'Prospects for the recovery of uranium from seawater', p. 68; see also J. G. Pina-Jordan, 'Measurement and modelling of uranium and strategic element sorption by amidoxime resins in natural seawater', unpublished MS Thesis, Texas A&M University, College Station, TX, 1985.
57 Calculations based on information contained in M. Benedict, T.H. Pigford, and H. W. Levi, *Nuclear Chemical Engineering*, 2nd edn (McGraw-Hill, NY, 1981), p. 708.
58 Letter: M. M. Miller, Department of Nuclear Engineering and Energy Laboratory, Massachusetts Institute of Technology, Cambridge, MA, 29 Jan. 1987.
59 Chowdhury and Stadig, 'Solar energy', pp. 42–3.
60 Office of Saline Water, DOI, *The A-B-Seas of Desalting*, p. 38; Office of Water Research and Technology, DOI, *The A-B-C of Desalting*, p. 29.
61 M. Tsezos and S. H. Noh, 'Extraction of uranium from sea water using biological origin adsorbents', *Canadian Journal of Chemical Engineering*, vol. 62, no. 4 (1984), pp. 559–61.
62 'Magnesium: uses in cars to spur structural consumption', *Modern Metals*, vol. 40, no. 12 (1985), p. 80; 'Twenty years hence – Plastics, aluminum, and mag?', *Materials Engineering*, vol. 100, no. 4 (1984), pp. 60–1.

Chapter ten

The United States Exclusive Economic Zone: the management challenge

Increased use of oceanic minerals will likely spawn jurisdictional and management problems between national governments and their subordinate political units and between management agencies within governments. It also will require vigorous national assessment and development programmes, guided by clearly defined policy objectives. The US EEZ is an excellent example of these new relationships and demands at work.

One objective of the Reagan Administration has been to reduce the US's dependence on foreign mineral imports (Table 10.1). The establishment of the 200-nmi EEZ was intended to help reduce this dependence. The EEZ added about 810 million ha to the territorial area administered by the US; estimates vary, however, depending upon one's data source and whether it includes waters surrounding some of the US Pacific Trust Territories which are no longer counted as US possessions.[1]

The US's EEZ Proclamation is, in reality, an extension of its 1976 Magnuson Fisheries and Conservation Act (MFCA). When this Act went into force on 1 March 1977, it established a 200-nmi fishery zone regulated by the US Coast Guard and the US National Marine Fisheries Service.

Table 10.1 United States net-import reliance for selected metallic minerals potentially available from the oceans, 1986

Mineral	%
Manganese	100
Platinum-group metals	98
Cobalt	92
Chromium	82
Nickel	78
Tin	77
Zinc	74
Silver	69
Vanadium[a]	54
Iron ore	37
Copper	27
Gold	21
Lead	20

Source: US Bureau of Mines, *Mineral Commodity Summaries, 1987* (DOI, Washington, DC, 1987), p. 2.
Note: a. The vanadium value is for 1984.

This act imposed numerous new demands upon the US Government, including the need for monitoring users, making scientific investigations and initiating international boundary negotiations. The EEZ continues to make these demands.

Boundary disputes

Implementation of the MFCA created twenty new international maritime boundaries (opposite and adjacent) for the US; negotiations to delimit these boundaries began soon after the MFCA came into force. The EEZ Proclamation, and the potential for mineral resources along many of these boundaries, has re-emphasised the urgency of establishing mutually recognised boundaries.

Canada

Prior to the establishment of the US EEZ, Canada and the US were negotiating four boundaries. One, in the Gulf of Maine (lying between New England and Nova Scotia), was settled by the ICJ through binding arbitration. When negotiations first began (1975), the dispute focused upon delimiting the Gulf of Maine's continental shelf boundary. Like the US, Canada in 1977 also established a 200-nmi fishery zone. The two zones overlapped within the Georges Bank area, an important fishing ground. At about the same time, geologists were waxing enthusiastic about this region's offshore petroleum potential. Canada had actually begun issuing exploration licences for Nova Scotian waters as early as 1964,[2] and drillers had brought in small oil and gas wells off Sable Island. Estimates in one US DOI study put the crude-oil potential of the Georges Bank area at about 200 million bbl and gas at about 139,000 million cubic metres.[3]

The oil and gas discoveries in the offshore of Nova Scotia and the USDOI petroleum potential estimates added fuel to the political fire coming from both Washington, DC, and Ottawa. After laborious negotiations, the two sides concluded a treaty to put the issue before the ICJ,[4] which gave its judgement on 12 October 1984, with the area being divided about equally (Figure 10.1).[5] By the time of the ICJ's decision, several dry exploratory wells along the US continental shelf had dampened the optimism of those who thought a petroleum bonanza might lie near New England's shore. In hindsight, based on the region's unproved petroleum potential, debate was overdrawn.

Of the other US-Canada boundary disputes, only one concerns an area with a major mineral resource potential, that is the Beaufort Sea, an area lying east of the US's giant Prudhoe Bay petroleum field. Debate centres on the seaward extension of the US-Canada land boundary at 141°W longitude, with the US opting for a median line, whereas Canada insists on a northward extension along the 141°W meridian. The locational

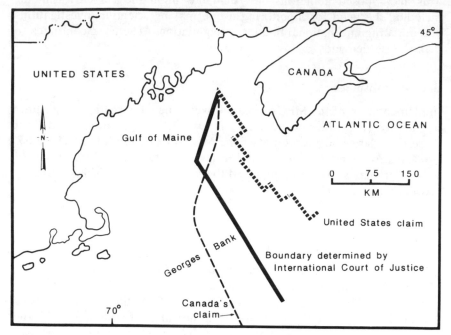

Figure 10.1   The International Court of Justice's Gulf of Maine boundary determination

Source: After H. R. Marshall, Jr, 'Disputed areas influence OCS leasing policy', *Offshore*, vol. 45, no. 5 (1985), p. 99. With permission.

difference of the two lines results in a significant wedge of offshore waters with considerable potential for petroleum.[6] Negotiations are still in progress (Figure 10.2).[7]

USSR

The US and the USSR are negotiating an important boundary in the Bering Sea, between Alaska and north-eastern Siberia. This boundary, established by the 1867 US-Russia Convention, passes through waters that, until the petroleum potential of the region was recognised, had an economic importance to the two states only for fishing.

Soviet and US negotiators are at odds concerning the Convention Line's exact position, because they use different methods of calculation. The US employs an arc of a great circle (a straight and shortest distance line on a globe), whereas the USSR uses a rhumb line (a straight line on a Mercator map). The difference in methods results in a lens-shaped area claimed by both sides (Figure 10.3).[8]

Efforts by the US to open the disputed region to petroleum exploration are now on hold. Only one lease sale (April 1984 in the Navarin Basin) has been held in areas immediately adjacent to the Convention Line. The

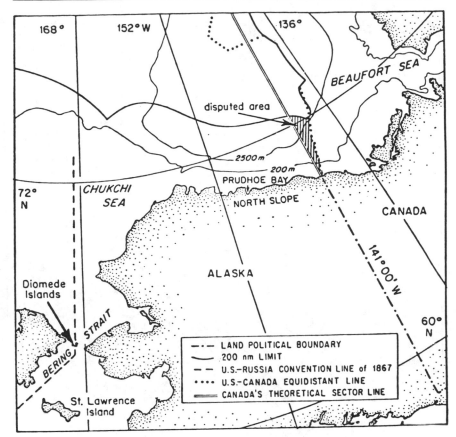

Figure 10.2 United States-Canada Beaufort Sea boundary dispute

Source: After R. Bowen and T. Hennessey, 'U.S. EEZ relations with Canada and Mexico', *Oceanus*, vol. 27, no. 4 (1984/85), p. 42. With permission.

highest bid for each block lying within 55.6km of the Convention Line was put in escrow for five years, by which time the DOI must have decided whether to award a given bid. If the federal government fails to accept an in-escrow bid (there are seventeen), then the concerned party may withdraw its money by providing notice within sixty days after the lapse of the five years. A similar programme is in effect for a lease sale held during 1984 in the Beaufort Sea boundary area;[9] four blocks from this sale are in escrow.[10]

Mexico and Pacific Island States

Recently geologists have suggested that deep waters of the Gulf of Mexico may hold significant quantities of petroleum. This situation should encourage the US Senate to act on a Treaty of Maritime Boundaries agreed upon by the US and Mexico in 1978. To date (September 1987), the Senate has taken no action.[11]

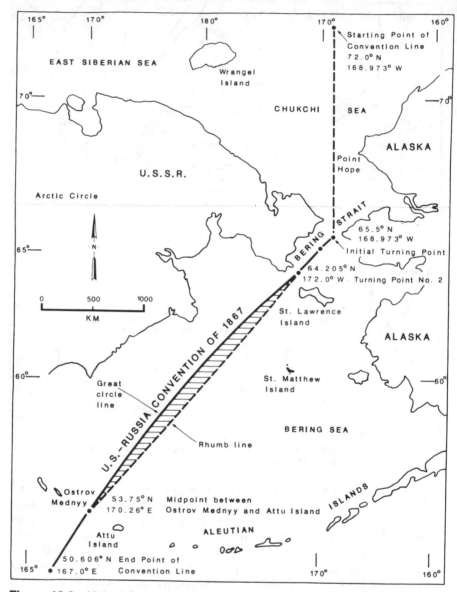

Figure 10.3 United States-USSR Bering Sea boundary dispute

Source: After O. Young, *Resource Management at the International Level* (Frances Pinter, London, 1977), p. 3. With permission.

Numerous treaty negotiations are under way to establish boundaries between US territories and island states in the south-western Pacific Ocean. A few have been settled, many have not. New Zealand and the Cook Islands have signed treaty agreements with the US, but boundaries between the US and Japan, Kiribati, Western Samoa, Tonga, The Marshall

Islands, and the Federated States of Micronesia must be established.[12] Given the rising interest of the US and of these countries in the region's ferromanganese-crust-laden seamounts, a vigorous push should be made to conclude treaties that precisely delimit these boundaries, otherwise industry may be leary of investing.

Managing the EEZ

When the US's many maritime boundary treaties (both new and renegotiated) have been agreed upon and ratified by the US Senate, federal agencies should be better able to move forward in identifying EEZ mineral resources and in implementing exploration and leasing programmes. Despite a lack of formal treaties, much work is already under way. The DOI's USGS, for example, is making detailed studies of twelve corridors that extend across the EEZ (Figure 10.4). These corridors were selected to provide a broad coverage of the continental margins' varied geology. The USGS has done shallow- and deep-water drilling and made seismic and side-scan sonar profiles.[13] From this information, USGS personnel are preparing a Continental Margin Map series (1: 1,000,000), with the first scheduled for release in late 1987.

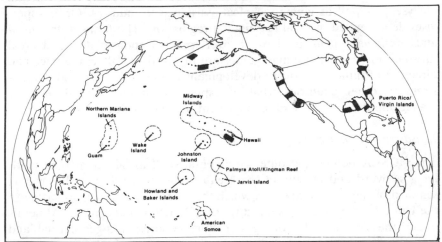

Figure 10.4 Representative corridors established for investigation of the EEZ's diverse geology

Source: D. L. Peck, 'The U.S. Geological Survey Program and Plans in the EEZ', in *Symposium Proceedings: A National Program for the Assessment and Development of the Mineral Resources of the United States Exclusive Economic Zone, 15–17 November 1983*, USGS Survey Circular 929 (USGS, Alexandria, VA, 1984), p. 81.

NOAA – through the National Ocean Service – in the DOC is also investigating the EEZ. This agency is preparing a series of atlases on selected areas. These volumes include mapped data and other information on living marine resources and ecosystems; physical oceanographic characteristics; environmental quality; and political boundaries and jurisdictions.[14] This effort, however, is clouded by sensitive-data-classification issues between NOAA and the US Navy; these problems, as of mid-1987, had not been resolved.

For the most part, marine scientific responsibilities of the DOI and DOC are complementary. But some overlaps occur because the wording of national legislation in the OCSLA and DSHMRA creates ambiguities in interpreting jurisdictional responsibilities of the two departments. DOI personnel in the USGS have a long-standing experience with, and the expertise to evaluate the geology and mineralisation of, the continental margin; the DOI's MMS is responsible for providing continental shelf mineral exploration licensing and for organising and holding competitive bidding mineral leasing sales. To date, this work has focused mainly on oil and gas. Thus, some critics have questioned the DOI's jurisdictional competence over other minerals, and they also note the absence of adequate data on OCS hard minerals. They ask: *How does industry bid and the DOI accept or reject bids fairly when little or no data exist on either side?*

NOAA's activities are housed in its Offices of Marine Minerals (OMM) and Minerals and Energy (OME). The OMM's responsibilities include the making of continental shelf resource evaluations and environmental analyses, as well as giving technical assistance to industry. OME supervises deep seabed mineral leasing and regulation. The terms *deep seabed* and *continental shelf* are keys to overlapping responsibilities and areal jurisdictions. How these ambiguities are resolved will affect the pace and direction of mineral resource development in the US's EEZ and demonstrate to other countries what to expect in marine mineral resource management problems and management agency rivalries.[15]

Jurisdictional ambiguities

The OCSLA defines the outer continental shelf as 'all submerged lands lying seaward and outside of the area of land beneath navigable waters seaward of the territorial sea in which the subsoil and seabed appertain to the United States and are subject to its jurisdiction and control'. Minerals under the Act include 'oil, gas, sulfur, geopressured-geothermal and associated resources, and all other minerals authorized by an act of Congress to be produced from public lands'.[16] In the MMS's interpretation of the law, the OCS 'extends at least to the seaward limit of the EEZ' and in some locations 'beyond the EEZ ... to the physical limit of the continental shelf'. The OCSLA's counterpart, DSHMRA, uses the definition of the continental shelf contained in the 1958 Geneva Convention on the Continental Shelf (GCCS) which defines it as

> the seabed and subsoil of the submarine areas adjacent to the coast but outside the area of the territorial sea, to a depth of 200 meters or beyond that limit to where the depth of the superjacent waters admits of the exploitation of the natural resources of the said areas.[17]

Finally, the UNCLOS III Convention defines the continental shelf as those 'sea-bed and subsoil ... areas that extend beyond' a state's

> territorial sea throughout the natural prolongation of its land territory to the outer edge of the continental margin, or to a distance of 200 nautical miles from the baselines from which the breadth of the territorial sea is measured where the outer edge of the continental margin does not extend up to that distance.

Under certain circumstances, however, the LOS Convention allows the EEZ to extend on to the continental shelf a distance of 350nmi (648km).[18]

In sum, the continental shelf could end and the deep seabed begin at varying distances offshore, depending upon which definition one uses. From this situation comes the offshore jurisdictional ambiguity for the DOI and the DOC. James Broadus and Porter Hoagland[19] and Robert McManus[20] have appropriately noted the potential for overlapping jurisdictions (Figure 10.5).

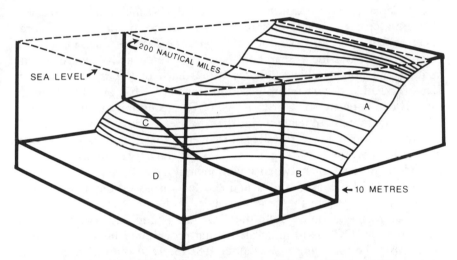

Figure 10.5 Hypothetical offshore mine sites within the 200-nmi EEZ

Source: After R. McManus, 'Legal status and 1983–1984 developments', in M. B. Hatem (ed.) *Marine Polymetallic Sulfides – A National Overview and Future Needs: Workshop Proceedings, 19–20 January 1983*, Maryland Sea Grant Publication no UM-SG-TS-83-04 (University of Maryland, College Park, MD, 1983), p. 85. With permission.
Note: A is on the outer continental shelf within the 200-nmi zone; B lies off the geological continental margin but within 200 nmi from shore; C sits on the shelf beyond 200 nmi; D is off the geological continental margin and beyond 200 nmi.

Situations portrayed in Figure 10.5 represent possible overlapping jurisdictions anywhere along the US's offshore, assuming the seaward limit of the geological continental shelf has been established, which it has not. In Area A, on the continental shelf and within the 200-nmi EEZ, the MMS (under the OCSLA) has an undisputed claim to control mineral leasing and mining. Area D, beyond the 200-nmi EEZ limit and off the continental shelf will be administered by NOAA (under the DSHMRA). Jurisdiction becomes fuzzy in areas B and C. Area B, although not on the continental shelf, lies within the 200-nmi zone and could, by the MMS's interpretation of the OCSLA, be placed under its control.[21] Under the DSHMRA (which uses the 1958 GCCS definition) Area B could be part of NOAA's jurisdiction. The location of Area C is on the continental shelf but beyond the 200-nmi EEZ. If the MMS interpretation of OCSLA and the 1958 GCCS definition are used (as well as the special circumstances provision in the LOS Convention) Area C comes under the control of the MMS. NOAA could, however, also claim jurisdiction over nodule deposits here (but not other minerals), given that nodules are assigned to it under the DSHMRA.[22] This confusing jurisdictional situation has recently become especially apparent in the offshore of the US's Pacific North-west.

The Gorda Ridge

Along Canada's British Columbia and the US's Washington, Oregon, and California coasts, a series of spreading-zone ridges occur – the Explorer on the north, the Gorda on the south, and the Juan de Fuca in between (Figure 10.6). Marine researchers have discovered active hydrothermal venting and polymetallic sulphides within each of the three systems.[23] Explorer Ridge occurs entirely within Canada's EEZ and the Juan de Fuca Ridge is partly so. Gorda Ridge (GR) lies entirely within the US's EEZ. Government officials would like to see industry explore the GR and to develop mineral resources discovered there, if they are exploitable. The government's eagerness to seek industry's involvement has helped create friction between governmental agencies (DOI and DOC) and between the coastal states and the federal government, as well as raising an outcry from the general public.

As early as 19 January 1982 the DOI had announced its intention of developing an OCS hard-minerals leasing programme on a case-by-case basis.[24] On the following 8 December 1982 the DOI published in the US *Federal Register* a 'Notice of Jurisdiction' over OCS minerals other than gas, oil, and sulphur. The notice included all OCS areas in general but specifically identified GR minerals.[25] And on 28 March 1983, only a little more than two weeks after President Reagan issued his 10 March 1983 EEZ Proclamation, the US DOI's MMS announced its intention to prepare a Draft Environmental Impact Statement (DEIS) and to hold a lease sale for the GR, an area encompassing more than 181,000km^2 of seabed.[26] In December 1983 the MMS released a DEIS for public comment [27] and set the leasing date for August 1984. The DEIS failed to satisfy industry, the general public, or the state governments of California and Oregon. Critics felt the DEIS had not adequately addressed the environmental impacts mining might have. They contended that not enough information was available to make specific statements or

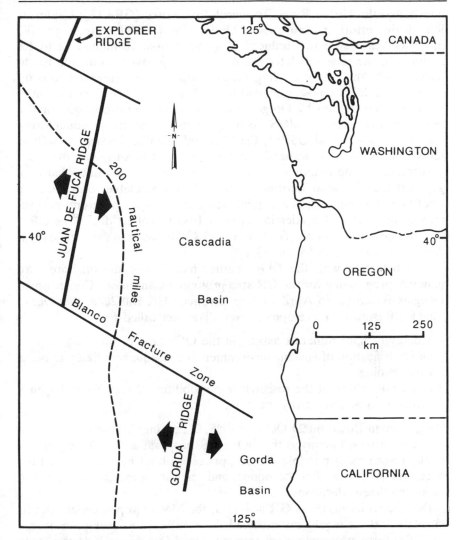

Figure 10.6 Gorda Ridge and Juan de Fuca Ridge spreading zones

Source: After B. A. McGregor and M. Lockwood, *Mapping and Research in the Exclusive Economic Zone* (USGS, Reston, VA, 1985), p. 25.

recommendations. The marine mining industry was not interested in any leasing and its members bluntly pointed out that the DOI did not even know if the GR actually had mineral potential. Some observers suggested that the DOI was in such haste because it wanted to put its jurisdictional stamp on an area not part of the geological continental shelf – an area usually administered by NOAA, within the DOC.[28]

As a result of this chorus of unhappy parties, the DEIS was withdrawn and in February 1984 the Secretary of the DOI and Oregon's Governor announced the establishment of a joint Federal-State Working Group, now called the Gorda Ridge Technical Task Force (GRTTF); California joined the effort in June 1984. Under the GRTTF agreement, the proposed lease area was reduced from more than 16 million ha to 1.6 million ha; the area was later trimmed to a little over 1.0 million ha, to include only the GR spreading centre proper. The agreement also included an indefinite postponement of the lease sale. The fifteen-member GRTTF is co-chaired by Oregon and California's state geologists and a representative from the MMS; its other members include personnel from the DOI's USGS and USBM, the Fish and Wildlife Service, as well as representatives from academic and private institutions and the state governments. The mandate of the GRTTF was to determine what data gaps existed and what information was already available. Based on the GRTTF's recommendations, joint research sub-groups from governmental agencies and academia began in 1985 to collect data,[29] an effort that has continued since. To date, the GRTTF has made no recommendation for a new EIS for the GR.[30]

Dissatisfaction with the DOI's earlier handling of the offshore hard minerals programme for the GR area prompted California's Congressman Douglas Bosco on 16 April 1985 to introduce HR Bill 2048, the 'Ocean Mineral Resources Development Act'. The Act called for

1 a delay of hard minerals leasing in the OCS area of the GR;
2 an identification of specific environmental consequences likely to occur from mining;
3 a clear definition of the research responsibilities of each federal agency involved in administering the area.[31]

Congressman Bosco on 24 October 1985, in hearings before the House of Representatives, portrayed the DOI's GR proposal as 'auctioning off ... public resources at possible fire sale prices' without having gathered the necessary environmental, economic, and scientific information needed to make intelligent decisions.[32]

Despite criticisms of its GR activities, the MMS has proceeded with its mandate – some would say assumed the mantle – to develop an overall OCS pre-lease prospecting programme. On 7 December 1984 the MMS published an announcement of proposed rule-making for OCS hard minerals exploration[33] and a month later (15 January 1985), it called for further recommendations of other areas for possible leasing on the OCS.[34] on 9 April 1986 it published an advance notice of proposed rule-making for post-lease operations, which were made available for public comment and federal agency response.[35] The MMS on 26 March 1987 published those questions directed to its 9 April 1986 proposal, as well as its responses.[36]

Under the MMS's proposed programme for obtaining pre-leasing exploration permits, decisions will be made on a case-by-case basis. Prospecting permits are not limited to US nationals, a provision that will likely stir debate within industry, the Congress and the general public.

The plan stipulates that issuance of a prospecting permit (set for a two-year period) does not commit the US Government to granting mining leases, a provision disliked by those in industry, because they could lose their exploration investments. To encourage industry's involvement, the DOI's programme provides protection for proprietary information of prospecting permittees, minimises fees, and eliminates bond requirements. On the other hand, in concert with long-standing federal policies, the programme provides no direct subsidies, and permits may be suspended or cancelled should a lessee's activities present 'a serious threat to life, property, mineral deposits, national security, or the environment'. Specific environmental protection and monitoring programmes will be required of lessees and the MMS will obtain baseline data as a guide for preparing EIS's for proposed sales.[37]

While the MMS has been working to implement an exploration licensing programme under its OCSLA mandate, bills have been introduced into the Congress in an effort to provide a separate system of regulations for OCS hard minerals. Industry is especially concerned for mandated competitive bidding required under the OCSLA,[38] which was formulated primarily for the leasing of natural gas, oil, and sulphur. Officials of several states have expressed similar concerns to the MMS. In a letter to the MMS, Jananne Sharpless, California's Secretary for Environmental Affairs, pointed out that:

> It is necessary to construct a statutory and regulatory framework that reflects the fact that hard minerals is a different and internally diverse industry compared to oil and gas, and one that is not yet economically viable offshore for all minerals and geographic locations.[39]

In the same vein, a letter to the MMS from Robert Grogan, Director of Alaska's Office of Management and Budget Division of Governmental Co-ordination, noted that:

> We continue to maintain ... that the Outer Continental Shelf Lands Act (OCSLA) should not be used as the legislative authority for marine mineral mining in the OCS. The OCSLA was developed primarily to regulate oil and gas development and does not address the specifics of [hard] mineral development, which has very different impacts.[40]

On 25 February 1987, in response to concerns voiced by a coalition of industry representatives, coastal states, and environmental groups, Congressman Mike Lowry from the State of Washington introduced HR 1260 'The National Seabed Hard Minerals Act'. The Bill is identical to HR 5464, which he introduced in the previous Congress on 15 August 1986. Lowry's Bill is designed to provide industry with clear guidelines of what to expect in exploration, leasing, and development programmes[41] and to allow exploration and commercial recovery in offshore shelves adjacent to the fifty states.[42] The Bill also calls for

1 assuring that companies investing in exploration have priority in leasing sales,
2 establishing state-federal task forces to guarantee that concerns of coastal states are heard,
3 setting aside undisturbed offshore 'reference areas' to serve as indicators of mining's environmental impacts.[43]

If passed into law, Lowry's Bill would require the DOI Secretary and the Administrator of NOAA to submit (within one year of the Bill's enactment) a special report to specified congressional committees. The report would specifically identify the two agencies' agreed-upon allocation of EEZ responsibilities, such as the MMS's retaining leasing and permitting authority and NOAA's having control of environmental research and monitoring. If this provision is fully met, it will have helped considerably to reduce NOAAs and the MMS's overlapping functions and jurisdictions.

What might be the consequence here or elsewhere of such confusion and apparent agency rivalry and jurisdictional ambiguity? Broadus and Hoagland feel that this situation may have both positive and negative aspects. On the negative side, it could (1) create added costs that delay development of a resource and (2) discourage industry from making investments. From the positive standpoint (1) private companies and environmental groups may have a greater influence in the resource use decision-making process through increased access to the competing agencies; (2) competing agencies may face greater accountability; and (3) more scientific expertise should be available for managing the resource.[44]

Past inertia and initiatives

The GR controversy is a sympton of the overall neglect by the US Government's executive and legislative branches to formulate and to implement a co-ordinated long-term oceans policy that can be translated by federal agencies into development and management programmes. A fundamental problem exacerbates this lack of direction, that is, authority for managing the oceans resides within a multitude of congressional committees, cabinet departments, special agencies, and executive offices, all of which are jealous of their prerogatives and do not want a 'super-agency' or 'czar of the oceans'.[45] In addition, budget competition among these various entities dissipates professional energies and dilutes the impact of implemented programmes.

In the past, many brief and aborted efforts were made to determine the potential of marine minerals and to establish a co-ordinated programme for their development. The USBM in 1963 organised a Marine Technology Center at Tiburon, California; its task was to determine the feasibility of ocean mining. And in 1968 a Public Land Law Review Commission study and another in 1975 by a Panel on Operational Safety in Marine Mining (POSMM) concluded that offshore hard minerals were potentially important but attempting to manage them under the OCSLA was not conducive to their development. The 1975 POSMM report also recommended that prototype hard-mineral mining in the OCS should begin, using representative areas to determine its feasibility and consequences.[46]

The year before the POSMM report, draft OCS hard-mineral regulations for leasing and mining were developed but never implemented. That same year, the USGS drew up a long-range marine mining programme, but it was not funded. This effort was followed by attempts 'to develop a prototype leasing program which would permit limited seabed mining operations with close environmental monitoring and control, but these were not carried to completion due to a lack of funds'. Three years later in November 1977, the Bureau of Land Management and the USGS called for the formation of an interagency task force to design OCS minerals policy leasing recommendations. The task force in 1979 reported that enough economic incentives and interest existed to allow commercial-scale hard-minerals mining in the OCS. But no regulations were developed,[47] the situation that existed when the MMS began its effort to open the GR for exploration and leasing.

Perhaps in response to the displeasure over the proposed development of the GR area and an admonishment from President Reagan to move forward in developing needed strategic minerals in the US's EEZ, the MMS in August 1983 established the Office of Strategic and International Minerals (OSIM). Since its establishment, planning efforts and co-operation with coastal states have become more co-ordinated and mutually beneficial, as demonstrated by the joint federal-state efforts in investigating the polymetallic sulphides of Hawaii and environmental and geological conditions of the GR off Oregon and California. These efforts have pointed the way for additional joint federal-state task forces of which there are now five, including agreements by the DOI with Georgia, with several Gulf Coast states and with North Carolina.[48]

Based upon the MMS's 15 January 1985 call (discussed previously) for recommendations of additional areas with hard-mineral resource potential, North Carolina responded by nominating Onslow Bay and indicated its desire to assist in developing a phosphorite leasing programme. OSIM followed up on North Carolina's interest by querying interested scientists, local phosphate firms, and North Carolina's Office of Marine Affairs as to the potential for developing the OCS's phosphorites. Subsequently North Carolina's Marine Science Council endorsed the idea and formed an *ad hoc* committee to develop an action plan, to include the university system, the state government, and the general public. The governor endorsed the *ad hoc* committee's proposal, and in March 1986 the DOI and North Carolina announced the formation of a joint task force to study offshore hard minerals. Its first job was to seek proposals for the preparation of an economic feasibility study to determine if the next step, the development of an EIS, is worthwhile.[49]

Future directions

As national, state, and local governments and industry become more interested in mineral exploitation within the EEZ, the need for more co-operative management efforts will arise, both for hard minerals and petroleum. Perhaps, the task-force concept can be taken a step further. Lewis Alexander and Susan Hanson, writing for the journal *Oceanus*, have pointed the way. They suggest that because the continental shelf's shallower seabed areas will be increasingly used for marine parks, aquaculture establishments, power plants, oil storage facilities and restaurants, marine management must be designed to accommodate multiple-users. They stress that both the unity and the diversity of geographic conditions of coastal regions can be better coped with if state governments are given more offshore responsibility, even beyond the traditional 5.5km limit. This responsibility could be focused within regional councils (similar to those now used in fishery management under the MFCA), which would help solve the dilemma of the need for certain ubiquitous regulations for the EEZ while simultaneously recognising regional differences. They recommend the establishment of eight councils, whose function would be to avoid multiple-user conflicts. Boundaries would conform to those of the MFCA councils, although these could be flexible, depending on local situations. Federally mandated guidelines would regulate the councils' actions, with the councils' members coming from federal marine-resource agencies, from state representatives and from the private sector. Initially council recommendations for management and conflict avoidance could be advisory. Once the functions and operational procedures are well established, an investigatory role might be added. National co-ordination of the councils could be performed by an agency in Washington, DC, as the Department of Transportation, DOI, or NOAA.[50]

Although the Alexander-Hanson proposal does not specifically identify marine mining as one of the councils' concerns, they note the importance of petroleum production in the offshore. The co-operative experience generated by the HA-JA and GRTTF could easily be applied in a broader context. The mineral industries, whether sand and gravel, shells, phosphorite or petroleum, should benefit from a regional multiple-user management programme. And certainly the coastal states would welcome an expanded role in offshore resource management; their concern was well illustrated at a conference of the Coastal States Organization (CSO) in early April 1987 which, in effect, made its own 'EEZ Proclamation'. The conference theme focused on how coastal states can expand their role in managing offshore resources,[51] a necessary change, as proclaimed by the CSO's director, R. Gary Magnuson, who sees the national government's policies as ineffective or as failures.[52]

Some coastal states are taking seriously the CSO's call for action in oceanic resources management. Oregon in May and June of 1987 enacted two important pieces of legislation. One item (Senate Bill 606) parallels concepts contained in the federal Lowry Bill. The Bill deals with hard mineral leasing and fee collection programmes and with public disclosure of information for areas within the state's territorial sea.[53] The other item (Senate Bill 630) established a comprehensive ocean management act that provides for the formation of an Oregon Ocean Resources Management Task Force and covers all oceanic resources out to 200 nmi.[54] The State of Washington is undertaking a similar, but less ambitious, look at oceanic resources management. In mid-1987 the Western Legislative Conference Ocean Resources Group was preparing materials as a guide for western state legislators in asserting state management interests in oceanic resources.[55]

Magnuson in 1985 recommended a programme (somewhat like that suggested by Alexander and Hanson) whereby federal legislation would establish regional EEZ ocean management authorities whose members would represent appropriate state and federal agencies. Their financial support would come from a portion of revenues generated by EEZ royalties and rents, as in petroleum production. Major decisions and a failure of agreement within the management authorities could be implemented or decided in Washington, DC.[56]

Paul Ryan, editor of the journal *Oceanus*, has suggested that the time may have come to consider establishing a 'separate federal agency to handle EEZ affairs' or to 'elect or appoint a governor for the area'. He noted, however, that such a plan would probably entail a change in the US Constitution. Whatever the form of the future administration of the EEZ, Ryan admonishes that 'there is an overriding need' to get on with the 'research, exploration, and survey of the US EEZ'.[57] Many Congressmen also feel the US should take stock of its ocean programmes and evaluate their future direction and potential.[58] No truly comprehensive examination of US ocean policy and programmes has been made since the work of the 1966 Stratton Commission whose report *Our Nation and the Sea* provided a strategic plan focused on environmental assessment, coastal-zone management, and resource management.[59]

Since 1983 several efforts to introduce legislation into the Congress to establish a national oceans marine policy commission similar to the Stratton Commission have failed, in part from a lack of support by the Reagan Administration. From its beginning in 1980, the Reagan Administration tried to gut numerous oceanic research and management activities. If not for the Congress's support, funding of state coastal zone management efforts and Sea Grant College programmes would have been eliminated.[60] Were the Reagan Administration to have its way, the 'O' could be removed from NOAA – the National Oceanic and Atmospheric Administration. In fairness to the Reagan Administration, however, one must admit that it has also sought to trim other federal agencies' functions that it viewed as competitive with private enterprise.

Indicative of recent events is the elimination of the National Advisory

Committee on Oceans and Atmosphere (NACOA). Established in 1971 and composed mainly of representatives from industry and academia, NACOA was the only national body attempting to give direction to US oceans research and management programmes. Through its annual reports (1972–87), it made recommendations to the President and the Congress on matters ranging from oceanic and atmospheric research needs to strategies for integrated and co-ordinated administrative programmes.[61] For example, in 1985–6, under the leadership of John Flipse, of Texas A&M University, NACOA assessed the roles and missions of NOAA. Its efforts were not well received by some members of the Congress, who criticised Reagan's NACOA appointments as presidential cronies that lacked the necessary expertise demanded under the legislative criteria.[62]

Long before NACOA published its report on NOAA (February 1987), the Congress moved to scuttle NACOA (by not funding it after 30 September 1986), perhaps anticipating that its report would not be flattering to NOAA's management efficiency and priorities or the Congress's support for this important agency. In its report NACOA stressed a positive view of NOAA's potential as the country's lead agency in oceanic and atmospheric research and resource management, if only it were adequately and consistently funded. And it expressed deep concern for the dilution of NOAA's effectiveness because of its mandate to function as a scientific, a managerial, and a service agency.[63]

On 23 April 1986 while NACOA was preparing its report on NOAA, a Bill, HR 4676, was introduced into the House of Representatives to eliminate NACAO. According to some critics, the Bill's sponsors were unhappy with the 'ideological makeup' of its membership.[64] Others see the Congress' action not as a matter of ideology but of gaining 'power' in directing oceanic programmes through the establishment of a second 'Stratton Commission', the very body whose recommendations led to the estabishment of NACOA. White House staff deny they helped in eliminating NACOA only to replace it with another 'Stratton Commission'. True to its principles, the Executive Branch has spoken against establishing a new oceans policy commission. Not so in the Congress.

On 19 February 1987 Senator Lowell Weicker (R-Connecticut) and Congressman Walter Jones (D-North Carolina), Chairman of the House Merchant Marine and Fisheries Committee, introduced Bills (S 562 and HR 1171) to form a National Oceans Policy Commission.[65] A hearing on the Jones Bill was held on 21 May 1987. Senators John Breaux (D-Louisiana) and Lowell Weicker; Jean-Michel Cousteau of the Cousteau Society; John Costlow of Duke University's Marine Laboratory; Edward Wolfe, Deputy Assistant Secretary of State; and John Carey, NOAA's Deputy Administrator, gave testimony. All witnesses supported the measure except Wolfe and Carey, who presented the Reagan Administration's 'opposition to ... what they see as an "unnecessary" commission', because it would be a redundancy of groups and agencies already co-ordinating and advising on oceans policy and programmes.[66] Despite a lack of support from the Executive Branch, on 9 July 1987, the House Merchant Marine and Fisheries Committee reported favourably on the Bill.[67]

The Weicker and Jones Bills are not identical but similar. Under the Jones Bill, the commission would have seventeen members, with fourteen nominated by the leadership of both the Senate and House of Representatives and then appointed by the President. The President would also select the chair and vice-chair. Nominees must represent a wide range of expertise in marine, local, and state government and national and international policies formulation. The Jones measure requires the commission to submit its report within two years and under the Weicker Bill, after eighteen months.[68]

Conclusions

Considering the US Government's professed eagerness to see its EEZ provide an assured supply of strategic minerals, its seems odd that the State Department has not made more progress in acquiring boundary delimitation treaties and that the Senate has not acted more expeditiously to approve those established, as with Mexico. With time, however, most boundary disputes must be resolved so that mining firms can obtain a clear title to seabed minerals.

The Congress is moving toward passing legislation that should help sort out ambiguities of federal agency mandates in managing the EEZ, and there is more co-operation between coastal states and federal agencies in managing mineral resources, even those lying beyond the 5.5km territorial sea limit. One observer suggests that the participating roles and rights of US states and territories in managing EEZ minerals should be 'spelled out in a comprehensive management regime'.[69] Passage of the Lowry Bill will meet this need. Whether special regional management councils will be established is difficult to predict, but the emerging success of several offshore mineral task forces, as betweeen the DOI and the states of Hawaii, Oregon, and California, North Carolina, Georgia, and the Gulf Coast States demonstrates the assets of co-operative management.[70]

It is still too early to say whether the US Government will succeed in establishing a co-ordinated oceans policy and development programme, but if out of the Jones and Weicker Bills before the Congress there comes a vigorous and independent commission that is not a 'Toy of the Congress' and which provides recommendations that can be translated into workable legislation, perhaps the US will finally achieve a focused oceans policy and development programme, one that includes minerals. We must wait and hope!

Notes

1 Office of Technology Assessment, US Congress, *Marine Minerals: Exploring our New Ocean Frontier* (USGPO, Washington, DC, 1987), p. 3.
2 F. C. F. Earney, 'The geopolitics of offshore petroleum', in H. E. Johansen, O. P. Matthews, and G. Rudzitis (eds), *Mineral Resources Development: Geopolitics, Economics, and Policy* (Westview Press, Boulder, CO, 1987), p. 59.
3 F. Rose, 'Fish – and perhaps oil – lie at the bottom of U.S. and Canada dispute', *Wall Street Journal* (13 April 1984), p. 30.
4 J. Cooper, 'Delimitation of the maritime boundary in the Gulf of Maine Area', *Ocean Development and International Law* vol. 16, no. 1 (1986), pp. 63–4.
5 For a detailed analysis see L. De Vorsey, 'Historical geography and the Canada–United States seaward boundary on Georges Bank', in G. Blake (ed.), *Maritime Boundaries and Ocean Resources* (Croom Helm, Beckenham, 1987), pp. 182–207.
6 F. C. F. Earney, 'The United States Exclusive Economic Zone: mineral resources', in G. Blake, (ed.) *Maritime Boundaries and Ocean Resources* (Croom Helm, Beckenham, 1987) p. 169.
7 Letter: M. O. C. Walker, Bureau of Oceans and International Environmental and Scientific Affairs, US Department of State, Washington, DC, 2 Sept. 1987.
8 H. R. Marshall Jr, 'Disputed areas influence OCS leasing policy', *Offshore*, vol. 45, no. 5 (1985), pp. 100–01; for an excellent review of the Bering Sea problem, see C. M. Antinori, 'The Bering Sea: A maritime delimitation dispute between the United States and the Soviet Union', *Ocean Development and International Law*, vol. 18, no. 1 (1987), pp. 1–47.
9 Marshall, 'Disputed areas influence OCS leasing policy', p. 101.
10 Memorandum: R. H. McMullin, Acting Regional Director, Minerals Management Service, Alaska Outer Continental Shelf Region, DOI, Anchorage, AK, to T. Giordano, Program Director, Office of Strategic and International Minerals, MMS, DOI, Long Beach, CA, 10 Sept. 1987.
11 Walker, Letter, 2 Sept. 1987.
12 Earney, 'The United States Exclusive Economic Zone', p. 174.
13 Using the Gloria side-scan sonar system (discussed in Chapter 5) developed by the UK's Institute of Oceanographic Sciences, an area the size of New Jersey is mappable in one day.
14 B. A. McGregor and M. Lockwood, *Mapping and Research in the Exclusive Economic Zone* (DOC, Washington, 1985), p. 36; Letter: M. Lockwood, Oceanographier, National Oceanic and Atmospheric Administration, DOC, Rockville, MD, 30 July 1987.
15 According to James Broadus and Porter Hoagland, of Woods Hole Oceanographic Institution, Colombia, France and the PRC already seem to have experienced marine minerals jurisdictional disputes among their governmental agencies. See J. M. Broadus and P. Hoagland III, 'Rivalry and coordination in marine hard minerals regulation', in *Exclusive Economic Zone Papers*, reprinted from *Oceans '84 Exclusive Economic Zone Symposium* (NOAA, DOC, Washington, DC, 1985), p. 56.
16 R. McManus, 'Legal status and 1983–1984 developments', in M. B. Hatem (ed.), *Marine Polymetallic Sulfides: A National Overview and Future Needs: Workshop Proceedings, 19–20 January 1983*, Maryland Sea Grant Publication no. UM-SG-TS-83-04 (University of Maryland, College Park, MD, 1983), p. 82.

17 'Convention on the continental shelf', Article 1, adopted 29 April 1958, UN Document A/Conf. 13/L. 53.

18 United Nations, *The Law of the Sea: Official Text of the United Nations Convention on the Law of the Sea* (UN, NY, 1983), pp. 27–8.

19 Broadus and Hoagland, 'Rivalry and coordination in marine hard minerals regulation', pp. 56–7.

20 McManus, 'Legal status and 1983–1984 developments', pp. 85–7.

21 A May 1985 legal opinion given by the Solicitor of the DOI states that 'the president's EEZ Proclamation extended US jurisdiction and MMS leasing authority to at least 200 miles offshore, or as far as the geologic OCS may extend. This authority does not apply to the OCS of US possessions and territories'. See 'Interior Department's proposed regulations for marine minerals mining on the OCS', *Oceans Policy News* (May 1987), p. 12.

22 McManus, 'Legal status and 1983–1984 developments', pp. 83–7; Broadus and Hoagland, 'Rivalry and coordination in marine hard minerals regulation', pp. 56–8; *The Ocean and the Future*, 'Hearings', testimony of J. M. Broadus before the Subcommittee on Oceanography of The Committee on Merchant Marine and Fisheries, House of Representatives, 99th Cong., 1st sess., 24 Oct. 1985 (USGPO, Washington, DC, 1986), pp. 164–5.

23 See J. Samson, 'Compilation of information on polymetallic sulfide deposits and occurrences off the West Coast of Canada', Draft Working paper (Canada Oil) and Gas Lands Administration, Energy, Mines and Resources Canada, Ottawa, Oct. 1985), pp. 3, 5–6, 24–5.

24 *Federal Register*, vol. 52, no. 58 (26 March 1987), p. 9,759.

25 *Federal Register*, vol. 47, no. 236 (8 Dec. 1982), p. 55,313.

26 *Federal Register*, vol. 48, no. 60 (28 March 1983), p. 12,840.

27 J. B. Smith, B. R. Holt, and R. G. Paul, 'The minerals management service's nonenergy leasing program for the Outer Continental Shelf/Exclusive Economic Zone', in *Minerals and Materials: A Bimonthly Survey* (April/May 1985), p. 42.

28 Some observers believe the DOI's position reflects only a desire to protect its authority under the OCSLA and does not seek to set final policy.

29 Smith, Holt, and Paul, 'The minerals management service's nonenergy leasing program'. p. 42.

30 Letter: G. McMurray, Marine Minerals co-ordinator, Department of Geology and Mineral Industries, State of Oregon, Portland, OR, 2 July 1987. For an excellent discussion of the GRTTF, see G. McMurray, 'The Gorda Ridge technical task force: a cooperative federal-state approach to offshore mining issues', *Marine Mining*, vol. 5, no. 4 (1986), pp. 467–75.

31 'Ocean Mineral Resources Development Act', H. R. 2048, House of Representatives, 99th Cong., 1st sess., 16 April 1985.

32 *The Ocean and the Future*, 'Hearings', testimony by Congressman D. B. Bosco, p. 88.

33 *Federal Register*, vol. 49, no. 237 (7 Dec. 1984), p. 47,871.

34 *Federal Register*, vol. 50, no. 10 (15 Jan. 1985), p. 2,264.

35 *Federal Register*, vol. 51, no. 68 (9 April 1986), p. 12,163.

36 *Federal Register*, vol. 52, no. 58 (26 March 1987), pp. 9,760–3.

37 ibid., pp. 9,758–62.

38 ibid., p. 9,758.

39 Mimeographed copies of Letters: J. Sharpless, Secretary for Environmental Affairs, State of California, Sacramento, CA, 19 June 1987 and 30 Sept. 1986, to the Minerals Management Service, DOI, provided by R. J. Bailey, OCS Co-ordinator, Department of Land Conservation and Development, State of Oregon, Portland, OR, 10 Aug. 1987.

40 Mimeographed copy of Letter: R. L. Grogan, Director, Office of

Management and Budget Division of Governmental Coordination, State of Alaska, Juneau, AK, 1 June 1987, to R. Stone, MMS, Long Beach, CA, provided by R. J. Bailey, OCS Co-ordinator, Department of Land Conservation and Development, State of Oregon, Portland, OR, Bailey, 10 Aug. 1987; on 5 July 1988 the MMS published final prospecting rules for marine minerals other than oil, gas and sulphur (see *Federal Register*, vol. 53, no. 128 (5 July 1988), pp. 25,242–60.

41 'Interior Department's proposed regulations', p. 12.
42 D. Moffitt, 'Ocean mining: a framework for North Carolina and other coastal states', *Legal Tides*, vol. 1, no. 4 (1986), p. 7.
43 'Interior Department's proposed regulations', pp. 12–13.
44 Broadus and Hoagland, 'Rivalry and coordination in marine hard minerals and regulation', pp. 59–60; *The Ocean and the Future*, p. 166.
45 For a good discussion of the difficulties in implementing integrated ocean management programmes, see S. A. Bleicher, 'Reflections on the failure of NOAA's ocean management office', *Coastal Zone Management Journal*, vol. 11, no. 4 (1984), pp. 353–67.
46 *Federal Register*, vol. 52, no. 58 (26 March 1987), p. 9,759.
47 ibid.
48 Telephone interview: R. G. Paul, Minerals Management Service, DOI, Long Beach, CA, 24 July 1987.
49 D. Moffitt, 'Ocean mining', pp. 6–7.
50 L. M. Alexander and L. C. Hanson, 'Regionalizing the US EEZ', *Oceanus*, vol. 27, no. 4 (1984/85), pp. 7–12.
51 See Center for the Study of Marine Policy in association with the Coastal States Organization, *Coastal States are Ocean States: Proceedings of a Conference Held at the Mayflower Hotel, Washington, D.C., 1–3 April 1987* (CSMP, University of Delaware, Newark, DL, 1987); see also '"Coastal States are Ocean States" Conference', *Oceans Policy News* (April 1987), p. 5.
52 R. G. Magnuson, 'The coastal state challenge', address presented to the North Carolina Coastal States Ocean Policy Conference, Raleigh, NC, 31 Oct. 1985.
53 State of Oregon, *Senate Bill 606*, 'A Bill for an Act Relating to hard mineral deposits in submersible and submerged lands', 64th Oregon Legislative Assembly – 1987 Regular Session, Salem, OR, 19 May 1987.
54 State of Oregon, *Senate Bill 630*, 'A Bill for an Act relating to ocean resources planning', 64th Oregon Legislative Assembly – 1987 Regular Session, Salem, OR, 24 June 1987.
55 Letter: R.J. Bailey, OCS Co-ordinator, Department of Land Conservation and Development, State of Oregon, Portland, OR, 10 Aug. 1987.
56 Magnuson, 'The coastal state challenge'.
57 P. R. Ryan, 'The Exclusive Economic Zone', *Oceanus*, vol. 27, no. 4 (1984/85), p. 4.
58 See, for example, D.K. Inouye, 'Resource development in the EEZ: lead or follow, the time has come', *Sea Technology*, vol. 28, no. 6 (1987), pp. 36–7.
59 Ryan, 'The Exclusive Economic Zone', p. 4.
60 Despite efforts by many Congressmen, funding for oceanic research has been declining. In fiscal year 1984 the US spent about $1,197 million on oceanographic research, whereas space programmes received $12,664 million. The gap grew in 1985 when only $1,477 million went to the ocean sector, with space receiving $15,236 million. In fiscal year 1986 the gap widened even more, with oceanic research receiving only $1,508 million and space programmes obtaining $16,721 million. For more detail, see Inouye, 'Resource development in the EEZ', and 'Federal ocean budgets for 1988, part I', *Sea Technology*, vol. 28, no. 2 (1987), pp. 30–4.
61 National Advisory Committee on Oceans and Atmosphere, *A Report to the*

President and the Congress, Fifteenth Annual Report (National Technical Information Center, DOC, Springfield, VA, 30 June 1986), p. 33.

62 Letter: J. E. Flipse, Associate Deputy Chancellor, College of Engineering, Texas A&M University System, College Station, TX, 5 Aug. 1987.

63 National Advisory Committee on Oceans and Atmosphere, *An Assessment of the Roles and missions of the National Oceanic and Atmospheric Administration* (Texas A&M University, College Station, TX, 1987).

64 'National Advisory Committee on Oceans and Atmosphere (NACOA)', *Oceans Policy News* (May 1986), p. 13.

65 'National Oceans/Marine Policy Commission', *Oceans Policy News* (March 1987), pp. 7–8.

66 'National Oceans Policy Commission', *Oceans Policy News* (June 1987), pp. 5–6.

67 'National Oceans Policy Commission', *Oceans Policy News* (Aug. 1987), p. 5.

68 'National Oceans/Marine Policy Commission', p. 7.

69 R. J. Bailey, 'Marine minerals in the Exclusive Economic Zone: implications for coastal states and territories', paper presented at the Western Legislative Conference, Pacific States/Territories Ocean Resource Group, San Francisco, CA, 28 Feb. 1987.

70 The task forces are specifically focusing upon: cobalt-rich ferromanganese crusts (Hawaii); polymetallic sulphides (Oregon and California); phosphorites (North Carolina); heavy minerals and phosphorites (Georgia); heavy minerals, sulphur, and sand and gravel (Alabama, Louisiana, Mississippi, and Texas). For details of activities of these Task Forces, see Minerals Management Service, DOI, *Briefing Book: OCS Marine Mining* (DOI, Washington, DC, July 1987).

Chapter eleven

Offshore petroleum frontiers

Of the world's relatively unexplored areas with hydrocarbon potential, the continental margins are the most important. Petroleum consultant, Michel Halbouty, in 1981 estimated the area of the world's prospective offshore petroleum-containing sedimentary basins at nearly 23 million km^2, about 31 per cent of the world's total petroliferus basins.[1] Estimates of the offshore basins' recoverable reserves vary. In 1982 one investigator, Karl Hinz, cited estimates for the offshore of between 874,000 million and 2.149 million million bbl of crude oil (45 per cent of the world's estimated total) and 170 and 175 million million m^3 of the world's exploitable gas supply, a proportion similar to that of crude oil.[2] When Hinz made these estimates, drillers had completed only six boreholes in Norway's waters north of 62°N.,[3] and Alaska's offshore was only beginning to be explored, as were several areas in Africa and south-east Asia. Hinz stressed that our 'present knowledge of the geological structure and development of the continental margins is still too full of gaps for realistic assessment of their hydrocarbon potential'.[4] Hinz's statement remains true today.

With an increasing world petroleum consumption, as well as anticipated improvements in exploration and production technology, the offshore should become an increasingly important part of the world's petroleum industry. In 1986 the offshore accounted for more than 24 per cent of the world's estimated total annual crude oil production (Table 11.1); offshore commercial gas production estimated at 350,000 million cubic metres[5] accounted for a roughly estimated 19 per cent of the world's total commercial production of 1.807 million million m^3.[6]

In 1986 the Middle East led all regions in total daily offshore crude oil production, followed closely by the Latin America/Caribbean and the North Sea regions – with each pumping more than 3 million bbl/d. The UK in 1986 surpassed all other states in offshore crude oil production, with an output of more than 2.2 million bbl/d. The US's daily production of offshore natural gas in 1986 totaled nearly 338 million m^3, more than twice the UK's and four times Norway's daily output, the world's second and third largest offshore gas producers (Table 11.2).

A few oil industry specialists in the early 1980s suggested that by the turn of the century the offshore might provide 50 per cent of the world's annual total crude oil production. From the perspective of 1987, this estimate seems too high. Several factors may slow expansion of offshore petroleum production. These include

Table 11.1 Worldwide offshore daily average crude oil production, 1986

Area/country	Thousands of bbl/d[a]	Area/country	Thousands of bbl/d[a]
Middle East		Malaysia	20.9
Saudi Arabia	1,107.0	Thailand	15.9
Egypt	589.6	Philippines	7.8
Iran	505.0	Japan	1.4
Dubai	330.6	China	1.0
Neutral Zone	266.0	Total	1,158.6[b]
Abu Dhabi	265.0		
Qatar	158.0	West Africa	
Ras al Khaimah	11.0	Nigeria	289.5
Sharjah	8.6	Angola/Cabinda	185.6
Total	3,240.9[b]	Cameroon	125.0
		Congo	115.0
Latin America/Caribbean		Gabon	105.0
Mexico	1,700.0	Ivory Coast	19.5
Venezuela	900.0	Zaire	16.6
Brazil	376.0	Ghana	00.3
Trinidad/Tobago	127.8	Total	856.5[b]
Peru	116.6		
Chile	11.3	Australia/New Zealand	
Total	3,231.7[b]	Australia	384.1
		New Zealand	14.0
		Total	398.1[b]
North Sea			
United Kingdom	2,236.6		
Norway	780.6	Union of Soviet Federated	
Denmark	55.1	Socialist Rebublics	165.0[b]
Netherlands	20.7		
Total	3,093.0[b]	Mediterranean	
		Spain	36.1
North America		Greece	27.2
United States	1,257.0	Italy	8.9
Total	1,257.0[b]	Tunisia	6.1
		Total	78.3[b]
South-east Asia/Far East			
India	621.0	Offshore total	13,479.1
Indonesia	391.9	World total offshore	
Brunei	98.7	and onshore	55,801.4
		Per cent offshore	24.2

Source: 'Worldwide offshore daily average oil production (000 b/d)', *Offshore*, vol. 47, no. 5 (1987), p. 52. With permission.
Notes: a. Rounded; b. Estimated.

1 increasingly hazardous and expensive offshore exploration and production frontiers;
2 present and near-term depressed market conditions;
3 governmental development, leasing and taxing policies;

245

4 unsettled geopolitical relations between many coastal states.

These topics form the core of this and subsequent chapters.

Exploration and production frontiers: technology and economics

As the world's easily accessible and giant crude oil and natural gas fields become depleted, the petroleum industry will push its oceanic exploration and production activities into ever more hostile environments and marginal fields.[7] Drilling rigs and production platforms are already working in deep-water and arctic regions (the main foci of this chapter), where they face prohibitive costs and experience hazards of isolation, wind, cold, and ice. The problems posed by these hazards are a function of (1) technological capabilities, that is, what is a hazard today may be routine tomorrow,

Table 11.2 Worldwide daily offshore gas production, 1986

Country	Million m^3/d	Country	Million m^3/d
United States	337.5	Thailand	9.7
United Kingdom	158.6	Abu Dhabi	8.9
Norway	85.3	Angola	7.7
Venezuela	42.8	India	6.8
Netherlands	41.9	Republic of Ireland	6.5
Australia	41.0	Brazil	6.1
Malaysia	40.8	Colombia	3.6
USSR	36.8	Egypt	3.1
Mexico	26.9	Japan	1.6
Brunei	23.1	Oman	1.0
Indonesia	18.9	Spain	0.7
Nigeria	13.5	Greece	0.1
Denmark	13.1	Congo	0.1
Italy	11.8	Trinidad	0.02
New Zealand	10.4		
		Total	958.03

Source: 'Offshore gas production (MMcfd)', *Offshore*, vol. 47, no. 5 (1987), p. 51. With permission.
Note: a. Converted from ft^3 to m^3 and rounded, with $35,3147ft^3$ equal to $1m^3$.

and (2) a company's incentive to develop needed technological capabilities, which is influenced by costs.

Arctic Regions

High latitude regions are the focus of much petroleum industry research, exploration and development.[8] Here, the costs are higher[9] and the penalties for errors greater than in less physically demanding regions. In polar regions engineers must cope with especially difficult hazards – short summers, severe weather conditions, and stationary and moving ice. Despite these difficulties, Argentina and Chile in the southern hemisphere and Canada, Norway, the USSR, and the US in the northern hemisphere are attempting to meet these challenges.

The US has an especially active exploration effort and nascent development programme in arctic waters. As of mid-1987 six leasing sales were scheduled for Alaskan waters – one in the Beaufort Sea for January 1988[10] and five in various basins within the Bering Sea for 1989 through 1991.[11] Exploration and development of resources in US arctic regions, however, will proceed only if world economic conditions allow it. Although development of a 40 million to 50 million bbl oil discovery in the Gulf of Mexico or off the California coast may be economic, it is not in Alaskan waters. According to a computer simulation study by the Congress's analytical arm, the Office of Technology Assessment (OTA), for an oil field to be economically viable in Alaska's offshore requires a recoverable reserve of 1,000 million to 2,000 million bbl. But this requirement will fluctuate, depending on world prices.[12]

Although costs for all offshore petroleum areas are high, they are especially crucial in arctic and deep-water frontiers. The OTA has developed cash-flow profiles for ten hypothetical nearshore fields (both large and small) within three Alaskan basins (Harrison Bay, Navarin, and Norton), the Gulf of Mexico and a deep-water area off California. The OTA's cost estimates are only 'approximations of those which may be encountered with actual projects in these offshore areas'. Other than taxes and royalties, four major cost sectors – exploration, development, operation, and transportation – come into play in calculating total cost estimates (Table 11.3).[13]

The OTA's analyses assumed a strike success ratio of 1:10 and that five delineation wells are drilled, if a strike is made. In the Gulf of Mexico region, however, only three delineation wells are assumed. Because of the great differences in environmental severity, in water depth and in transport needs, the several areas' costs differ radically. Exploration costs for the Navarin basin are more than ten times those for the Gulf of Mexico. On the other hand, the exploration costs for the deep-water area off California are only 8 per cent less than the Norton basin. Development costs for a large field in the Navarin basin are sixty-five times greater than for the Gulf of Mexico.

Unfortunately Alaska's offshore has not been productive of major discoveries recently. Consequently estimates for Alaskan waters have been reduced significantly. In 1981 the USGS released a report that put Alaska's estimated crude oil resources (undiscovered and recoverable) at 12,200 million bbl and natural gas resources at 1.8 million million m³; the MMS in 1985 provided the OTA with revised data which set crude oil reserves at 3,300 million and natural gas at 0.4 million million m³. These reductions represented a 73 per cent decline for crude oil and a 78 per cent decline for natural gas.[14] An important part of these resources occurs in the Beaufort Sea, an area that well illustrates the problems of working in arctic environments, especially with ice.

The most significant engineering problem and production cost in the Arctic is ice. Polar pack ice, ice floes, icebergs, and land-fast ice must be

Table 11.3 Computer-simulated comparative United States offshore exploration and development costs[a]

Area	Water depth (m)[d]	Field size (mill. of bbl)	Exploration cost ($ million)	Development cost ($ million)	Operating cost ($ mill/yr)	Transportation cost ($/bbl)	Production lead-times (yr)
Gulf of Mexico							
Small field	120	15	78	105	7	0.00	2
Large field	120	50	78	168	12	0.00	2
California							
Deepwater							
Small field	1,000	150	400	450	16	2.50	10
Large field	1,000	300	400	900	24	2.00	10
Norton Basin[b]							
Small field	15	250	435	1,038	72	6.50	8
Large field	15	500	435	2,076	102	5.00	9
Harrison Bay[c]							
Small field	15	1,000	720	3,162	120	12.50	12
Large field	15	2,000	720	6,324	168	10.00	12
Navarin Basin[b]							
Small field	135	1,000	825	5,460	132	6.50	11
Large field	135	2,000	825	10,920	240	5.00	11

Source: Office of Technology Assessment, US Congress, *Oil and Gas Technologies for the Arctic and Deepwater* (OTA, Washington, DC, 1985), p. 119.
Notes: a. Costs refer to total, undiscounted outlays in 1984 US $; b. Bering Sea region; c. Beaufort Sea region; d. converted from ft to m and rounded.

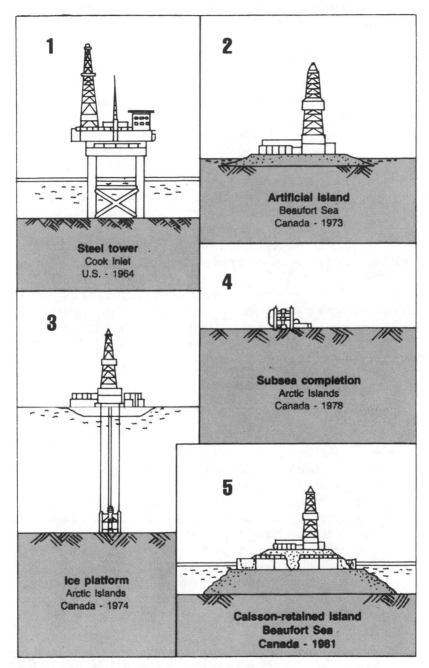

Figure 11.1 Generalised evolution of petroleum drilling and producing bases in arctic regions

Source: After L. A. LeBlanc, 'Operators probe for least-cost production', in *Harsh Environment and Deepwater Handbook* (PennWell Publishing, Tulsa, OK, 1985), p. 51. With permission.

managed before successful drilling and development can occur. Development of aerial satellite technology will enhance the oil industry's ability to cope with ice problems. US satellite data gathering programmes for ice conditions, however, are presently underfunded, and industry must pay most of its own ice surveillance costs.[15]

To avoid problems of icebergs and shifting ice packs and pressure ridges, drillers must have a stable base from which to operate (Figure 11.1). Traditionally drilling for and producing petroleum in arctic sea-ice environments required the building of artificial gravel or ice islands. To build a gravel island of perhaps 120m² in only 6m of water can require some 230,000m³ of aggregates which must often be obtained onshore at a considerable distance from the drilling site.[16] Imperial Oil Limited of Canada built the world's first Arctic Ocean artificial island (dubbed Immerk) in the Beaufort Sea during the summers of 1972 and 1973. Dredges extracted gravel from the seabed and built a work area of about 90m diameter with an above-sea-level height of 4.5m. Built at a cost of about US$5 million, Immerk was a success.[17] By 1979 a total of fifteen artificial islands had been constructed,[18] mostly in Canada's southern Beaufort Sea.[19]

In the mid-1970s oilmen began experimenting with using ice platforms as a drilling base; ice is a cheaper building material than aggregates, and permitting agencies prefer ice islands because, when summer arrives, they melt and do not alter the local environment, as do gravel islands that are also expensive to dismantle.[20] Union Oil Company, during the winter of 1976–7, constructed an ice island (275m diamter) in the Beaufort Sea in waters 2.5m to 3.0m deep, off the Colville River. Engineers constructed a containment ring of snow on the sea ice, pumped sea-water into the enclosure, and allowed it to freeze. Several pumping-filling-freezing cycles increased the weight of the enclosed ice-containment area, causing it to sink and to fuse with the seabed.[21] An ice-free moat constructed (and maintained all winter) around the island, isolated the island from moving sea ice.

From their experience with constructing flooded islands, engineers developed techniques of building islands by using spray ice. Construction crews pump water from below the ice sheet and then use movable and adjustable nozzles to produce ice platforms with varying properties. The spray, thrown as high as 60m into the air, freezes before landing on the ice sheet. As the spraying proceeds, the weight of the newly formed ice depresses the ice sheet until it becomes bonded with the seabed. Additional spraying builds up a mound (freeboard) of protective ice surrounding the island (Figure 11.2).[22]

In Canada's arctic-island region, drillers have used floating spray-ice platforms to drill in waters as deep as 300m. An area with a 300-m diameter and a thickness of about 6m will support drilling operations, even though the ice platform is not bonded with the seabed. Sea-ice surrounding these islands is relatively stationary; few major shifts in the

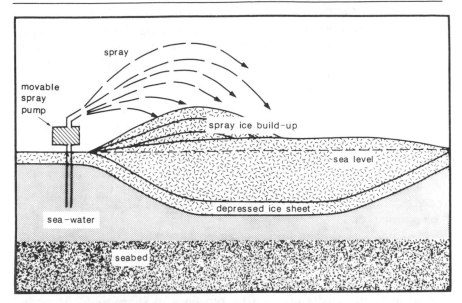

Figure 11.2 Spray-ice construction, forming a grounded island

Source: After R. Goff, 'Ice Islands may aid Beaufort development', in *Harsh Environment and Deepwater Handbook*, (PennWell Publishing, Tulsa, OK, 1985), p. 123. With permission.

ice occur. Spray-ice, however, is very porous and less dense than regular sea-ice. Consequently it must be monitored carefully, because differential settling of the ice island is a potential hazard, especially during drilling. For good ice to form, the temperature during spraying should be no higher than -21°C.[23] Drillers also employ natural sea ice as a working platform, by constructing a 20-cm insulating 'sandwich' drilling pad of timber, polyurethane foam, and polyethylene plastic.[24]

Oil companies working in arctic environments during the 1970s and early 1980s 'depended upon sheer size, bulk ... or gravity to withstand the environmental forces and provide a sufficient safety factor'. Until recently, designs were premised on stable petroleum prices and a good profit margin. Mistakes then could be smoothed over by anticipated petroleum price rises, but no longer. Depressed prices and undependable profits of the mid-1980s, demand less costly, more integrated and flexible systems in arctic regions.[25]

Research engineers now put greater effort into developing more mobile and self-contained drilling platforms. One such unit is Exxon's Concrete Island Drilling System (CIDS). Designed by Global Marine Development Corporation and built in Japan, CIDS began operating off Alaska's north-western coast in the late summer of 1984. CIDS's deck holds a five-storey building that provides office, living, dining, recreation, laundry, and hospital space, as well as a control room and a rooftop helipad. A base of approximately 800 square metres[26] anchored to the seabed by twenty-four watertight compartments filled with heated sea-water (to prevent freezing) holds CIDS in place.[27]

Operators of CIDS use a large ice-spray freeboard, which some say is unnecessary with this big, well-designed drilling rig. They also constantly monitor the surrounding ice for stresses that might potentially shift the rig. Two radar scanners sweep the adjacent 'icescapes' to check for changes in their position. Pressure panels, distributed at several sites in the surrounding ice, measure wind and current forces exerted on the freeboard, and special sensors built into CIDS record the magnitude of the ice pack's stress on its base. Other safety sensors monitor for hydrogen sulphide, a lethal gas that poses a hazard during drilling. Most rig personnel spend a two-week shift onboard and then two weeks off, as far away as the California sun,[28] a scarce commodity in the arctic winter. As of late 1987, however, no one was commuting to work on CIDS; it has been 'stacked' (idled) since early 1986,[29] because of the depressed petroleum market.

If exploratory drilling brings in an economically productive well, the oilmen must harness and regulate the well for production. Because production platforms are expensive to build and maintain, producers attempt to minimise their use. Sub-seabed completion systems, whereby wellhead control units are tied by pipelines to shore-based or central production platform storage areas, are one method of reducing the need

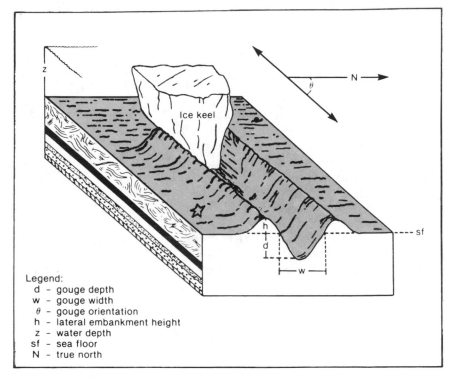

Legend:
 d – gouge depth
 w – gouge width
 θ – gouge orientation
 h – lateral embankment height
 z – water depth
 sf – sea floor
 N – true north

Figure 11.3 Schematic diagram of ice-keel gouging in shallow offshore area

Source: Office of Technology Assessment, US Congress, *Oil and Gas Technologies for the Arctic and Deepwater* (OTA, Washington, DC, 1985), p. 61.

for production platforms. This technique, under development since the mid-1970s and now used in many offshore oil and gas fields, can provide savings in arctic frontier areas. Ice, however, presents difficult problems.

Moving pack ice comes under tremendous differential stress, causing it to buckle into surface ridges and underlying keels. These keels can gouge into the seabed, cutting pipelines and shearing off sub-seabed completion (wellhead) units. Before producers can safely use sub-seabed completion in ice floe- and iceberg-prone regions, they must know the maximum water depth at which keel gouging occurs and how deep into the seabed the keels cut (Figure 11.3). In deep water, submarines can provide upward-looking sonar profiles, but in areas less than 100m, there are no adequate methods. Geologists recently have studied the Beaufort Sea for sedimentational and biological filling of gouges (relative to post-glacial seabed isostatic uplift) in an effort to determine the maximum water depths at which gouging has occurred. They concluded that within the last few hundred years ice gouging has occurred at water depths as great at 64m.[30]

One method of protecting subsea completion units from ice-ridge and iceberg keels is to construct 'glory holes' (excavated depressions) that put the completion unit below the predicted scour line (Figure 11.4). Glory holes are costly to construct, may be infilled by bottom-current sediments, and can cause the stalling of an iceberg. Another technique involves construction of a silo protected by a cover deflector. Engineers can design the upper part of the silo and flowlines to break free if hit by an iceberg or ice-ridge keel, allowing the wellhead or blowout preventor

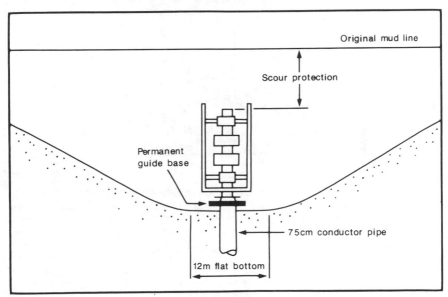

Figure 11.4 Typical 'glory hole' oil well completion on the continental shelf

Source: After R. M. Oglesbee and L. G. Kuhlman, 'Weather, depths impact subsea well projects', in *Harsh Environment and Deepwater Handbook* (PennWell Publishing, Tulsa, OK, 1985), p. 63. With permission.

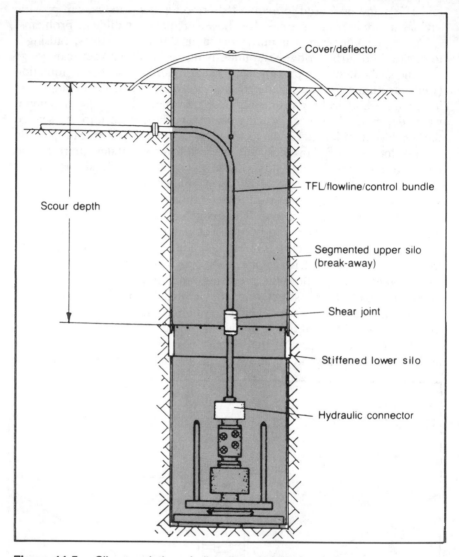

Cover/deflector

TFL/flowline/control bundle

Scour depth

Segmented upper silo
(break-away)

Shear joint

Stiffened lower silo

Hydraulic connector

Figure 11.5 Silo completion of oil well on the continental shelf
Source: R. M. Oglesbee and L. G. Kuhlman, 'Weather, depths impact subsea well projects', in *Harsh Environment and Deepwater Handbook* (PennWell Publishing, Tulsa, OK, 1985), p. 63. With permission.

to remain in place (Figure 11.5). Silo techniques are less advanced than glory-hole techniques.[31]

Petroleum producers have made good progress in coping with ice in high latitude and polar regions, but much remains to be done before ice becomes a truly routine environmental component. The same is true of another offshore challenge, deep water.

Figure 11.6 Jackup drilling rig

Source: Courtesy Rauma-Repola Metal Industries. With permission.
Note: The USSR has purchased this jackup model for use in arctic waters. The rig has 140-m legs and
can drill to a depth of 6,500m in 100m of water.

Deep-water regions

Offshore exploration drilling technologies have evolved from pier-based wells of the later 1890s that tapped seaward extensions of Southern California's Summerland field. During 1938 in 8-m waters off Louisiana, drillers struck oil in what was to become the famous Creole field. By 1946 wildcatters had drilled nine offshore wells along the Gulf Coast – five off Louisiana and four off Texas. In 1947 Kerr-McGee used the world's first mobile drilling rig. Engineers built the rig from Second World War Navy surplus vehicles – a landing craft (for a platform) and two barges (for flotation).[32] During succeeding decades, increasingly sophisticated systems have pushed deep-water exploration and production into more difficult frontiers.

Since 1965 petroleum companies' record drilling depths have gone from a modest 190m to 2,120m of water.[33] By late 1987 Brazil had pushed actual production to a water depth of 413m.[34]

Figure 11.7   Semi-submersible drilling rig
Source: Courtesy Rauma-Repola Metal Industries. With permission.
Note: The rig is designed for operation in harsh environmental conditions, although it is not specially outfitted for arctic conditions. The rig weighs 13,700 tonnes; the height to the main deck is 35m and the length is 78m.

Some of the most critical technology needs for deep water are improvements in (1) structural designs, welding techniques, and seabed platform foundation engineering; (2) methods of riser and pipeline installation, maintenance, and repair; (3) drilling, well control, and completion; and (4) diving and navigation support activities. Petroleum industry engineers believe these technology needs can be met and that, eventually, waters as deep as 2,400m will be exploited,[35] although riser technology will be a major challenge (especially in water depths greater than 3,000m).[36]

Drilling rigs

Jackups are the drilling workhorses of relatively shallow waters, usually less than 125m (Figure 11.6). A jackup has legs that rest on the seabed and a self-elevating work pad. Overall, the jackup is the least expensive rig to build and to operate, and while on location, it provides a highly stable workbase. When under tow, however, the legs are normally elevated above the water (to reduce drag), making the rig unstable.

Semi-submersibles and drill ships are capable of working in deeper waters than the jackups. Self-propelled, they can readily move from one location to another, an excellent asset during periods of exceptionally severe weather or with the approach of ice floes or icebergs. They are also dynamically positioned while drilling. Thrusters on every side help to maintain these units accurately on site. The semi-submersibles are supported by pontoons (with an adjustable bouyancy). The deck provides not only for drilling activities but also living quarters and helipads (Figure 11.7). Drill ships, the most mobile and capable of working in ultra-deep areas, are now drilling in waters of more than 2,000m (Figure 11.8).

Production platforms

Drill ships and semi-submersibles can also function as production platforms, but most production systems are either bottom-based steel (See Figure 11.12 – hybrid system) or submersible concrete platforms of various designs. In future, producers expect to have a capability of using concrete submersibles in waters of 250m, as will occur in Norway's Troll field in the North Sea (Figure 11.9). These units are very stable and can provide storage space for petroleum.

Brazil

One of the world's leading areas for deep-water research and technology is offshore Brazil. In early 1988 Brazil held the water-depth record (492m) for petroleum production. Petrobrás – Brazil's state oil company – plans to move into much deeper waters in the near future. It is working jointly with private domestic firms in a research programme to develop production wells in waters deeper than 2,000m. A Petrobrás research team, the Technological Capability Program for Deep Water Oil Exploitation (Procap), hopes to see developed a diverless wet Christmas tree, an important piece of equipment for deep water, because divers cannot

Figure 11.8 Ice-class drill ship, the *Valentin Shashin*

Source: Courtesy Rauma-Repola Metal Industries. With permission.

Note: The Rauma-Repola Mäntyluoto Works in Finland have built and delivered to the USSR several drill ships of this type, the UL1.

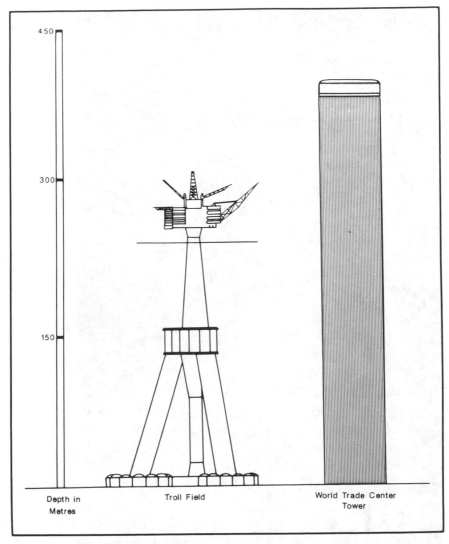

450

300

150

Depth in
Metres

Troll Field

World Trade Center
Tower

Figure 11.9 Schematic view of a potential production platform for Norway's
Troll field

Source: After Office of Technology Assessment, US Congress, *Oil and Gas Technologies for the Arctic
and Deepwater* (OTA, Washington, DC, 1985), p. 48. With permission, Shell Oil Company.

safely work for long periods at depths greater than 500m. Procap and
research groups elsewhere are continuing to improve deep diving equipment (Figure 11.10) and remotely operated vehicles (ROVs), which
can perform many shallow-water (Figure 11.11) and deep-water tasks.[37]
Recently researchers have focused on reducing the dependence of offshore operators on tethered underwater work vehicles. Several recent
symposia demonstrated that unmanned and untethered (autonomous)

259

Figure 11.10   An atmospheric diving suit

Source: Courtesy OSEL GROUP. With permission.
Note: Divers working in this suit do so at a surface atmospheric pressure. Its depth rating is 700m and it has a forward speed of more than 0.8m/sec. The system provides capabilities for various mechanical tasks and video observation.

Figure 11.11 The *MiniRover* MK II

Source: Courtesy Deepsea Systems International, Inc. With permission.
Note: The *MiniRover* MK1 is designed for relatively shallow-water work and is capable of dives to 258m. It is useful for under-ice survey and inspection jobs. Its length is 66cm (without skids), it weighs 21–25kg in air, travels 2.2–2.4 knots/hr in still water, and has umbilical cords of up to 912m.

submersibles are likely to become an integral part of the industry, once guidance and control and power systems are refined. These units will perform programmed missions 'totally internal to the submersible'.[38] A technology with a significant potential but which needs more intensive research is that of manned one-atmosphere underwater work stations. These dry systems allow workers to remain submerged for extended periods at sea-level pressures. Some small units are already used for a few seabed tasks, such as wellhead completions and maintenance and pipeline connections. For example, construction engineers used one-atmosphere work chambers to install subsea completion units in Brazil's Garoupa field.[39]

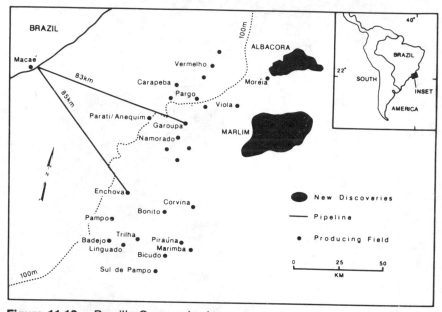

Figure 11.12   Brazil's Campos basin
Source: After 'Campos: 60% of output and new strikes keep coming', *Petrobrás News*. no. 106 (Sept. 1987), p. 5. With permission.

Brazil's main offshore petroleum fields occur in the Campos basin, discovered in 1974[40] (Figure 11.12). The Campos basis has thirty-seven named oil or gas fields with more than twenty already producing. More will come on stream in the near future. Recent exploratory drilling identified two new major fields, the Marlim in 850m of water and the Albacora at 425m. Geologists estimate the Marlim field's gas reserves at 100,000 million m^3 and Albacora's at 100,000 million to 150,000 million m^3;[41] crude oil reserves are put at 2,000 million bbl for the Marlim field and 500 million bbl for the Albacora.[42] These two fields – even with waters deeper than 500m excluded[43] – have doubled Brazil's proved oil reserves (2,250 million bbl), for a total of nearly 5,000 million bbl;[44] some

66 per cent of the country's proved reserves are located in the offshore.[45] The productivity of exploration wells in these fields has been an important stimulus for Brazil's effort to plunge into deeper waters with their exploration rigs. According to Brazilian researchers, the most efficient production systems for these deep waters will be floating units, such as tankers and semi-submersibles. Floating production platforms evolved from Petrobrás's use of drilling rigs as temporary production platforms. These worked so well that Brazilian construction yards are now building semi-submersible production platforms capable of working in waters of more than 350m. A hybrid system could use a fixed platform in combination with a semi-submersible (Figure 11.13). Petrobrás's managers are also looking at compliant or flexible guyed-towers (Figure 11.14) and tension-leg platforms (Figure 11.15) as Campos basin production units.[46]

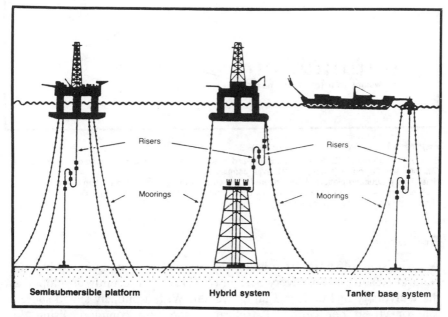

Figure 11.13 Typical deepwater floating production systems, ships, and semi-submersibles

Source: J. Redden, 'Brazil sees giant possibilities', *Offshore*, vol. 47, no. 2 (1987), p. 30. With permission.
Note: The semi-submersibles are in combination with bottom-based steel-supported units; these systems can regulate and store seabed petroleum output.

New technologies

Petroleum producers are always alert for new, innovative technologies that can reduce costs and environmental impacts. This topic is too broad to examine here in detail, except to illustrate a few examples of state-of-the-art technologies now used in exploration, production, and environmental control.

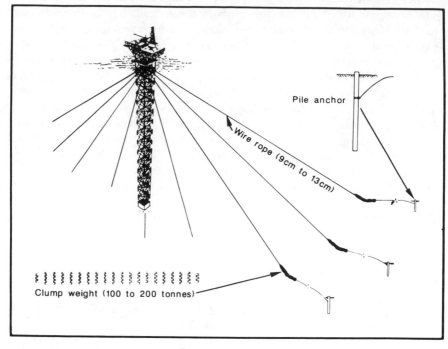

Figure 11.14 A flexible guyed-tower

Source: L. D. Power, L. D. Finn, R. W. Beck, and J. M. Hulett, 'Tests recognize guyed tower potential',
Offshore, vol. 38, no. 5 (1978), p. 221. With permission.
Note: This production platform is engineered to provide for movement of the system in response to
applied forces such as storm waves. Undulations of heavy clump weights help maintain the equilibrium
of the platform, as waves pass by.

Exploration

A promising offshore petroleum exploration and mapping technique,
Natural Resources Discovery System (NRDS), was recently developed at
the University of Lund, in Lund, Sweden. With a publicly stated success
rate of 70–80 per cent,[47] PetroScan – the firm commercialising the
technique – is presently seeking a patent in the US.[48] Work with the
NRD method began in the early 1980s, and by the mid-1980s after some
fifty prognoses (survey-forecasts), it was a proved success. By July 1987
PetroScan had completed major prognoses off the coasts of seven African
states, four off Middle Eastern states, eight in the US's Gulf Coast region,
and fifteen in waters off Norway, Sweden, Denmark, and the UK.[49]

PetroScan technicians employ orbiting satellite altimeter data from
GEOS 3 and SEASAT (corrected for meteorological, tidal, current and
seabed topographical conditions) to map variations in the height of the
ocean's surface. The accuracy of measurement is within a few centimetres
of amplitude (Figure 11.16).[50] Short-distance undulations of the surface
occur in association with petroleum deposits (Figure 11.17). Where the

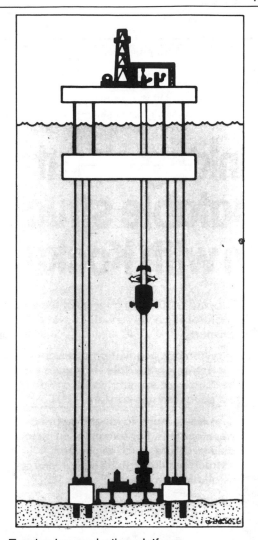

Figure 11.15 Tension-leg production platform
Source: 'Production progresses in 2 U.K. fields', *Offshore*, vol. 38, no. 5 (1978), p. 232.
Note: Tethers, tied to a seabed template, and buoyant columns allow compliance while the platform's excess flotation holds it on station.

surface height is relatively shallow, there is a good likelihood that seabed petroleum deposits are present, because hydrocarbons are less dense than rock. Adjacent, surrounding rocks have a greater gravitational force than the petroleum and, therefore, increase the water mass above them. A depression of 25cm in the water mass is a good indicator that a significant volume of petroleum is present, whereas a bulge of 5cm or more is indicative of little petroleum potential.[51] The technique provides a horizontal enhancement and is usable at any ocean depth (where ice cover is absent) and has the best resolution when there are at least 50

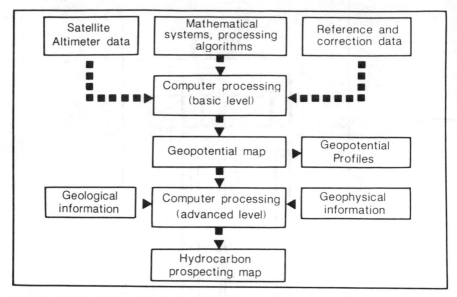

Figure 11.16 Schematic diagram of PetroScan's Natural Resources Discovery System for data processing

Source: After L. A. LeBlanc, 'Advanced technology', *Offshore* vol. 47, no. 4 (1987), p. 21. With permission.

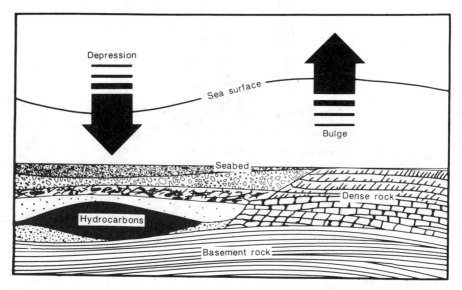

Figure 11.17 Schematic ocean surface profile determined from processing data obtained by PetroScan's Natural Resources Discovery System

Source: L. A. LeBlanc, 'Advanced technology', *Offshore*, vol. 47, no. 4 (1987), p. 21. With permission. Lower water-surface elevations indicate the presence of petroleum in the seabed, whereas higher elevations indicate the absence of petroleum.

million bbl of oil present within depths of 5km.[52] After positive anomalies have been mapped, traditional seismic techniques provide a vertical enhancement that help identify specific structures and traps.

Because the NRDS method is usable in all offshore areas and is cheaper than traditional seismic methods, it should be especially helpful in high-cost frontier areas.[53] Some prognoses have been done for as little as US $15/km^2.[54] If PetroScan is given the co-ordinates of the desired survey area, it can do a prognosis of an area as large as 40,000km^2,[55] and it has a nearly 100 per cent accuracy for predicting 'where not to prospect for oil and gas'.[56]

Production and environmental control

Many critics of the offshore petroleum industry accuse it of carelessly spilling crude oil into the sea. Some severe accidents have occurred, as in the North Sea Ekofisk Complex's Bravo platform blowout in 1977 and the Gulf of Campeche's Ixtoc I blowout in 1979. The Bravo blowout lasted only nine days, the Ixtoc I blowout nine months.[57] These spills are regrettable and did damage the environment, but they should be put into the proper perspective.

A 1985 joint report published by the US's National Academy of Sciences, the National Academy of Engineering and the Institute of Medicine provided estimates of various sources of marine hydrocarbon pollution. Between 12.4 million and 64.1 million bbl of crude oil enters the oceans annually, from both natural and societal sources. Excluding the oil production process itself, human sources include (among others) ships' bilge water; tanker accidents; loading terminal spillage; refinery, municipal, industrial emission and surface-runoff wastes; and ocean dumping. Natural sources involve sediment erosion and marine seeps. The world's marine seeps annually give off an estimated 146,000 to 14.6 million bbl of oil. The best estimate puts the seepage at 1,457,000bbl/yr. Because the range of the natural seep estimate is so wide, the best estimate value is not reliable. Offshore petroleum production activities discharge 291,000 to 437,000bbl of oil annually.[58]

At least one oil producer, Atlantic Richfield Company (ARCO), along with several partners (Mobile, Western LNG, and Phillips Oil Company), has pointed the way for what may be done not only to reduce natural hydrocarbon seep pollution but also to add to their petroleum production profits. In September 1982 ARCO placed the world's first two permanent containment structures over petroleum seeps off Coal Oil Point, 16km west of Santa Barbara, California (see Figure 13.7). ARCO spent several years obtaining permits and $8 million in designing and constructing the containment structures (Figures 11.18 and 11.19). The structures were designed to withstand earthquakes, 14-m tsunamis, and 100-year storms. Sixteen 25-tonne concrete blocks were positioned on the perimeter of the cylinder after it was put in place. If the seeps should stop emitting crude

Figure 11.18 Seep containment systems under construction by Kaiser Steel at Napa, California

Source: Courtesy ARCO Oil and Gas Company. With permission.

Note: Each unit weighs 350 tonnes.

oil and gas (as they intermittently do),[59] the structures can be moved to a new site. Once the gas is piped onshore, a processing facility scrubs and cleans it for sale to local businesses and homes. The captured oil collects in the caps' separators. A workboat visits the site periodically to pick up and transfer the oil to a nearby ARCO platform from which it is piped onshore to a processing plant.[60]

Coal Oil Point
Seep Containment Project

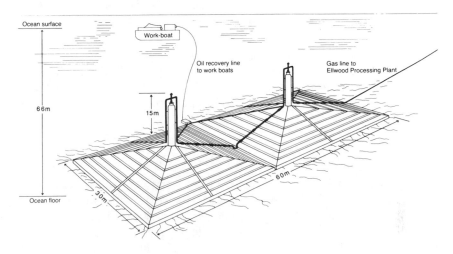

Seep Containment Structure

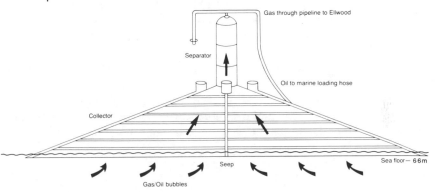

Figure 11.19 Schematic layout for ARCO's Coal Oil Point seep containment system (in place 2.8km offshore)

Source: Courtesy ARCO Oil and Gas Company. With permission.

The containment structures have captured more gas than the engineers had expected. During the first ten months of operation, some 28,000 m^3/d of gas were captured instead of the anticipated 14,000m^3/d. The anually captured gas[61] is sufficient to supply about 400 homes for an entire year.[62]

ARCO agreed to capture (during 1982–7) a certain amount of reactive gas that turns into photochemical smog when exposed to the sun. Only 15 per cent of the captured gas is classified as reactive; the remainder is methane. By capturing the gas, ARCO earns and banks pollution credits. These credits offset emissions created by the company's new onshore facilities, or they can be sold to other companies without an opportunity to earn them.[63] Beginning in 1988, however, hydrocarbon emission credits remaining in ARCO's bank will be depreciated over the following five years until no pollution tradeoffs remain.[64]

There are more than 1,460 seeps in the Coal Oil Point area, 50 of which emit large amounts of hydrocarbons. These seeps will be active for thousands of years. A conservative estimate puts their oil seepage alone at 100bbl/d, which (totalled annually) is 33,000bbl more than were released during the 1969 Santa Barbara blowout. ARCO and other oil companies have considered developing additional seep containment sites, but most of these seeps are not nucleated enough to allow an economically efficient capture programme.[65]

Conclusions

Geologists do not yet have enough data to determine with precision how much economically exploitable petroleum the world's continental margins hold. It is only during the last decade that exploration has been pushed into polar and deep-water areas. Increasingly, improved technologies for working with ice make year-round activities feasible in arctic seas. Drilling rig and production systems designs now allow petroleum development programmes in waters deeper than 400m, and producers expect, in the not too distant future, to produce seabed petroleum from waters deeper than 2,000m. New technologies such as PetroScan's NRD exploration methods and ARCO's Coal Oil Point natural-seep production systems exemplify the engineering challenges and economic investments of the offshore petroleum industry.

In the decades ahead, those in industry will push into increasingly difficult areas. They will make high-latitude and deep-water petroleum regions an integral part of our economic oceanic resource base.

Notes

1 M. T. Halbouty, 'World petroleum reserves and resources with special reference to developing countries', in *Petroleum Exploration Strategies in Development Countries: Proceedings of a United Nations Meeting Held in The Hague 16–20 March 1981* (UN Natural Resources and Energy Division, Department of Technical Co-operation for Development, published in co-operation with the UN by Graham & Troutman, London, 1982), p. 4.
2 K. Hinz, 'Assessment of the hydrocarbon potential of the continental margins', in F. Bender (ed.), *New Paths to Mineral Exploration: Proceedings of the Third International Symposium, Held in Hannover, Federal Republic of*

Germany at the Federal Institute of Geosciences and Mineral Resources, 27–29 October 1982 (E. Schweizerbart'sche, Verlagsbuchhandlung, Stuttgart, FRG, 1983), p. 55. The *Oil and Gas Journal* in late 1986 gave a total (onshore and offshore) world *proved* crude oil reserve of just over 697,000 million bbl and a total (onshore and offshore) world *proved* natural gas reserve of just under 103 million million m^3; data for the USSR were reported as 'explored reserves', including proved, probable, and some possible reserves; data for Canadian gas included proved reserves, and some probable reserves. See 'Worldwide oil and gas at a glance', *Oil and Gas Journal*, vol. 84, nos 51/52 (1986), pp. 36–7. *Proved* reserves are 'those quantities which geological and engineering information indicate with reasonable certainty can be recovered in the future from known reservoirs under existing economic and operating conditions'. See British Petroleum Company plc, *BP Statistical Review of World Energy* (BP, London, 1987), pp. 2, 21.

3 E. Bergsager, 'First drilling in Norwegian Sea off Norway yields encouraging results', *Oil and Gas Journal*, vol. 79 no. 23 (1981), pp. 115, 118, 120, 124.

4 Hinz, 'Assessment of the hydrocarbon potential of the continental margins', p. 75.

5 'Offshore gas production (MMcfd)', *Offshore*, vol. 47, no. 5 (1987), p. 51.

6 G. V. Hough, 'World survey – natural gas: steady progress is maintained', *Petroleum Economist*, vol. 54, no. 8 (1987), p. 295.

7 For an excellent review of state-of-the-art offshore technology, see D. A. Fee and J. O'Dea, *Technology for Developing Marginal Offshore Oilfields* (Elsevier Applied Science Publishers, London, 1986).

8 A good résumé of recent arctic activities is contained in N. D. Xuong, 'A brief history of the search for arctic offshore oil', *Oil and Enterprise: The Oil and Gas Review*, no. 29 (Dec. 1985), pp. 14–19.

9 X. Boy de la Tour and D. Champlon, 'Economic Aspects of Hydrocarbons production in arctic areas', *Oil and Enterprise: The Oil and Gas Review*, no. 29 (Dec. 1985), pp. 47–52.

10 Minerals Management Service, DOI, *Proposed Notice of Sale: Oil and Gas Lease Sale No. 97, Outer Continental Shelf Beaufort Seu (January 1988)* (DOI, Washington, DC, 1987).

11 Memorandum: R. H. McMullin, Acting Regional Director, Minerals Management Service, Alaska Outer Continental Shelf Region, DOI, Anchorage, AK, to T. Giordano, Program Director, Office of Strategic and International Minerals, MMS, DOI, Long Beach, CA, 10 Sept. 1987.

12 Office of Technology Assessment, US Congress, *Oil and Gas Technologies for the Arctic and Deepwater* (USGPO, Washington, DC, 1985), pp. 8–9.

13 ibid., pp. 117–18.

14 ibid., p. 31; these data are not directly comparable because the USGS included data from state waters, whereas the MMS data included only federal waters. See L. W. Cooke, *Estimates of Undiscovered, Economically Recoverable Oil and Gas Resources for the Outer Continental Shelf as of July 1984* (MMS, DOI, Washington, DC, 1985), p. 21.

15 Office of Technology Assessment, *Oil and Gas Technologies*, p. 16.

16 D. Holmstrom, 'CIDS: island in the ice', *EXXON USA*, vol. 24, no. 3 (1985), p. 10.

17 J. G. Riley, 'How Imperial built first arctic island', *Petroleum Engineer International*, vol. 46, no. 1 (1974), pp. 25–8; see also J. G. Riley, 'The construction of artificial islands in the Beaufort Sca', *Journal of Petroleum Technology*, vol. 28, no. 4 (1976), p. 366.

18 Letter: G. W. Kalyniuk, Support Manager, ESSO Resources Canada, Edmonton, Alberta, 13 March 1979.

19 For a detailed look at Canada's Beaufort Sea petroleum industry activities, see M. B. Todd, 'Development of Beaufort Sea hydrocarbons', *The Musk-Ox*, no. 32 (1983), pp. 22–43.

20 R. Goff, 'Ice islands may aid Beaufort development', in *Harsh Environment and Deepwater Handbook* (PennWell Publishing, Tulsa, OK, 1985), pp. 123–4.

21 F. C. F. Earney, *Petroleum and Hard Minerals from the Sea* (Edward Arnold, London, 1980), p. 160.

22 Goff, 'Ice islands may aid Beaufort development', p. 123.

23 ibid., pp. 123–4.

24 G. Ives, 'Ice platform concept proven for arctic offshore drilling', *Petroleum Engineer International*, vol. 46, no. 6 (1974), p. 14.

25 L. A. LeBlanc, 'Operators probe for least-cost production', in *Harsh Environment and Deepwater Handbook* (PennWell Publishing, Tulsa, OK, 1985), pp. 50–1.

26 S. B. Wetmore 'Arctic offshore drilling: a new challenge', in *Harsh Environment and Deepwater Handbook* (PennWell Publishing, Tulsa, OK, 1985), p. 146.

27 D. Holmstrom, 'CIDS: island in the ice', pp. 10–11.

28 ibid., pp. 11–14.

29 Letter: J. B. Davis, Co-ordinator, Public Affairs Department, EXXON Company, USA, Houston, TX, 8 Oct. 1987.

30 E. Reimnitz, P. Barnes, and L. Phillips, 'Evidence of 60 meter deep arctic pressure-ridge keels', in *Harsh Environment and Deepwater Handbook* (PennWell Publishing, Tulsa, OK, 1985), pp. 41–7.

31 R. M. Oglesbee and L. G. Kuhlman, 'Weather, depths impact subsea well projects', in *Harsh Environment and Deepwater Handbook*, (PennWell Publishing, Tulsa, OK, 1985), pp. 62–63.

32 Earney, *Petroleum and Hard Minerals from the Sea*, p. 92.

33 Office of Technology Assessment, *Oil and Gas Technologies* p. 49.

34 'Campos: 60% of output and new strikes keep coming', *Petrobrás News*, no. 106 (Sept. 1987), p. 4.

35 Office of Technology Assessment, *Oil and Gas Technologies*, pp. 73–4.

36 J. P. Riva Jr, *World Petroleum Resources and Reserves* (Westview Press, Boulder, CO, 1983), especially pp. 61–2. A riser is a generic term for a series of tubular systems connecting a seabed termination with a facility at the sea surface.

37 'Campos', pp. 4–5.

38 R. W. Corell, 'Research more focused for autonomous underwater vehicles', *Sea Technology*, vol. 28, no. 8 (1987), pp. 41–5.

39 M. E. Jones, *Deepwater Oil Production and Manned Underwater Structures* (Graham & Troutman, London, 1981), pp. 69–74.

40 E. Taylor, 'Brazil pumps $7 billion into Campos offshore fields', *Offshore*, vol. 46, no. 11 (1986), p. 62.

41 'New offshore fields boost Brazilian gas reserves', *Offshore*, vol. 46, no. 11 (1986), p. 66; 'Brazil continues to hit oil in 13-year-old Campos play', *Offshore*, vol. 47, no. 11 (1987), pp. 48, 50–1.

42 'Brazil slates deepwater development', *Oil and Gas Journal*, vol. 84, no. 22 (1986), p. 32.

43 W. Freire, 'Taking the plunge into deepwater petroleum exploration', *OPEC Bulletin*, vol. 17, no. 10 (Dec. 1986–Jan. 1987), pp. 12–13.

44 J. Redden, 'Brazil sees giant possibilities', *Offshore*, vol. 47, no. 2 (1987), p. 29.

45 Freire, 'Taking the plunge into deepwater petroleum exploration', p. 13.

46 Redden, 'Brazil sees giant possibilities', p. 30.

47 PetroScan, AB, the NRD Method for Location of Offshore Hydrocarbon Deposits (PetroScan, Göteborg, Sweden, 1985), n.p.
48 'Hydrocarbons correlate with sea heights', *World Oil*, vol. 204, no. 4 (1987), p. 39.
49 PetroScan AB, 'Company information', mimeographed data provided to the author, 29 Sept. 1986, Göteborg, Sweden, n.p.
50 PetroScan AB, *The NRD-method: What You May Not Realize about the Waves of the Sea* (PetroScan, Göteborg, Sweden, 1987) n.p.; see also 'New analytical technique boosts odds in exploration', *Ocean Industry*, vol. 20, no. 5 (1985), p. 23.
51 L. A. LeBlanc, 'Hydrocarbon deposits affect sea level', *Offshore*, vol. 47, no. 4 (1987), p. 21.
52 'NRD used to chart Norwegian hydrocarbons', *Offshore*, vol. 46, no. 4 (1986), p. 68; Letter: B. Lundgren, President, PetroScan AB, Göteborg, Sweden, 5 Nov. 1987.
53 'Hydrocarbons correlate with sea heights', p. 40.
54 LeBlanc, 'Hydrocarbon deposits affect sea level', p. 21.
55 Interview: F. Wassén, Manager, PetroScan, Göteborg, Sweden, 29 Sept. 1986.
56 'PetroScan, 'Company information', n.p.
57 National Academy of Sciences, *Oil in the Sea: Inputs, Fates, and Effects*, (National Academy Press, Washington, DC, 1985), pp. 567–8.
58 ibid., p. 82. Data are converted from tonnes, with 1 tonne equal to 7.285bbl of petroleum.
59 The rate of flow varies from day to day, depending upon tidal levels and the presence of storms (lower barometric pressures evidently cause an increased flow); see R.B. Spies, 'Natural submarine petroleum seeps', *Oceanus*, vol. 26, no. 3 (1983), p. 26.
60 ARCO Oil and Gas Company, Atlantic Richfield Company, *Santa Barbara Channel Seep Containment Project* (ARCO, Los Angeles, CA, 1985), pp. 6, 10.
61 R. Sollen 'Air pollution trapped on ocean floor', *Oceanus*, vol. 26, no. 3 (1983), p. 23.
62 ARCO Oil and Gas Company, *Santa Barbara Channel Seep Containment Project*, p. 9.
63 Sollen, 'Air pollution trapped on ocean floor', p. 27.
64 ARCO Oil and Gas Company, *Santa Barbara Channel Seep Containment Project*, p. 10.
65 ibid., pp. 2, 12.

Chapter twelve

The United Kingdom and Norway: offshore petroleum development policies

National governments play an especially important role in offshore petroleum exploitation, because initially they own most of the resource base. They also establish policies for developing these resources. Government policies are, in turn, influenced by internal and external political pressures and economic forces not under their control. Furthermore, in contrast to the petroleum industry, governments usually have only a limited expertise and technology for developing petroleum resources. These realities, as Øystein Noreng noted in his incisive volume *The Oil Industry and Government Strategy in the North Sea*, put industry and government in a 'bargaining situation'.[1]

Like all mineral producers, offshore petroleum enterprises respond negatively and positively to national policies in licensing, governmental participation, production regulations, and taxation. Because of special hazards and expenses associated with offshore petroleum production, each of these policy sectors can affect a company's interest in negotiating or bidding for exploration or production licences and can influence its decisions on exploration and development programme timing, products produced, and production rates.

National governments also respond to the offshore producers, noting their willingness to co-operate with governmental management and regulatory agencies and to contribute to the overall national economy. Governments also react to internal political pressures exerted by other industries and by the general public who may harbour misgivings about the impacts of large-scale offshore oil production whose control may come largely from foreign interests. Two states – the UK and Norway – well illustrate these relationships that are also common to many other petroleum-producing countries.

In 1962 when the UK and Norway were first approached by international oil companies interested in the North Sea's prospects, both states were relatively uninformed about how to manage these powerful firms. On the other hand, the oil companies were not well versed in investing and working under relatively rigorous controls managed by strong governments. Both the UK and Norway had to cope with several special characteristics of the petroleum industry that make it difficult to control.

These characteristics include an exceptionally capital intensive structure; a high rate of profit; a vertical integration; a low elasticity in demand for its products; and a large cash flow. The industry is also difficult to enter,[2] not only because of the large capital investments needed but also because of the highly specialised technicians required.

Although both Norway and the UK have been eager to develop their off-shore petroleum resources (Figure 12.1), they have approached the task somewhat differently, in part because of their differences in population size (Norway 4.2 million vs the UK's 57 million), overall energy self-sufficiency,

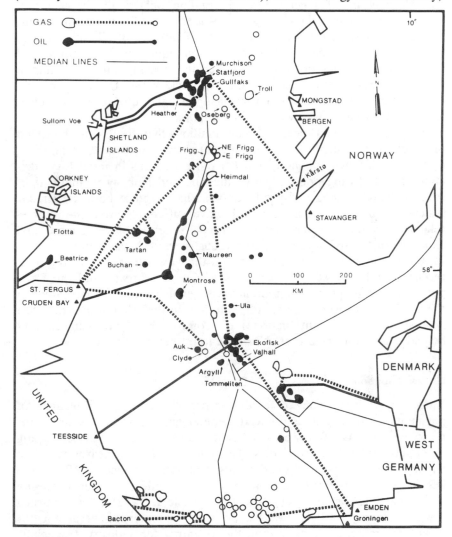

Figure 12.1 Selected United Kingdom and Norwegian North Sea oil and gas fields and pipelines

Source: After Bergen Bank, *Petroleum Activities in Norway 1987*, (BB, Bergen, April 1987), p. 14. With permission.

petroleum dependence, and industrial and institutional structures. The relative impacts of major petroleum developments were
expected to be (and have been) greater on Norway's smaller population
and less industrialised economy. The UK's problem was to cope with
both an overall energy-production deficit and an increasing dependence
on petroleum.[3] For example, during 1976, even after a decade of oil
activities, the UK produced only 86 million bbl of oil, whereas Norway
produced 100 million bbl. Oil consumption in Norway was 66 million bbl;
the UK consumed nearly 670 million bbl. A further comparison shows
that Norway in 1976 generated 154 million bbl of oil equivalent in
hydropower, whereas the UK's output was 8 million bbl of oil equivalent.
And finally, although the UK mined 509 million bbl of oil equivalent in
coal, it consumed 530 million bbl of oil equivalent. Norway's coal
production (on Svalbard) about equalled its consumption of 3.7 million
bbl of oil equivalent.[4] These data make clear the UK's urgency in
developing its energy base and in reducing its balance-of-payments deficit
associated, in large part, with energy imports.

Overall, Norway has been (1) more cautious than the UK in the speed
of developing its offshore petroleum; (2) more conservative in allowing
foreign oil-firm participation; and (3) more rigorous in regulatory, taxation, and licensing policies. Both the UK and Norway maintain tight
control over licensing and production in the offshore; both use similar
systems of allocating licences; both have suffered petroleum-industry-
related economic growing pains; both have sought to develop a national
expertise in the offshore petroleum industry; both have been concerned
for the petroleum industry's impacts on their overall economy; and until
recently, both had state oil companies. Finally, Norway and the UK's
petroleum development programmes were evolving during a period when
the OPEC states were forging a 'revolution' in the relationships between
host states and the international oil companies, a circumstance that
helped shape both countries' North Sea petroleum policies.[5]

United Kingdom

Early on, with its concern focused especially on a better balance of
payments account and a reduced dependence on foreign sources for
petroleum, the UK followed a policy of 'full speed ahead' in developing
its offshore petroleum resources. The UK went from a nearly complete
dependence on imports in the early 1960s to self-sufficiency in 1980,[6] and
became a net exporter during the 1980s, with crude oil output increasing
by eleven times between 1976 and 1986 (Table 12.1). The UK's journey
toward autarchy was not always smooth for the national government, for
local and regional areas within the country, for industry (especially
manufacturing) or for the international and domestic oil companies
operating in UK waters. The UK's autarchy may be short-lived, with
production peaking in the 1980s and then declining in the 1990s. Crude

oil production in 1986 was more than 2.6 million bbl/d. By 1991 production will have declined to about 1.6 million bbl/d. Although output should be substantial in the 1990s, production will be a function not only of the recovery of crude oil prices but also of new discoveries and of leasing, production, and taxation policies.[7]

Table 12.1 United Kingdom oil production, 1976–86

Year	Production (Thousands of bbl/d)
1976	240
1977	765
1978	1,095
1979	1,600
1980	1,650
1981	1,835
1982	2,125
1983	2,360
1984	2,580
1985	2,655
1986	2,665

Source: British Petroleum Company plc, *BP Statistical Review of World Energy* (BP, London, June 1987), p. 5.

Licensing

The Continental Shelf Act of 1964 vested all UK North Sea waters and seabed in the state. And it required that all licence applicants be resident citizens (that is maintain a subsidiary in the UK). This provision assured that the international oil companies would pay taxes to the UK Government and would operate within the British court system. Specific licensing requirements are left in the hands of the Ministry of Power (now the Department of Energy). The initial licensing criteria developed were only loosely stated but stressed that preference would be given to those foreign firms that (1) had already contributed to continental shelf resource development and the country's overall fuel economy; (2) were willing to include domestic companies in their work programmes;[8] and (3) showed a desire to expedite their exploration and development efforts. From the beginning of its licensing rounds, the UK has assured expeditious development of its licensed blocks. It requires that within six years after a given round, lessees must surrender 50 per cent of their leased area.[9]

Discretionary allocation vs competitive bidding

Considerable debate has been generated within the UK (as elsewhere) about the merits of different licensing methods. From the beginning of its offshore licensing programme, the UK has based its policies on a discretionary allocation system, that is direct negotiations by an oil company with the government. As Mabro *et al.* noted, in an astute analysis titled *The Market for North Sea Crude*, discretionary licensing is 'a powerful

277

and effective weapon in the hands of the Government and can be used to ensure that both licenses and would-be licensees cooperate with Government policy'.[10]

The UK held its first licensing round in 1964, when the government offered 960 exploration blocks with an average size of 250km.[2] Round two came in 1965 and round three in 1969–70.[11] During these early rounds,[12] the government (bent on an accelerated development effort) based its final allocations mainly on the strength of the work programmes submitted by the prospective licensees,[13] an approach that, in effect, became a form of competitive bidding, with the government actually informing applicants that to receive a given licence would require a more vigorous work programme. The requirement of detailed work programmes helped to screen out companies that sought licences merely for speculation.[14] In the third round, an applicant's willingness to allow the participation of the UK's nationalised Gas Council and National Coal Board was also taken into account, another form of quasi-competitive bidding.[15]

By the end of the fourth round (1971–2), 2,655 blocks had been offered, and licences issued totalled 245 within 863 blocks.[16] Only during the fourth, eighth, and ninth rounds did the government directly offer any licences under competitive cash tenders. Among the blocks offered in each round, a minority of fifteen was designated as cash tenders; all fifteen blocks were licensed in the fourth round, seven in the eighth round, and thirteen in the ninth round.[17] The fifth (1976–7) and sixth (1978–9) rounds were based on discretionary allocations. In the seventh round (1980–1), oilmen were permitted to identify unlicensed blocks of their choice in a precisely defined area within the North Sea.[18] These experiments were successful, but the government continues to use the discretionary system, including its recently completed tenth round in May 1987 when 51 blocks (of 127 offered) were awarded.[19]

Colin Robinson, in an article published in *Lloyds Bank Review*, explained the government's affinity for the discretionary system as a function of politicians and civil servants liking it, 'presumably because it gives them control over the oil companies'. He further noted that the oil companies also seem 'content with the discretionary system though that is possible because they fear that a licence auction regime would be additional to existing taxation instead of replacing it'.[20] Danny Hann suggested, in the *Scottish Journal of Political Economy*, that both government and oil industry bureaucrats prefer the discretionary approach because it is more complex than an auctioning system, which helps justify their positions, a situation that also applies to complex taxation systems.[21] From a practical standpoint, a discretionary allocation programme also allows the government more easily, and legitimately, to award licences to domestic firms that might otherwise be outbid by larger and better financed international oil companies. This consideration is probably of greater importance to the government than the accelerated work programmes encouraged under the discretionary system.[22] Nevertheless, during the UK's first four rounds, foreign oil companies (especially US firms) never accounted for less than 62.5 per cent of the licences awarded.[23]

Robinson feels the auctioning system, although not perfect, would (1) provide the national treasury with more revenue, because as discoveries are made, the ante would rise through competition, and (2) favour low-cost operators who can outbid inefficient operators.[24] Kenneth Dam takes a similar stand. He feels that auctioning is more likely to place 'scarce rights or resources in the hands of those firms that can most efficiently exploit them' and it helps to avoid problems associated with an arbitrarily set size of licensing blocks. Dam also favours the auctioning of licences because it helps retain 'the economic rent for the state', whereas discretionary allocations transfer 'the economic rent to the licensee', with economic rent defined as the difference between overall revenues and production, management, and capital costs.[25] This situation is, of course, premised on the assumption that no collusion occurs, which was probably most common in the early years of licensing, because of the oil firms' scramble to obtain reserves outside the Middle East.[26]

Block size

An important element in the ability of oil firms to maximise production efficiency and for the government to ensure the efficient use of the petroleum resource is the size of the blocks assigned. If these are too small, fields may overlap the licence areas of several competing firms. Unless adjustments can be made, less than optimal pumping strategies may be used, which – in the long term – may reduce a field's maximum petroleum yield. Adjustments have, in fact, occurred through reallocations of portions or entire surrendered blocks.

Those in industry consider the $250km^2$ average block size used in early UK licensing rounds as too small,[27] although interested parties could reduce this problem by negotiating shared production and management controls with neighbouring licensees. Until 1975 an assignee could transfer *control* of a licence to another party, but the *licence* itself could not be transferred without permission from the Ministry of Power. In 1975 under the Petroleum and Submarine Pipe-line Act, the government also stopped control reallocations among assignees, unless approved by the Ministry of Power.[28] Permission to transfer licences must now be obtained from the Secretary of State for Energy.

Royalties and taxes

Since 1964 when the UK Government extended its 1934 Petroleum Production Act (PPA) to include offshore operations, the offshore taxation system has been changed several times.[29] Often, these changes were dictated more by political and bureaucratic concerns than by their economic implications for the offshore petroleum industry.[30] Nevertheless, as suggested by MacKay and Mackay as early as 1975, these changes also probably reflect the difficulties in meeting the needs of different producers operating fields of significantly different sizes. They note that 'the trick is to ensure that the state captures the economic rent without discouraging ... exploration and production.' These goals are 'extremely difficult to achieve and any system of taxation which applied uniform conditions will certainly fail to meet one of these objectives'.[31]

The current tax regime consists of three tiers (Table 12.2) – a royalty, a petroleum revenue tax (PRT), and a corporation tax (CT). The PPA provides for Crown-assessed royalties on all oil and gas produced offshore. Royalty rates depend on the concessionary round in which a company received a given licence. As of 1986, for licences issued in the first four rounds, the royalty was 12.5 per cent of the petroleum's wellhead production value; for licences granted in the fifth and later rounds, the royalty was 12.5 per cent of the landed-value. Royalties are payable in cash or kind. Most oil field payments are in kind. Exceptions (as of 1984) were Argyll, Auk, Beatrice, Buchan, Heather, Maureen, Montrose, and Tartan.[32]

Table 12.2 United Kingdom Government revenue from the North Sea (Millions of £)

	Royalties	PRT[a]	SPD[b]	CT[c]	Total
Total 1970–71 to					
1975–76	65	–	–	23	88
1976–77	71	–	–	10	81
1977–78	228	–	–	10	238
1978–79	289	183	–	90	562
1979–80	628	1,436	–	270	2,334
1980–81	992	2,410	–	480	3,882
1981–82	1,396	2,390	2,025	650	6,461
1982–83	1,643	3,274	2,395	460	7,772
1983–84	1,900	6,100	–	900	8,900
Total	7,212	15,793	4,420	2,893	30,318

Source: P. Sanderson 'North Sea oil and the UK economy: boon or bane?' *Barclays Review*, vol. 59, no. 4 (1984), p. 84.
Notes: a. Petroleum Revenue Tax; b. Supplementary Petroleum Duty; c. Corporation Tax.

The PRT (established in 1974) is the UK's most important petroleum levy; it is a profits tax, applied field by field. Thus a firm with investments in six fields has six separate PRT assessments, none of which can be charged against another to offset losses in a given field. Assessable profit for a field is the 'total value of oil produced *less* royalties, allowable operating and capital expenditure ... and losses from the field that have been carried forward'.[33] Interest costs to finance the capital costs for field development are not deductible.[34] In 1986 the PRT rate of assessable profits was 75 per cent, more than a two-thirds increase over its 1974 rate of 45 per cent.[35] The tax is collected twice each year. To assist operators, the government permits a specified amount of tax-free petroleum production.[36]

All firms operating in the UK pay the CT. This tax applies to an oil company's profits as a whole and not to individual oil fields, as with the PRT. In contrast to the increased PRT rate, the CT rate has been reduced from 52 per cent in 1982–3 to 35 per cent in 1986–7.[37]

The oil price collapse in early 1986 caused North Sea operators both to defer development projects and to seek tax relief. Although both large and small producers sought changes in the UK's tax structure, they lacked consensus about specific changes needed.[38] The UK Offshore Operators Association (UKOOA), representing the major operators, negotiated with the government during late 1986 and early 1987. These negotiations resulted in the 1987 national budget's providing two important changes in the PRT. One change 'introduces a new cross-field allowance whereby the investor' may 'elect to offset up to 10 per cent of new field development costs against PRT income from other fields'. This option may be used for a six-month PRT period 'up to the time when field payback is attained' but is 'restricted to new field developments from 17 March 1987, excluding the Southern gas basin'. A second provision allows the deduction of research expenditure for 'general' North Sea exploitation (that is research not directed to a specific field) against 'any PRT income after three years'.[39]

According to Alexander Kemp and David Rose, in an article published by the *Petroleum Economist*, the new-field development-cost provision – by speeding up relief on capital expenditure – is directed toward stimulating investments in new fields. This option, however, does not allow an uplift on an 'expenditure deducted against other PRT income'. Operators gain the most when a 'new field still does not pay PRT after the cross-field allowance has been fully used'. Kemp and Rose note that:

> This is a function of (a) development costs, (b) the size of the field, and (c) the oil price: the higher are development costs, the smaller the field, and the lower the oil price, the more likely it is that the field will not be subject to PRT.
>
> The gains from the new allowance are thus broadly targeted on fields of low profitability though there might be exceptions with a large field where PRT could still be payable at low levels of profitability.[40]

When a small field does not pay PRT, the investors' 'position is improved at all discount rates'.[41] This improvement may help attract investors in North Sea ventures, despite relatively depressed petroleum prices. Potential investors in small fields need help, because the government's tax regime has made many of these fields economically unattractive. Indeed, the government's fiscal policies have served, if inadvertently, as a major 'depletion tool'.[42] But according to Hann, both 'the government and ... industry' have (at times) mistakenly equated 'small fields with low profit fields'.[43]

Production and marketing controls

Under the UK's PPA, the state (through the Secretary of State for Energy) may set oil field production limits, a provision that irks oil producers. In addition, the producers' production plans can be rejected or modified, if the programme seems 'contrary to good oilfield practice' or if they are not in the 'national interest'.[44] Although the government, through consultation with a company, can set production limits on producing fields, it has not done so up to 1987, and it has consistently reassured industry it will not do so, except in an extreme emergency.[45] The Energy Secretary also has the power to delay a new field's production start-up. This provision was used on one occasion – when the Clyde field's start-up was delayed for two years.[46]

An especially irritating operational restraint concerns a UK requirement that, before export, all offshore crude must be brought onshore in the UK. This policy provides employment for British trade union workers, but it also causes transportation and marketing inefficiences. Until 1982 the government also required the landing of natural gas on British territory before it could be sold. Once landed, only the British Gas Corporation (BGC) was allowed to buy the gas, which it purchased at an artificially low price. This prescribed marketing constraint encouraged the petroleum companies to put most of their exploration efforts and investments into the UK's northern North Sea region where the chance of crude oil discoveries is better. In 1982 Parliament ended the BGC's monopsony on North Sea gas when it passed the Oil and Gas Enterprise Act (OGEA), which was supplemented by the Gas Act of 1986. Under OGEA, the BGC is forced to compete for gas supplies, a change that may have increased the price for consumers. This change has also forced the BGC to reduce costs, because earlier inefficiencies can no longer be externalised via low prices paid to offshore gas producers.[47] On the other hand, OGEA's introduction of common carriage provisions may have helped restrain price increases for industrial users.[48]

Still another component of control that once irritated the private commercial oil firms was a decade of operating with the UK's national petroleum-industry watchdog, the British National Oil Corporation (BNOC) created, in large measure, as a response to public and political pressure to have the government 'recoup the ... wealth that has been cavalierly turned over to international oil companies', which had struck it rich in the North Sea at the same time oil prices quadrupled in response to the OPEC and Arab states' precipitated crisis of 1973–4.[49]

BNOC

With the onset of the petroleum supply shortfall in 1973–4, the Conservative Government sought help from multinational oil corporations holding UK concessions. The government wanted these firms to cover the shortfall by diverting to the UK some of their production from other areas.

The companies did not co-operate, but rather distributed the shortfall, more or less, equally among their many customers. Soon thereafter, when the UK's continental shelf resources began to come on stream, the government decided to provide itself with an assured petroleum supply. Both the Labour and Conservative parties had suggested that national security (strategic) concerns demanded some form of state control over national petroleum production. Despite a heated opposition by the Tories, in 1975 the UK Parliament (under a Labour Government) passed the Petroleum and Submarine Pipeline Act (PSPA), which established BNOC, an organisation destined for a decade of difficulties and controversy.[50]

Robert Mabro *et al.* labelled the 'strategic' argument for the establishment of BNOC as specious, because the state in an emergency can 'take immediate and full control' of the distribution of petroleum produced in its waters, although this action would require that international energy agreements within the EEC and IEA be met. Other reasons for establishing BNOC centred on the Labour Party's desire for direct involvement in administering and operating this important resource sector, and also the asset of having precise and adequate information that is acquired best when government personnel participate as members of operating committees.[51]

Once in place, BNOC faced hard bargaining to bring the industry under governmental control.[52] It found that moving from statute to functioning entity was not an easy road. Although he did not necessarily have BNOC in mind, Kenneth Dam in 1975 bluntly outlined the difficulties faced by national oil companies when he said

> more than an official signature on a statute ... is required to create a national oil company that is more than a financial shell. To create a national oil company that can actually extract oil from the ground requires the cooperation, voluntary or coerced, of private companies.[53]

By January 1977 BNOC had acquired participation agreements with Conoco, Gulf, Tricentrol, and Ranger, and by mid-1978, nearly all petroleum firms had agreed to the programme.[54] But according to Brent Nelsen, these agreements were little more than paper transactions.[55] Under this programme, BNOC could also buy a partnership (up to 51 per cent) in all leases issued during the first four rounds and automatically received 51 per cent control in all leases, beginning with the fifth round in 1977.[56]

Because BNOC personnel sat on the operating committees, were privy to the companies' private information, and acted as advisers to the national government, the oil firms complained of a conflict of interest. They felt at a disadvantage when seeking exploration and production licences, because the government was fully informed about their affairs and because BNOC received leases in its own right. This situation led to a diminution of oil company activity in UK waters.[57]

The companies did not have long to fret. A new Conservative

Government came to office in 1979, and BNOC's powers were curtailed, although not quickly enough for some critics who had long wanted the agency dismantled. In 1982 BNOC's production activities were eliminated when Parliament passed the OGEA. The Act transferred BNOC's business, exploration, and production functions to its subsidiary, Britoil. Soon thereafter, Britoil was reorganised and made 51 per cent privately owned.[58] Then, in 1985, after only a decade of operation, BNOC itself was abolished.

Robert Mabro, Director of the Oxford Institute for Energy Studies, spoke against this action. He noted that the decision, based on a simplistic free-market ideology and a lack of understanding of the UK's economic interest in oil specifically and the economics of oil in general, was a mistake. He further argued that the national government should not allow the market alone to set oil prices, rather it should be able to intervene, and because it could not now do so, the next 'oil crisis' would be worse for the UK.[59] In reality, as Mabro *et al.* appropriately pointed out in their *Market for North Sea Crude* analysis, one reason for BNOC's demise was its failure to maintain crude oil purchase and sales prices advantageous to the UK.[60]

Although not an originally mandated role of BNOC, one of its primary functions came to be the setting of prices for the purchase and sale of participation crude, as well as the sale of in-kind crude. BNOC faced difficulties at both ends of the petroleum stream. It could neither control the volume of crude it received nor adequately hold back on sales. The latter problem was especially severe because BNOC had no facilities for storage when market prices were inadequate. All the while, BNOC had also to attempt to set purchase prices that did not trigger price crises in the world market and cause trading losses for the government. BNOC may have been destined for failure from the beginning, because it was 'denied the means to become an efficient trader (the necessary discretion over the acquisition of oil and the availability of storage facilities)' at the same time the national government was 'denying itself the means of reconciling BNOC's objectives through production controls and supply intervention'. In sum, the task set for BNOC was too ambitious and it fell victim to what Mabro *et al.* identified as not only an ideological struggle but an 'ambivalence of official UK views on the price of oil and on the role that the UK should play on the world petroleum market'.[61]

Petroleum – boon or bane?

Paul Sanderson stressed that 'the exploitation of North Sea oil ... has been of enormous consequence for the UK economy'. Government revenue from North Sea petroleum for the years 1970–84 totalled £30,000 million (see Table 12.2). The value of North Sea oil and gas sales and services during this same period totalled nearly £19,000 million, and the petroleum industry's contribution to the UK's gross domestic product (when purchases of goods and services, interest, profits and dividends due abroad are subtracted) amounted to about £13,600 million.[62]

Besides the direct financial gains to the UK's economy, many indirect benefits accrued from North Sea petroleum. For example, its revenues allowed the government to reduce borrowing, a benefit to the taxpayer. In addition, petroleum contributed to an improvement in sterling's real exchange rate during the late 1970s and early 1980s, a situation that enhanced the country's terms of trade – that is, the UK could pay for a given amount of imports with fewer exports, 'a benefit to consumers'.[63]

The petroleum industry has also contributed to important structural changes in the UK's economy. In part, because of the rise in the UK's exchange rate, the price of its exported manufactured goods has increased. The UK, once a net exporter of manufactures, is now a net importer. According to Sanderson, some economists argue that an increase in the real exchange rate is a 'means by which a necessary transfer of resources from the traded to the non-traded sector occurs'; therefore, this transfer 'should not be resisted', because it allows the UK to make more efficient use of its resources, that is 'to export oil rather than manufactures'. Sanderson also noted that investing petroleum revenues overseas can help reduce a rise in the exchange rate, which will create 'a surplus on current account, alleviating the pressure on manufacturing'. But unless all petroleum revenues are 'invested abroad and the earnings reinvested' some structural change will occur. On the other hand, if in the past all petroleum revenues had been sent abroad, 'no gain from oil would ever have been realised'. In sum, Sanderson felt that although the petroleum industry has caused structural change in the UK's economy, especially in the manufacturing sector, overall economic benefits have outweighed economic costs, as in the easing of the balance-of-payments problem and helping to reduce the inflation rate.[64] Kemp, Hallwood, and Wood, however, have noted that although

> oil revenues provide a substantial increase in the level of resources available to the country, ... it is the performance of the non-oil economy which will determine whether those resources can provide a contribution to expansion or a rentier's cushion against decline.[65]

Depending on one's viewpoint, oil may be looked upon as a boon or a bane. Scotland exemplifies this perceptual dichotomy. Long plagued by a persistently high unemployment rate and a net emmigration of its people, Scotland in the 1960s and early 1970s experienced a significant influx of workers from outside the area, both from other UK regions and foreign countries, although the latter source of immigration was not as great as was often assumed.[66]

The surge of capital and inflow of workers, along with their associated demands for services, housing, and schools, placed a strain on both large and small communities, such as Aberdeen on Scotland's eastern shore and Lerwick in the Shetlands. From 1974 to 1981 workers employed in

Scotland in firms wholly related to North Sea petroleum went from 13,000 to 50,000.[67] If indirect employment is added in, the total (as of 1976) was estimated at perhaps 68,000 workers.[68] Aberdeen experienced two-thirds of this growth. The total capital investment in the North Sea's offshore and onshore petroleum industry between 1965 and 1982 is estimated at some £35,000 million (1982 prices).[69] Many Shetlanders feared the environmental, social, and economic impacts such massive capital infusions might have on their rural lifestyle and pace of life. Another of their fears was justified – unfulfilled hopes. For example, the Sullom Voe oil terminal in northern Shetland, whose need and potential environmental impacts caused heated debate locally, regionally, and nationally, is not now (and may never be) used to capacity, given current and likely future petroleum market conditions. This situation is a bitter pill for the islands' 16,000 inhabitants, because the Shetland County Council sought (and obtained) national parliamentary legislation that gave it the right to levy a tax on every barrel of oil passing through Sullom Voe.[70]

Thus many Scots feel as if they have been robbed of the benefits of their offshore petroleum. Abandonment of all platform construction yards on Scotland's west coast (Kishorn, Ardyne Point, and Hunterston) and greatly reduced activity at Nigg on Cromarty Firth exemplify the problem. And in some areas, under-utilised housing and schools stand as testimony to the oil industry's contribution to a boom and bust economic cycle, an episode not soon forgotten by a people long suspicious of the motivations of the industrial and urban south of the UK.[71]

Prospects

The UK in late 1986 was estimated to have reserves of 5,300 million bbl of crude oil and 0.6 million million m^3 of natural gas.[72] In 1983 (based on production rates for that year) the UK had oil reserves enough to last another twelve to forty-six years, depending on the reserve estimate used.[73] G. C. Band stressed that the UK cannot take future petroleum reserves for granted. These will depend on the UK's maintaining an 'effective ... licensing policy, vigorous exploration activity, cost-effective technology, and a combined energy-pricing and fiscal climate that will encourage the development of smaller fields',[74] and in frontier areas as in the Rockall Trough to the west of Shetland, areas already offered for leasing and now being explored. Further efforts will be needed in already explored areas, as in the Irish Sea, English Channel, and the Firths of Clyde and Forth. As geological information for inshore areas becomes available, it may assist in locating onshore areas with petroleum potential.

Given that the UK's overall petroleum production has already peaked, the government's rejection in 1985 of a negotiated plan to import major supplies of gas available from Norway's Sleipner field compromised long-term national energy needs for short- and medium-term expediencies. Gas field depletion will be unnecessarily hastened and the country's overall dependence on imports (by the turn of the century) increased, all so that present domestic suppliers can capture markets and the balance-of-payments account can be improved.[75]

Because of the current short-term volatility of the oil market and its long-term price uncertainties, predicting future depletion is difficult. But because petroleum is a wasting (depleting) asset, the UK's economic structure and social conditions must face further adjustments. A diminution of petroleum exports will reduce the current account and 'require an expansion of the UK traded goods sector, which should be encouraged by a fall in the real exchange rate and a consequent improvement in the competitive position of UK industry'. The decline in the petroleum industry, however, should not be so abrupt as was its growth, and a 'return to pre-oil trading patterns based on ... a surplus of manufactures as oil production declines ... should be possible', although these will not be the same industries as existed before the oil era.[76]

Norway

Petroleum firms began exploratory drilling in Norway's North Sea sector during 1966. Initially they had little success. Then in 1969 drillers struck oil in an area that was to become the Ekofisk complex. In 1970 with Ekofisk's success assured and the public's attention focused on the potential mineral wealth of Norway's offshore, the national government began a serious reassessment of the significance of petroleum for the country's future.[77] Out of this reassessment came a tightening of controls on the international oil companies and a demand for greater direct state participation in the industry. This heightened concern resulted in the state's assuming a greater share in a privately owned firm, Norsk Hydro, with the state's share increasing from 48 per cent to 51 per cent.[78] Two years later (1972) the Storting, Norway's national parliament, established a wholly state-owned petroleum company, Den Norske Stats Oljeselskap (Statoil).

By 1974, with oil revenues beginning to flow into the Norwegian economy, a major concern for the government had become how to keep national and local economies from 'overheating'. But the anticipated flood of revenue did not materialise because of a subsequent general world recession and increasingly expensive exploration and production costs encountered by the petroleum industry as its activities moved northward in the North Sea. By 1977 the government found itself deep in debt, partially as a consequence of borrowing and large expenditures made attempting to buffer Norway's economy from the world recession, a period during which the country experienced a concurrent inflation in wages and prices and a decline in its export markets. Despite mandatory wage and price controls during late 1978 and throughout 1979, inflation continued into 1980 and 1981. A sharp increase in petroleum prices in 1979 and associated increases in revenues helped save the situation. It also demonstrated that the Norwegian economy had become intricately tied to the petroleum economy, 'a situation that economic policy a few years earlier explicitly sought to avoid'.[79] By 1981 the gross value of petroleum production (including extraction and processing) was approximately 25 per cent greater than for the country's entire manufacturing industry sector; in 1985 petroleum's value was 31 per cent greater.[80] Nevertheless, in contrast to the rapid pace of development in the UK's offshore petroleum programmes, Norway has sought a slowly paced effort. In the mid-1980s this approach has been relaxed, if only slightly, mainly because its national oil companies and the Norwegian platform construction and service industries have matured and are in need of new projects.[81] In late 1987, however, the government was once again attempting to slow the pace of development.

Licensing

In its licensing programme, Norway uses bilateral negotiations between the Ministry of Petroleum and Energy and domestic and foreign oil companies; no bidding system is used. The government began awarding petroleum exploration and development licences in 1965. In the first round, the government awarded seventy-eight blocks, the largest number given so far in the twelve rounds now completed. These blocks were awarded under twenty-two licences, with a preference given to those applicants that included Norwegian companies as partners.[82] The government has also given preference to companies willing to purchase Norwegian services and equipment,[83] often at inflated prices. This policy has resulted in adding to inflation within Norway. Some critics in Norway also feel that, if continued, this policy may lead to uncontrolled and hidden subsidies to foreign firms and create industrial enterprises without a true economic viability. These same critics suggest that what Norway really needs is a greater transfer of 'technology, managerial know-how, and research and development... through cooperative ventures between' Norwegian firms and foreign oil companies.[84] Foreign oil firms have also acquired leases in Norway's offshore only through state supervision and participation. A widespread consensus exists among political parties in Norway that 'the impact of the petroleum industry on the Norwegian economy is too great to allow private enterprise to have complete control over the depletion of offshore resources'.[85]

Norway, through careful licensing, has maintained control over the location and intensity of petroleum industry activities. This control allows the government to protect the country's national economy, marine and coastal environments, and strategic interests, as in the far north. The Norwegians have been highly concerned with the environmental dangers posed by the petroleum industry to fishing and ecological conditions, especially in areas north of 62° N.[86] To ensure these concerns are monitored, the Storting in 1972 established the Norwegian Petroleum

Directorate, which makes detailed evaluations of which blocks should be opened to exploration and development.[87]

Not until 1979, during the fifth round of licensing, did the government allow awards north of 62° N.[88] When drillers first began operating north of 62°, they were restricted to the summer season, because of the government's fear of blowouts and weather-related accidents that might occur during the dark of winter. In the 1983–4 winter season, however,

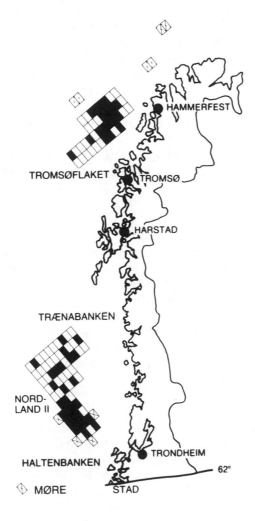

■ Blocks awarded
⊠ Blocks awarded North of Stad
   under the 11th round (Part I)

Figure 12.2 Location of blocks awarded north of Stad, Norway, by the Ministry of Petroleum and Energy

Source: Bergen Bank, *Petroleum Activities in Norway 1987* (BB, Bergen, Norway, April 1987), p. 46. With permission.

drillers operating on Haltenbanken were allowed to continue working. Now, with more than two decades of experience and an improved technology, companies may operate year-round, even in the North Cape region.[89]

Norway in late 1987 and early 1988 was in the midst of a twelfth concessionary round.[90] In March 1987 companies were allowed for the first time to nominate blocks for licensing. The government in early 1987 also offered several 'strategic' blocks in addition to those included within the normal licensing round. Eighteen firms made bids on several blocks in the western Barents Sea north and north-east of Finnmark. Six blocks were awarded (Figure 12.2). Within a decade or two, the Barents Sea region may be producing gas for export, which will provide revenues for a further reduction of Norway's external debt. So far, however, drillers have discovered only small amounts of gas, which are not large enough to exploit at current prices. Should large deposits be discovered, development will not begin until the late 1990s,[91] and finding markets for this gas may present a problem. According to Arild Holt-Jensen of the Geografisk Institutt, in Bergen, output is likely to be sent by ship to North American markets.[92]

Royalties and taxes

Offshore petroleum revenue in Norway comes from four sources: (1) an ordinary, two-part corporate income tax – state and municipal, (2) a special petroleum tax (SPT), (3) a capital tax, and (4) a royalty.[93]

Both the state income and municipal taxes go to the state's coffers, but are calculated separately. The state corporate income tax 'is 27.8 per cent of net income less dividends' and various deductions for royalties, depreciation, operating costs, and interest paid during construction phases of operation. The municipal corporate income tax is 23 per cent of net income. Norway began collecting SPT in 1975, after the crude oil price surge of 1973–4 and the oil companies' windfall profits.[94] The SPT is 30 per cent and is applied to 'the net income for municipal tax, less a tax-free allowance or "uplift", and with no deduction for dividends or for losses in activities other than petroleum production and pipeline transportation'. Capital tax is levied at 0.3 per cent 'on book values (as depreciated) on tangible, financial net wealth'. It is not deductible in the assessment of other taxes. Operators pay royalties only on projects approved before 1 January 1986. Producers using licences awarded before 1972 pay royalties of 10 per cent. For licences awarded after 1972, producers pay 8–16 per cent for oil and 12.5 per cent for other products.[95] Revenue authorities calculate royalties as a percentage of the gross value of production at the well-head; payments are made in cash or in kind. Since March 1974 all royalties have been paid in kind.[96]

Critics of Norway's taxation system complain that 'less profitable fields are relatively highly taxed'. In contrast, the UK's system makes distinctions between low- and high-profit fields. The consequence for Norway has been that only a few small fields have been developed, as in satellites of Statfjord and Ekofisk.[97] Thus with the added burden of depressed oil prices in 1985 and 1986, Norway felt compelled to make tax adjustments if firms were to continue developing their concessions.[98] Consequently the government passed reform tax legislation in the autumn of 1986.[99] Under the new legislation, depreciation write-offs on new expenditures in fixed assets may begin as of the first production.[100] The government also established a 6.6 per cent tax uplift for the first fifteen years of a new field's active life and for all old fields during a six-year phase-out period.[101] These changes have helped somewhat to renew interest in Norway's offshore; during 1987, however, the Norwegian Government was under pressure to make further tax concessions, but it refused to do so.[102]

Statoil

To protect the country's petroleum industry interests, Norway in 1972 established its state oil company – Statoil. The Norwegian Government has allowed Statoil to function as a powerful force in Norway's petroleum industry. Until the end of 1985 it automatically received a 50 per cent share in each licence granted and it could gradually claim (depending on the productivity of a given field) up to 80 per cent.[103] Until recently, commercial oil partners had to pay Statoil's share of exploration costs. In sum, Statoil has not had to face may 'real' financial risks (although it would probably disagree) and it has had no shareholders to answer to when it makes decisions.[104]

The Tommeliten takeover

It is not surprising, therefore, that Statoil and private oil firms often differ in viewpoint, a situation vividly illustrated in recent events associated with the small Tommeliten oil and gas field situated near the Ekofisk complex. Statoil rode roughshod over the three partners – Phillips, Norsk Fina, and Norsk Agip. The partners considered the Tommeliten field as potentially too economically marginal (relative to taxes and market prices) to proceed with development. Statoil's management dismissed these concerns and, on its own, declared the field commercial. Statoil gave the partners an ultimatum either to develop the field or be prepared to sell it to Statoil. And if they chose neither option, then the state would expropriate the Tommeliten, without payment.[105]

Phillips had requested to be made operator of the field. When its request was denied, the company relinquished its 25.87 per cent share.[106] Norsk Agip and Norsk Fina declined to pick up Phillips' share. Consequently Statoil became operator of the field and now holds 70.64 per cent ownership, with Norsk Fina owning 20.23 per cent and Norsk Agip 9.13 per cent. The project was scheduled for development in four stages, with production to begin in June 1988; it should produce for twenty years.[107]

How could Statoil act so arbitrarily? Statoil officials knew they had Phillips in a Catch-22 situation. Phillips needed Tommeliten's gas sales to help replace lost revenues from gas used for reinjection during subsidence control within the Ekofisk field. And yet, if it had gone ahead with development, the Norwegian Government and Statoil could have argued that the tax burden on marginal fields is not too high, after all. In the view of the *Petroleum Economist* editors, this approach to private and national 'co-operation' neither builds rapport nor instils confidence in potential investors, and it smacks of an OPEC mentality, whereby one views the petroleum as a one-time resource 'from which the maximum rent must be extracted at the point of production'. It also shows a lack of understanding that the 'real value of oil comes from the employment and technology created in exploiting it'[108] (Figure 12.3). Although no longer a partner, Phillips agreed (in May 1987) to purchase Tommeliten's gas for sale in western European markets.[109]

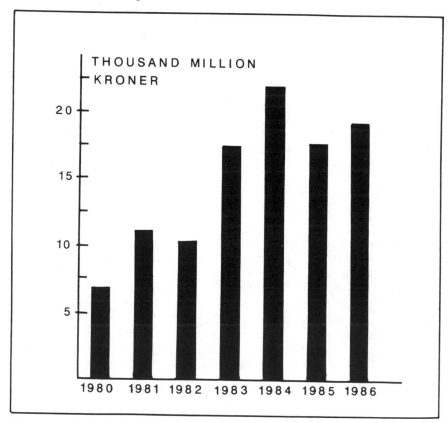

Figure 12.3 Norway's accrued investments in oil and gas production and pipeline transport (in constant 1980 prices)

Source: H. Skånland, 'Economic perspectives', *Norges Bank Economic Bulletin*, vol. 58, no. 1 (1987), p. 6. With permission. Data for average exchange rates in 1980 provided by Letter: P. Thomas and S. Kristiansen, Norges Bank, Oslo, Norway, 9 Oct. 1987.
Note: In 1980 the British £ averaged £1 to 11.4927 kroner and the US$ averaged $1 to 4.9394 kroner.

The reining in of Statoil

Statoil's recent high-handed actions seem to justify the Norwegian Government's re-examination of Statoil's role in Norway's offshore petroleum industry. Like many state oil companies, it had become (after only a decade) a strong entity that functioned as if it were responsible for setting its own policies, rather than being responsible to the government. By the mid-1980s Statoil had a tremendous cash flow (US$2,300 million in only the first six months of 1985)[110] and influence; today, although Statoil's profits have dropped, [111] it still has a large cash flow and some 7,000 employees. Although for the past six years Statoil has been having its wings clipped, the Tommeliten episode illustrates that it is still a force to be reckoned with.

The reining in of Statoil began in 1981 with the election of a Conservative Coalition Government. This was the period when the flow of Statfjord oil, gas, and cash surged.[112] Instead of letting most of this money funnel through Statoil, the government diverted it directly into the Ministry of Finance. In 1985, the government further reduced Statoil's power. In ninth-round leasing, Statoil's ownership share remained at 50 per cent in each block awarded, but its voting rights were 'reduced to 20 per cent, with the state holding the remaining 50 per cent directly'. The government has also reduced Statoil's share in some established fields, as in Gullfaks where its control was cut from 85 per cent to 15 per cent.[113] In late 1987, Statoil was again under close scrutiny. It under-estimated (by 50 per cent) construction costs of an expansion and modernisation of its Mongstad oil terminal and refinery located north of Bergen. Kåre Willoch, the former Conservative Party prime minister, called for an official investigation of the cost overrun and a consideration of whether Statoil should continue as a 100-per-cent state-owned company. This situation is an embarrassment to Willoch himself, given that he was head of the government when the Mongstad project was approved in 1984.[114] On the other hand, his prolonged opposition to the project (despite a powerful lobbying effort by Statoil), would seem to vindicate him.

Statoil looks abroad

Although the government has clipped Statoil's domestic wings, it has decided to allow Statoil to expand its activities outside Norway. This policy will allow Statoil to market the large supply of crude and gas that it will receive by 1990 from new fields coming on stream.[115]

In 1985 Statoil purchased Svenske Esso (Sweden) and in mid-1986 negotiated a deal with Exxon Corporation to purchase Dansk Esso (Denmark) with its some 450 petrol stations. Collectively Statoil's total ownership of petrol stations in Norway, Sweden, and Denmark is 1,600.[116] Also in mid-1986, the MPE assented to a Statoil request to buy into Swedegas (the Swedish state gas company), now being privatised by 40 per cent. This venture should help Statoil sell large amounts of gas that will likely enter the western European market within the next decade. In anticipation of these markets, Statoil is assessing potential locations for gas storage on the continent, such as abandoned salt mines or the Netherlands' huge Groningen gas field 'which is expected to have been emptied around the turn of the century'.[117]

On the wings of new infusions of capital coming from foreign investments, new oil and gas fields coming on stream and its new mandate for controlling gas exports, Statoil (despite the Mongstad scandal) may soar once again.[118]

Norway, the United States, and OPEC

Although as members of NATO they are military allies, the US and Norway have divergent economic and political interests regarding Norway's petroleum industry policies. This divergence exists in part, because Norway is a net exporter of petroleum, whereas the US is a net importer. The two states differ in their preference for oil and gas prices, which has recently resulted in Norway's becoming sympathetic to the Organization of Petroleum Exporting Countries' (OPEC) objective of restricting output to maintain higher prices. The US and Norway also differ in their attitudes about the appropriate pace of development of Norway's reserves.

Policy differences became apparent during the 1973–4 international oil crisis, when OPEC raised prices and restricted output at the same time the Arab states embargoed exports to selected world states. The US was eager for Norway to join the western states' International Energy Agency (IEA), organised as a counterforce to OPEC,[119] whose purpose is to distribute available oil supplies among its members during major supply shortfalls.[120] Membership in the IEA presented Oslo with internal political problems similar to the European Economic Community issue;[121] thus Norway joined the IEA, but under an 'associate status' whereby it reserved the right to make its own decisions when the IEA invokes emergency measures. Norway's reservations did not please the US. On the other hand, Norway showed little inclination to join with OPEC's Arab members in squeezing oil consumers and punishing friends of Israel. This situation continued for more than a decade.

Blue-eyed Arabs – fact or fiction?

By the mid-1980s world oil markets had become glutted and OPEC was in disarray, without discipline among its members. Despite a sharp drop in oil prices in late 1985, Norway was still unwilling to co-operate in OPEC's call for production cut-backs .[122] As prices continued to slide in the spring of 1986 and after a change in government, Norway indicated it might be willing to co-operate with OPEC, if doing so would stabilise prices, but Oslo remained hesitant, because OPEC could not seem to get its own house in order.[123] Although not following OPEC's production cut-back actions directly, Norway in late 1986 reduced exports by 10 per cent and stored its excess production,[124] in effect, a production cut-back. Significantly the announcement came on 9 September, the same day the UK's Secretary of State for Energy, Peter Walker, and Norway's Arne Øien were meeting and only a day before Prime Minister Margaret Thatcher was to visit Norway. OPEC took great delight in publicising this situation, because of Mrs Thatcher's repeated rejections of OPEC's suggestions to have the UK join its production-control programme.[125]

In January 1987 Norway announced a six months' 7.5 per cent (about 80,000 bbl/d) cut in the overall production growth rate, to begin 1 February 1987. In mid-year officials extended this programme to the end of 1987.[126] Despite this curbing of growth in output, Norway's national petroleum analysts expected production in 1987 would exceed 1986's output, because the Gullfaks field had come on stream, and by 1 December, this had happened[127]

The government expected the decision to cut back the rate of growth in oil production to bring rumblings of discontent from western European consumers and Norway's offshore producers. They were not disappointed, with some criticism coming from, perhaps, unexpected quarters. Norway's state oil company, Statoil, announced that this policy change 'could compromise Norway's reputation as a reliable producer'.[128] Arvid Frihagen, an authority on Norwegian petroleum legislation and a professor of public and international law at the University of Bergen, questioned the government's legal right to impose production cuts on foreign oil companies. He insisted that 'it is far from clear that Norway's petroleum legislation gives access to regulate the companies' oil production'.[128] The oil companies probably agreed. British Petroleum (BP) strongly voiced its displeasure and a spokesman for Shell pronounced Norway's action as 'support for OPEC' and 'incompatible with democratic institutions'. Nevertheless, the oil companies gave in, a precedent characterised by Paul Dempsey, an editor for *World Oil*, as a surrender of the right to 'control ... the most sensitive part of their balance sheet'.[130]

Through its cut-back programme Norway is co-operating with OPEC, but Øien has bluntly told OPEC leaders that its policy will continue only so long as OPEC controls overproduction among its members. Øien and OPEC's oil ministers met in mid-September 1987 in the UK to discuss the world market situation.[131] Such co-operation, according to some critics, means the Norwegians must now carry the appelation 'blue-eyed Arabs of the North'. This characterisation is unfair and overdrawn.

Between January and August 1986, Norway's North Sea oil industry lost 4,000 jobs. At mid-year, forty of about ninety drilling rigs previously active off Norway's coast were unused.[132] The impact of this loss has rippled throughout the entire country. Should the state sit idly by and hope for the best? Surely not! Furthermore, the Norwegians have never used their petroleum as a political weapon by capriciously excluding certain states' access to Norway's resources, as have the Arab states to theirs. The Norwegians, however, might be justified in casting a few stones of their own, considering the petropolitical pressures put upon them by allies.

From the mid-1970s through to the early 1980s the US pushed Norway to pump more oil and gas and to allow more foreign companies to participate in the development of its petroleum resources. Pumping more oil and gas into the market can lower prices, whereas it is to Norway's advantage to have higher prices. For example, the Norwegian Government in mid-1986 expected that if the oil prices then current (US$10/bbl) were to continue throughout 1987, Norway's oil revenues would decline from the US$7,000 million accrued in 1984 to only US$670 million in 1987, a tenth of 1984's revenues.[133] Fortunately for Norway, prices during 1987 climbed toward the US$17 to $20/bbl range, although they fell back again (to US$15 to $16/bbl) near the close of 1987.[134] Even though the US Government desires lower prices for its petroleum consumers, these conditions damage its own oil industry. On the other hand, the US welcomes the effects lower prices can have on the USSR's petroleum export revenues and thereby its overall economy. Furthermore, some political-economic analysts argue that low petroleum prices weaken the Arab states and, in effect, strengthen Israel,[135] because economic growth in the Arab states is slowed. Charles Doran, on the other hand, from the perspective of a decade ago, suggested that higher petroleum prices encourage the Arab states to focus their energies on investing petro-dollars rather than on destroying Israel.[136] In sum, this issue is moot.[137]

The gas gap

An excellent example of US pressure on Norway occurred after the USSR invaded Afghanistan. To punish the USSR, the US embargoed sales of petroleum pipeline and pumping equipment needed to transport natural gas to western Europe from the Soviets' Urengoy field in north-central Siberia. Convinced that western Europe would perhaps have future gas shortages and might become too dependent on the USSR for its energy supplies, US Government officials strongly encouraged Norway to fill the anticipated gap.

Even though Norway in the mid-1980s moved to develop its huge Troll oil and gas field[138] (estimated to contain 1.3 million million m^3 of gas), Washington, DC, under the Reagan Administration, saw Norway as dragging its feet. The US position was as much ideological as it was concerned for secure gas supplies for Europe. When in early June 1986 Norway and European mainland countries reached agreement on the first phase of the Troll field's development, observers in Washington, DC, breathed an 'ideological sigh' of relief. As Jonathan Stern so caustically put it, the Reagan Administration was satisfied that Europe had "finally 'got its act together" on the security of natural gas supplies'.[139]

Some of Norway's current and potential natural gas clients may question this perception. Ruhrgas, for example, has told the Norwegians (whose petroleum fields have suffered three labour-dispute shutdowns [140] between 1981 and 1986) that the USSR has been a far more reliable supplier of gas than has Norway.[141] Statoil's effort to acquire storage areas – as at Groningen – is wise, because these can serve as buffers during interruptions. A Groningen location would be especially useful, because its centrality to markets will allow the movement of gas not only to continental Europe but also to the UK, if interruptions occur like that suffered in April 1986 when a strike in Norway's Frigg field sector spread to its UK counterpart. A Groningen storage site could also allow the UK to store gas on the continent in case of disruptions in UK supplies.[142]

Prospects

Despite Washington DC's concern for the security of Europe's future energy supplies and a desire for a more rapid development of Norway's petroleum output, Oslo will not be dictated to in its policy-making decisions, especially so given market conditions of 1986–7.[143]

If petroleum prices decline significantly in future, there may be a reduced demand for Norway's gas. As of 1986 Norway was the world's

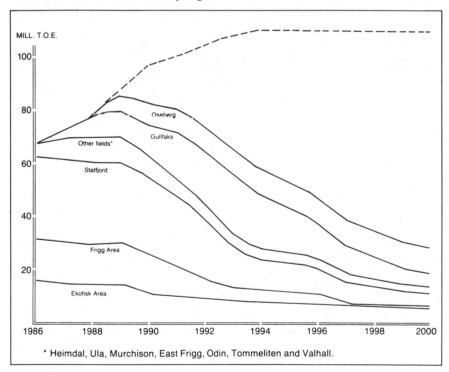

* Heimdal, Ula, Murchison, East Frigg, Odin, Tommeliten and Valhall.

Figure 12.4 Projected Norwegian petroleum production 1986–2000

Source: Bergen Bank, *Petroleum Activities in Norway 1987* (BB, Bergen, Norway, April 1987), p. 10. With permission.
Note: Overall output is expected to increase until about 1994, then remain relatively stable until beyond the year 2000.

third largest gas exporter.[144] Martin Saeter believes that, within a matter of decades, 'Norway will probably be the sole gas-exporting country in Western Europe,' which will put it in a strong bargaining position.[145] On the other hand, an integration of Norway's gas production and markets, its anticipated crude oil and gas production (Figure 12.4), proved reserves and proximity to large markets in Europe, give it considerable geo-political leverage that should continue well into the next century.

Thus the US must learn to live with the reality that when the world sees Norway (a small, NATO-allied state) as both sympathetic to OPEC and practising (what is for the US) a capitalistic heresy of state-ownership of a portion of an important industry like oil, it diminishes the US's inter-national stature. Furthermore, considering Norway's long-standing support of NATO, to accuse Norway of being a less-than-loyal ally whether in matters of defence or in general foreign political questions is grossly unfair. Is not each state entitled to be master of its own mineral storehouse?

Although Norway somewhat accelerated development of its petroleum industry during the early and mid-1980s, its future policy is likely to parallel that of the past – a slow to moderate pace. In late 1987 Norway's Ministry of Petroleum and Energy was putting pressure on lessees to postpone or reduce annual investments in new field developments. The ministry fears that some 95 per cent of Norway's proved oil reserves will have been committed for development prior to 1995, with 75 per cent of these reserves being depleted by the year 2000. Saga (for its Snorre field), Shell (Draugen), Conoco (Heidrun), Norsk Hydro (Brage) planned to submit new work programmes during September–December 1987. But the government has warned the operators that the Storting during 1987 is unlikely to approve new development plans and may, in fact, establish a staggered development ('queuing') programme. Saga has warned that 'unless the Snorre project is approved before the summer of 1988, it will have to be postponed for at least five years on economic grounds.[146]

Many Norwegians have argued (and some continue to argue) that the country is going forward too quickly in its petroleum development pro-grammes and that, during times of depressed markets, the petroleum may be more valuable left under the sea.[147] More importantly they fear potential economic and social dislocations associated with too rapid a development of the industry in a country with only 4.2 million people. According to Mallakh, Noreng, and Poulson, the direct demand for labour by the petroleum industry has been viewed as a 'minor problem but the demand for labor resulting from the domestic use of oil revenues was seen as a major problem'.[148] And, indeed, some labour and industrial dislocations in other sectors of the economy have occurred.[149] But because of massive governmental subsidies (Figure 12.5) made possible by petroleum revenues, the primary industries, and labour-intensive manufacturing industries seem not as severely affected as many parties anticipated.[150] The question is, of course, from where will these subsidies come when the oil spigot runs dry?

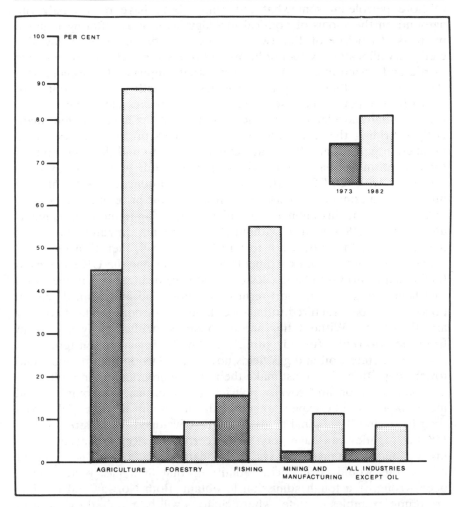

Figure 12.5 Per cent of factor income received from the state by specific sectors of the Norwegian economy in 1973 and 1982

Source: A. Utne, 'Norway: gains and strains in an oil economy', *EFTA Bulletin*, vol. 24, no. 4 (1983), p. 9. With permission.
Note: Factor income is defined as income to the production factors – labour and capital. Governmental financial support of factor income for all industries (other than petroleum) tripled from about 3 per cent in 1973 to nearly 9 per cent in 1982.

Conclusions

Both Norway and the UK have much at stake in the future of offshore oil and gas. But Norway's well-developed hydro-electric power industry allows more room to manoeuvre in her policies, whereas the UK's heavy dependence on petroleum energy somewhat limits her options, if she is to minimise petroleum imports.[151]

Although Norway and the UK have approached their development of offshore petroleum somewhat differently, both have occasionally run aground on the shoals of political ideology, as well as differing regional interests. Criticism of the two countries' continuing use of the discretionary allocation system of licensing (as based merely on bureaucratic inertia and vested interests) seems somewhat simplistic. In essence, both states must find that this system provides them with the most control of their offshore petroleum resources, even if this means sacrificing some of the potentially available economic rent. Both states have profited financially by riding the 'coat-tails' of OPEC via its pricing and controlled production policies and both have experimented to find the right policies for maximising private gain and the public good, as in Norway's co-operation with OPEC. Although Norway's production restraints are intended as a temporary measure, if once again the price of a barrel of oil falls far enough, this anomaly in policy could become more permanent, and even the UK – with or without the Iron Lady, Margaret Thatcher – may be forced to bend, if not to OPEC's siren call, then to necessity.

During the process of experimentation, Norway and the UK have given (and taken from) OPEC a few lessons, and have demonstated to the world that licensing, taxation, short-term production, and long-term depletion policies must be structured to balance the needs of both the host country and the lessee. Without this accommodation, neither party can profit from the *oljeeventyr* (oil adventure), as the Norwegians often term it.

As these states' oil and gas fields now in production mature and trend toward depletion, they must make their leasing programmes more flexible and their taxation and royalty policies less rigorous, less complex, and more stable, if they expect to attain the maximum lifetime yield from their fields and if marginal fields are to be developed. Hann stressed that UK tax regimes can have very negative consequences for incentives to invest, but to change the system fundamentally means 'government bureaucrats would be implicitly admitting' that their past policies have been wrong,[152] a tough admission to obtain. Both Norway and the UK are prime examples of states where changes will be required by the turn of the century. Perhaps they can then gain a little more time to prepare for what Jens Christian Hansen has called 'life after oil', a time 'no Norwegian [Briton] dares to talk about'.[153]

Notes

1 Ø. Noreng, *The Oil Industry and Government Strategy in the North Sea* (Croom Helm, London, 1980), p. 19.
2 ibid., pp. 15–18.
3 I. Kuczynski, *British Offshore Oil and Gas Policy* (Garland Publishing, NY, 1982), pp. 6-11–6-12.
4 British Petroleum Company plc, *BP Statistical Review of World Energy* (BP, London, June 1987), pp. 4, 7, 26–7, 30; where appropriate, data are converted from millions of tonnes of oil equivalent, with 7.285 bbl of oil equal to 1 tonne.

5 For a useful comparison of Norway and the UK's policies in the 1960s and 1970s, see T. C. Daintith and I. W. Gault, 'Oil and gas: national regimes', in C. M. Mason (ed.), *The Effective Management of Resources* (Frances Pinter, London, 1979), pp. 53–70; for a discussion of the significance of OPEC to the emergence of the North Sea as an oil-producing region, see P. R. Odell, 'The economic background to North Sea oil and gas development', in M. Saeter and I. Smart (eds), *Political Implications of North Sea Oil and Gas* (Universitetsforlaget, Oslo, 1975), pp. 51–63.

6 P. Sanderson, 'North Sea oil and the UK economy: boon or bane?', *Barclays Review*, vol. 59, no. 4 (1984), p. 84.

7 'UK seen as net crude oil importer in 1991', *Oil and Gas Journal*, vol. 85, no. 29 (1987), p. 31.

8 K. W. Dam, *Oil Resources: Who Gets What How?* (University of Chicago Press, Chicago, IL, 1976), pp. 24–5.

9 See 'Regulation and taxation of United Kingdom oil and gas exploration and production' (Department of Energy, London, 1987), p. 3 (mimeographed); supplied by Letter: C. Figg, Oil and Gas Division, Department of Energy, London, UK, 24 November 1987.

10 R. Mabro *et al.*, *The Market for North Sea Crude* (Oxford University Press, Oxford, 1986), p. 94.

11 D. C. Watt, 'Britain and North Sea Oil: policies past and present', *Political Quarterly*, vol. 47, no. 3 (1976), pp. 378–9, 382–3. Block sizes may vary, because they are defined by 1° of latitude and 1° of longitude, so that they become smaller to the north.

12 For an excellent critique of the early licensing rounds, examine K. Chapman, *North Sea Oil and Gas: A Geographical Perspective* (David & Charles, London, 1976), especially pp. 40–142.

13 D. I. MacKay and G. A. Mackay, *The Political Economy of North Sea Oil* (Westview Press, Boulder, CO, 1975), pp. 25, 28.

14 T. I. Williams, *A History of the British Gas Industry* (Oxford University Press, Oxford, 1981), pp. 159–60.

15 Dam, *Oil Resources*, pp. 28–9.

16 Mabro *et al.*, *Market for North Sea Crude*, p. 94.

17 'Regulation and taxation, p. 3.

18 Mabro *et al.*, *Market for North Sea Crude*, p. 94.

19 'Riding out the storm: United Kingdom', *World Oil*, vol. 205, no. 2 (1987), p. 48.

20 C. Robinson, 'The errors of North Sea policy', *Lloyds Bank Review* (July 1981), pp. 21–2.

21 D. Hann, 'Political and bureaucratic pressures on UK oil taxation policy', *Scottish Journal of Political Economy*, vol. 32, no. 3 (1985), p. 285.

22 Dam, *Oil Resources*, pp. 40, 42. The UK Government has also intervened directly in an effort to assist its domestic offshore supply industry to compete with established, multinational oil-corporation-based industry. Government personnel have audited the multinationals' supply purchasing activities, an action that also helps break their monopoly of oil field development information; see M. Jenkin, *British Industry and the North Sea: State Intervention in a Developing Industrial Sector* (Macmillan, London, 1981), especially pp. 212–14.

23 MacKay and Mackay, *Political Economy of North Sea Oil*, p. 25.

24 Robinson, 'Errors of North Sea policy', pp. 21–2.

25 Dam, *Oil Resources*, pp. 5–7, 49.

26 Robinson, 'Errors of North Sea policy', pp. 21–2; C. Rowland and D. Hann, *The Economics of North Sea Oil Taxation* (St Martin's Press, NY, 1987), p. 13.

27 Chapman, *North Sea Oil and Gas*, pp. 85–7.
28 Dam, *Oil Resources*, pp. 48–9.
29 J. C. B. Cooper, 'The oil industry of Scotland', *Scottish Bankers Magazine*, vol. 76, no. 303 (1984), p. 88.
30 Hann, 'Political and bureaucratic pressures on UK oil taxation policy', pp. 283–4.
31 MacKay and Mackay, *Political Economy of North Sea Oil*, p. 37.
32 Mabro *et al.*, *Market for North Sea Crude*, pp. 111–13.
33 ibid., p. 114.
34 Letter: M. J. Woodage, Oil and Gas Division, Department of Energy, London, UK, 30 October 1987.
35 A. C. Ross, 'The United Kingdom's experience with North Sea oil and gas', *IDS Bulletin*, vol. 17, no. 4 (1986), pp. 42–7; according to M. P. Devereux and C. N. Morris, there is considerable agreement among tax specialists that a simple flow-of-fund CT alone, instead of the PRT and CT combination, would better meet the government's objectives and reduce distortions in company investment, financing and output decisions; see M. P. Devereux and C. N. Morris, *North Sea Oil Taxation: The Development of the North Sea Tax System*, Report Series no. 6 (Institute for Fiscal Studies, London, 1983), p. 62. C. Rowland and D. Hann see the government's objectives as receiving an early tax return, a 'fair' overall field lifetime tax take and convincing the public that it is 'in charge' of the oil sector; see Rowland and Hann, *The Economics of North Sea Oil Taxation*, pp. 108–9.
36 For a detailed discussion of the PRT, see D. Bland, *UK Oil Taxation* (Oyez-Longman, London, 1984), pp. 31–122.
37 Marbo *et al.*, *Market for North Sea Crude*, pp. 117–18.
38 M. Quinlan, 'Companies press for urgent action', *Petroleum Economist*, vol. 53, no. 6 (1986), pp. 205–6.
39 A. G. Kemp and D. Rose, 'Impact of budget proposals', *Petroleum Economist*, vol. 54, no. 5 (1987), p. 169; see also 'Oil taxation', *Finance Act 1987*, chap. 16, pt V (HMSO, London, 15 May 1987), pp. 45–7.
40 ibid.
41 ibid., p. 170.
42 G. Manners, 'North Sea oil: benefits, costs and uncertainties', *Geoforum*, vol. 15, no. 1 (1984), p. 60; see also E. L. Lynk and M. G. Webb, 'An unintended consequence of the taxation system for UK North Sea oil', *Scottish Journal of Political Economy*, vol. 33, no. 1 (1986), pp. 58–73.
43 Hann, 'Political and bureaucratic pressures on UK oil taxation policy', p. 284.
44 C. Robinson, 'Oil depletion policy in the United Kingdom', *Three Banks Review*, no. 135 (September 1982), p. 5. F. Atkinson and S. Hall have suggested that the UK's 'soundest depletion policy would seem to be one of preventing production from rising much above the self–sufficiency level while encouraging exploration and production over the longer term'; see F. Atkinson and S. Hall, *Oil and the British Economy* (Croom Helm, Beckenham, 1983), p. 9.
45 Mabro *et al.*, *Market for North Sea Crude*, pp. 95–8.
46 Manners, 'North Sea oil', p. 60.
47 C. Robinson, 'Errors of North Sea policy', pp. 22–31. J. Carter, a London-based international energy consultant, claims that the BGC's accounting system does not allow an easy determination of profitable and unprofitable sectors of its activities; see J. Carter, 'The changing structure of energy industries in the United Kingdom', in J. M. Hollander, H. Brooks, and D. Sternlight (eds), *Annual Review of Energy*, vol. 11 (Annual Reviews, Palo Alto, CA, 1986), pp. 451–69.
48 Woodage, Letter, 30 Oct. 1987.

49 Dam, *Oil Resources*, p. 9.
50 Mabro *et al.*, *Market for North Sea Crude*, pp. 99–100.
51 ibid., pp. 100–1.
52 G. Corti and F. Frazer, *The Nation's Oil: A Story of Control* (Graham & Troutman, London, 1983), especially pp. 120–48.
53 Dam, *Oil Resources*, p. 4.
54 F. C. F. Earney, *Petroleum and Hard Minerals from the Sea* (Edward Arnold, London, 1980), p. 75.
55 Letter: B. Nelsen, Fulbright Scholar, Norsk Utenrikspolitik Institutt, Oslo, Norway, 6 November 1987.
56 Earney, *Petroleum and Hard Minerals from the Sea*, p. 75.
57 Mabro *et al.*, *Market for North Sea Crude*, p. 101.
58 'Britain', *OPEC Bulletin*, vol. 14, no. 8 (1983), p. 35.
59 See R. Mabro, 'Significance of abolishing British National Oil Corporation', *Financial Times* (21 March 1985), n.p., as quoted in *OPEC Bulletin*, vol. 16, no. 3 (1985), pp. 33, 79.
60 Mabro *et al.*, *Market for North Sea Crude*, p. 107.
61 ibid., pp. 105–8; for an examination of BNOC as an ideological entity and how political expediency may have overshadowed economic efficiency in its operation, see D. Hann, 'The process of government and UK oil participation policy', *Energy Policy*, vol. 14, no. 3 (1986), pp. 253–61.
62 Sanderson, 'North Sea oil and the UK economy', p. 84.
63 ibid., pp. 85, 89.
64 ibid., pp. 86–7, 89; for a detailed examination of the relationships of the UK's offshore petroleum industry to the country's industrial structure see Atkinson and Hall, *Oil and the British Economy*, pp. 83–100.
65 A. G. Kemp, C. P. Hallwood, and P. W. Wood, 'The benefits of North Sea oil', *Energy Policy*, vol. 11, no. 1 (1983), p. 130.
66 See A. Marr, *Foreign Nationals in North-East Scotland*, North Sea Study Occasional Paper no. 7 (Department of Political Economy, University of Aberdeen, Scotland, 1975).
67 W. Elkan and R. E. D. Bishop, 'North Sea oil: responses to employment opportunities', *Energy Economics*, vol. 7, no. 2 (1985), pp. 127–33.
68 H. D. Smith, A. Hogg, and A. M. Hutcheson, 'Scotland and offshore oil: the developing impact', *Scottish Geographical Magazine*, vol. 92, no. 2 (1976), pp. 74–91.
69 In 1986 K. Chapman estimated that, as of 1985, those employed in North Sea oil-related work in Scotland totalled 63,500, excluding engineering firms that sell the oil industry part of their output; see K. Chapman, 'There's too much oil!', *Geographical Magazine*, vol. 58, no. 8 (1986), p. 378.
70 P. L. Baldwin and M. F. Baldwin, *Onshore Planning for Offshore Oil: Lessons from Scotland* (Conservation Foundation, Washington, DC, 1975), p. 112.
71 For a more detailed discussion of the impact of petroleum on Scotland, examine I. R. Manners, *North Sea Oil and Environmental Planning: The United Kingdom Experience* (University of Texas Press, Austin, TX, 1982); T. M. Lewis and I. H. McNicoll, *North Sea Oil and Scotland's Economic Prospects* (Croom Helm, London, 1978).
72 British Petroleum Company plc, *BP Statistical Review of World Energy* (BP, London, June 1987), pp. 2, 21.
73 Sanderson, 'North Sea oil and the UK economy', p. 89.
74 G. C. Band, 'UK North Sea production prospects to the year 2000', *Journal of Petroleum Technology*, vol. 39, no. 1 (1987), p. 64.
75 J. Stern, 'After Sleipner: a policy for UK gas supplies', *Energy Policy*, vol. 14, no. 1 (1986), pp. 9–14.

76 Sanderson, 'North Sea oil and the UK economy', pp. 88–9.
77 B. S. Aamo, 'Norwegian oil policy: basic objectives', in M. Saeter and
   I. Smart (eds), *The Political Implications of North Sea Oil and Gas*
   (Universitetsforlaget, Oslo, 1975), pp. 81–92.
78 Noreng, *Oil Industry and Government Strategy in the North Sea*, p. 44.
79 R. El Mallakh, Ø. Noreng, and B. W. Poulson, *Petroleum and Economic
   Development: The Cases of Mexico and Norway* (D. C. Heath, Lexington,
   MA 1984), pp. 86–8, 112.
80 Central Bureau of Statistics, *Norges Offisielle Statistikk B 690, Statistical
   Yearbook of Norway 1987* (CBS, Oslo, 1987), pp. 234–5, and *Statistikk B
   388, Statistical Yearbook of Norway 1983*, pp. 134–5.
81 Letter: A. Holt-Jensen, Geografisk Institutt, Universitetet i Bergen, Bergen,
   Norway, 9 November 1987.
82 Mabro *et. al.*, *Market for North Sea Crude*, p. 135.
83 F. C. F. Earney, 'Norway's offshore petroleum industry', *Resources Policy*,
   vol. 8, no. 2 (1982), p. 136.
84 El Malllakh, Noreng, and Poulson, *Petroleum and Economic Development*,
   p. 120.
85 Mabro *et al.*, *Market for North Sea Crude*, p. 135.
86 Ministry of Industry and Crafts, *Petroleum Exploration North of 62° N*,
   Report no. 91 to the Storting (1975–6)(MIC, Oslo, 1976); for a helpful study
   of the potential environmental effects of petroleum production in the North
   Sea, examine D. W. Fischer, *North Sea Oil: An Environment Interface*
   (Universitetsforlaget, Bergen, Norway, 1981).
87 Norwegian Petroleum Directorate, *Petroleum Outlook 1986* (Stavanger,
   1985), pp. 1–8.
88 Earney, 'Norway's offshore petroleum industry', p. 136.
89 Bergen Bank, *Petroleum Activities in Norway 1987* (BB, Bergen, Norway,
   April 1987), p. 38.
90 ibid., pp. 39, 41.
91 'Beyond the Norwegian 62nd Parallel: ever further north', *Oil and
   Enterprise: The Oil and Gas Review*, no. 25 (August 1985), p. 39.
92 Holt-Jensen, Letter, 9 November 1987.
93 Bergen Bank, *Petroleum Activities in Norway*, p. 43.
94 Mabro *et al.*, *Market for North Sea Crude*, p. 149.
95 Bergen Bank, *Petroleum Activities in Norway 1987*, pp. 42–3.
96 Mabro *et al.*, *Market for North Sea Crude*, p. 149.
97 ibid., p. 155; for a helpful discussion of the effects of taxation on petroleum
   exploration in Norway, see S. D. Flåm and G. Stensland, 'Exploration and
   taxation: some normative issues', *Energy Economics*, vol. 7, no. 4 (1985),
   pp. 237–40.
98 Bergen Bank, *Petroleum Activities in Norway 1987*, p. 42.
99 'Tax reforms for oil companies on Norwegian Shelf', *Norinform*, no. 28
   (2 September 1986), p. 2.
100 Mabro *et al.*, *Markets for North Sea Crude*, p. 156.
101 'Norway's tax plan sparks opposition', *Oil and Gas Journal*, vol. 84, no. 30
    (1986), p. 42.
102 'Norway makes no changes in oil taxation', *Norinform*, no. 14 (5 May 1987),
    p. 6.
103 By 31 December 1984 Statoil had held nearly 42 per cent of Norway's total
    originally recoverable oil reserves in fields on stream and under development;
    see S. Myklebust, B. Stenseth, and P. E. Søbye (eds), *North Sea Oil and Gas
    Yearbook: Comparative Statistical Analysis, 1986* (Central Bureau of
    Statistics of Norway and Universitetsforlaget AS, Stavanger, 1987), p. 68.
104 'State oil power in Norway', *Petroleum Economist*, vol. 53, no. 2 (1986), p. 38.

105 ibid., p. 38.
106 'Phillips gives up field stake', *Petroleum Economist*, vol. 53, no. 4 (1987), p. 143.
107 'Tommeliten on go', *Norwegian Oil Review*, vol. 12, no. 6 (1986), p. 55.
108 'State oil power in Norway', p. 38. MacKay and Mackay stressed this point (relative to the UK) more than a decade ago; see MacKay and Mackay, *Political Economy of Oil*, p. 141.
109 'Tommeliten on go', p. 55.
110 'Statoil buoyant', *Norinform*, no. 28 (17 September 1985), p. 5.
111 'Statoil profits drop', *Norinform*, no. 10 (17 March 1987), p. 3.
112 'Statfjord C on Stream', *Norinform*, no. 22 (9 July 1985) pp. 2–3.
113 Mabro *et al.*, *Market for North Sea Crude*, p. 145.
114 'Statoil overspend stirs political strife', *Norinform*, no. 31 (6 October 1987), pp. 1–2. The controversy resulted in the replacement of Statoil's head, Arve Johnson, and its board; see 'Harald Norvik new Statoil chief', *Norinform*, no. 1 (12 January 1988). p. 1.
115 Mabro *et al.*, *Market for North Sea Crude*, p. 145.
116 'Statoil buys up Dansk Esso A/S', *Norinform*, no. 15 (6 May 1986), p. 8.
117 'Statoil plans gigantic gas storage facilities on continent', *Norinform*, no. 33 (7 October 1986), p. 4.
118 For an outstanding critique of the role of Statoil and its relationship to the state in the development of the offshore petroleum industry, see M. G. Visher and S. O. Remøe, 'A case study of a cuckoo nestling: the role of the state in the Norwegian oil sector', *Politics and Society*, vol. 13, no. 4 (1984), pp. 321–41.
119 F. Laursen, 'Security aspects of Danish and Norwegian law of the sea policies', *Ocean Development and International Law*, vol. 18, no. 2 (1987), p. 216.
120 For a discussion of the IEA's objectives, consult 'International Energy Agency (IEA) Ministerial Communiqué', *OECD Observer*, vol. 36 (September 1985), pp. 33–7.
121 J. C. Ausland, *Norway, Oil and Foreign Policy* (Westview Press, Boulder, CO, 1979) pp. 32–4.
122 'Norway will not cut oil production yet', *Norinform*, no. 34 (29 October 1985), p. 3.
123 'Door to OPEC ajar', *Norinform*, no. 17 (20 May 1986), p. 1.
124 'Cut in Norwegian oil exports', *Norinform*, no. 30 (16 September 1986), p. 1.
125 'Norway trims oil output', *OPEC Bulletin*, vol. 17, no. 8 (1987), p. 18. Some energy specialists recommend that the UK shoulder part of the responsibility for stabilising oil prices by cooperating with OPEC; see J. Duggan, 'Now is the time for the United Kingdom to support OPEC', *OPEC Bulletin*, vol. 18, no. 1 (1987), p. 4.
126 R. Vielvoye, 'UK production slippage', *Oil and Gas Journal*, vol. 85, no. 30 (1987), p. 28.
127 'Norway curbs oil production'. *Norinform*, no. 2 (20 January 1987), p. 1.
128 P. Dempsey, 'Norwegian initiative', *World Oil*, vol. 204, no. 5 (1987), p. 23.
129 'Will legal problems foil Norway's oil initiative?', *Norinform*, no. 29 (9 September 1986), p. 1.
130 Dempsey, 'Norwegian initiative', p. 23.
131 'Norway would like early meeting with OPEC', *Norinform*, no. 26 (1 September 1987), p. 2.
132 'Four thousand jobs lost in the North Sea', *Norinform*, no. 26 (19 August 1986), p. 2.
133 'Oil revenues may be decimated'. *Norinform*, no. 25 (15 July 1986), p. 2; 'Oil price fall threatens state revenue', *Norinform*, no. 3 (28 January 1986), p. 1.

134 In 1986 the break-even point for large, existing fields was $12 to $15/bbl, but at this price (then and in 1987) little new investment could occur 'and a major part of the present oil and gas reserves would become uneconomic'; see Ø. Noreng, 'Energy policy and prospects in Norway', in J. M. Hollander, H. Brooks, and D. Sternlight (eds), *Annual Review of Energy*, vol. 11 (Annual Reviews, Palo Alto, CA, 1986), pp. 394–5.

135 'The American dimension and Norwegian oil policy', *Norwegian Oil Review*, vol. 12, no. 6 (1986) p. 10.

136 C. F. Doran, *Myth, Oil and Politics: Introduction to the Political Economy of Petroleum* (The Free Press, NY, 1977), pp. 42, 44–5.

137 For an examination of the Arab viewpoint, see A. al-Sowayegh, *Arab Petropolitics* (St Martin's Press, NY, 1984).

138 'Troll Field – the gateway to a new oil era in Norway', *Norinform*, no. 2 (18 January 1983), p. 1; 'Troll contract – Norway's biggest ever', *Norinform*, no. 20 (10 June 1986), pp. 1–2.

139 J. P. Stern, 'Norwegian Troll Gas: the consequences for Britain, continental Europe and energy security', *World Today*, vol. 43, no. 1 (1987), pp. 1, 3.

140 See, for example, 'North Sea strike affects Norway's treasury', *Norinform*, no. 13 (22 April 1987), p. 7.

141 R. Vielvoye, 'Norway's Labor dispute', *Oil and Gas Journal*, vol. 84, no. 15 (1986), p. 59.

142 Stern, 'Norwegian Troll gas', p. 3.

143 For a good analysis of the varying perceptions of Norwegians and Norway's allies relative to Norway's petroleum policies, see W. Østreng, 'Norwegian petroleum policy and ocean management: the need to consider foreign interests', *Cooperation and Conflict*, vol. 17, no. 2 (1982), pp. 117–38.

144 Ø. Noreng, 'Energy policy and prospects in Norway', p. 395.

145 M. Saeter, 'Natural gas: new dimensions of Norwegian foreign policy', *Cooperation and Conflict*, vol. 17, no. 2 (1982), pp. 139–50.

146 'Norway edges towards staggered field developments', *OPEC Bulletin*, vol. 18, no. 8 (1987), pp. 57–8.

147 A decade ago, K. Chapman suggested that hydrocarbons are much more important to us as industrial raw materials than as fuels and should not be squandered unnecessarily; see K. Chapman, 'North Sea hydrocarbons too precious to burn', *Geographical Magazine*, vol. 50, no. 5 (1978), pp. 492, 494, 498.

148 El Mallakh, Noreng, and Poulson, *Petroleum and Economic Development*, p. 89.

149 A. Holt-Jensen, 'Norway and the sea: the shifting importance of marine resources through Norwegian history', *GeoJournal*, vol. 10, no. 4 (1985), pp. 398–9; A. Utne, 'Norway: gains and strains in an oil economy', *EFTA Bulletin*, vol. 24, no. 4 (1983), pp. 9–10.

150 By 1981, some 20–25 per cent of the total employment in Norway was subsidised by the state; see Visher and Remøe, 'A case study of a cuckoo nestling', p. 234. Governmental policies have worked to discourage labour's migration from one location, industry, and employer to another, so much so that it may have been a detriment to the country; see W. Galenson, *A Welfare State Strikes Oil: The Norwegian Experience* (University Press of America, Lanham, MD, 1986), pp. 43–7.

151 See T. Lind and G. A. Mackay, *Norwegian Oil Policies* (C. Hurst, London, 1980), pp. 6–7.

152 Hann, 'Political and bureaucratic pressures on UK oil taxation policy', pp. 292, 294.

153 J. C. Hansen, 'Regional policy in an oil economy: the case of Norway', *Geoforum*, vol. 14, no. 4 (1983), p. 361.

Chapter thirteen

The geopolitics of offshore petroleum confrontation, conflict and co-operation

A general acceptance of the enclosure of the world's offshore for EEZs established by a majority of coastal states presents many challenges – challenges of confrontation, conflict, and co-operation. As the use of oceanic minerals expands in coming years, these challenges will emerge internationally, regionally, and locally. Indeed, they are already doing so.

Geopolitics pervades mineral industry analyses at all areal scales, as in the international LOS deep seabed mining debates and the US's inter-agency EEZ management disputes. Mineral geopolitics concerns the politicisation of economic relationships and the use of legal and economic power, even military force, to acquire or to deny access to mineral resources. Offshore petroleum activities vividly exemplify these processes.[1] Successes or failures in resolving offshore petroleum geopolitical disputes can have a major impact on whether and how these resources are developed. Furthermore, the locus of many contemporary offshore boundary problems associated with important fishing areas, transportation routes, and strategic military zones coincides with that of proved or potential offshore petroleum resources. This chapter examines the inter-relationships of these variables as they occur in various areas of the world's oceans.

International geopolitics

The geopolitics of offshore minerals at the international level often overshadows national and local events, with disputes sometimes de-generating from confrontation to conflict. On the brighter side, states are also initiating peaceful efforts to discover, develop, and manage offshore petroleum resources via joint-development programmes. The People's Republic of China (PRC or China) and its neighbours exemplify both approaches – the bellicose and the conciliatory.

The PRC and its neighbours

China is committed to developing its offshore petroleum resources, initially estimated to range from a modest 30,000 million bbl[2] to a more optimistic 112,000 million bbl.[3] After a decade of exploration, with only a limited success, many geologists within and outside the PRC would likely place that country's reserves somewhere below the mid-point of these estimates. But many areas remain relatively unexplored, in part, because of the overriding problem of disputed offshore boundaries, a situation

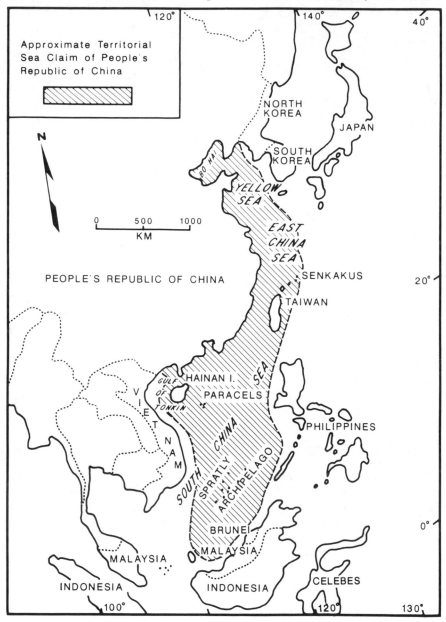

Figure 13.1 Approximate territorial sea claim of the People's Republic of China

Source: Author.

that may make it difficult to attract capital. Paul Yuan noted that 'only when political stability and peace are reasonably assured for a considerable period of time' can 'there be any meaningful, long-term commitment of foreign capital in the development of petroleum resources in China's offshore seabeds'.[4]

According to the PRC, Chinese peoples (since at least 200 BC)[5] have had a long-standing maritime political and commercial influence in waters adjacent to their southern and eastern shores. Thus the PRC views these waters as Chinese. Vietnam makes similar claims. Both states' viewpoints have become more apparent as the petroleum potential of these waters has been demonstrated. The PRC claims South China Sea waters to within only about 40km of the shores of Brunei; the Philippines' Luzon and Palawan Islands; and Malaysia's Sarawak and Sabah provinces. Beijing claims all waters surrounding Taiwan, because it considers the island an integral part of the PRC. Taiwan, Vietnam, and the Philippines also lay claim to numerous island groups within the South China Sea. With the growing interest of these littoral neighbours in providing petroleum for their expanding economies and burgeoning populations, the potential for confrontation, conflict, and co-operation increases (Figure 13.1).

Yellow and East China Seas

Based on Beijing's public posturing, tensions in the Yellow and East China Seas region occur because of

1 differing interpretations of how the outer limits of territorial sovereignty should be determined;
2 Japan and South Korea's co-operation in joint offshore petroleum exploration activities near and in disputed waters;
3 the defection of Taiwan from mainland China;
4 the continued presence of the western-supported government of South Korea.[6]

The PRC and South Korea are the main protagonists in the Yellow Sea disputes. Because neither state (as of mid-1987) has declared an EEZ, their arguments for establishing offshore boundaries hinge on traditional principles. South Korea claims Yellow Sea waters out to a median line, a claim the PRC rejects. Beijing opts (unofficially) for the principle of natural prolongation based on silt deposition by China's river systems, as the Hwang Ho. Adherence to this principle puts the PRC's outer limit well to the east of a median line.[7] Even though South Korea has granted numerous exploration concessions within its claimed area, Beijing and Seoul have not pursued boundary negotiations. Indeed, the PRC has never published a formal document on its claim and the basis for it;[8] to do so would confirm the legitimacy of the South Korean Government.[9] In the Korean Strait, lying between Japan and South Korea, Seoul also applies the median-line principle, but to the south-west (in the north-eastern East

China Sea), it uses the natural-prolongation principle; Japan insists upon a median line in this area, and uses some very small islands (Danjo Gūnto and Tori Shima) as the base from which the median line is measured. The South Koreans and Japanese have temporarily solved their boundary agreement impasse by forming a Japan-Korea Joint Development Zone (JDZ), including waters within and adjacent to their disputed areas. Because South Korea uses both the median-line and natural-prolongation principles, the PRC ridicules its claims as opportunistic, noting that South Korea uses whatever principle gives it the greatest advantage.[10] No immediate delimitation of precise boundaries in this region is likely because drilling programmes here have had little success, making the boundary claims – for the time being – relatively unimportant. This statement cannot be made for waters farther to the south in the East China Sea, as in the vicinity of the Senkaku (Tiao-yü T'ai) Islands.

Although ownership of the Senkaku Islands has long been disputed, the issue became more intense after a 1968 UN-sponsored research effort showed good prospects for petroleum resources in the region.[11]

The disputants (Japan, South Korea, and PRC) use different premises for making their claims. Taiwan contends that the uninhabited eight-island Senkaku group is part of Taiwan's continental shelf and that the 1,000-m-deep Okinawa Trough separates Japan's continental shelf from the Senkakus. Taiwan also notes that Japan, from 1895 to the Second World War, included the Senkakus as part of its Formosa (Taiwan) Prefecture. Beijing claims Taiwan and its possessions are an integral part of China's territory. Both Taiwan and the PRC argue that the Senkakus are part of the natural prolongation of their continental shelf area. They also draw upon history, insisting that Chinese fishermen and merchantmen, as early as the fifteenth century, used the Senkakus as a haven from storms. Beijing also insists that, under the 1952 Sino-Japanese peace treaty, all territories ceded by China to Japan in the 1895 Treaty of Shimonoseki were to be returned to the PRC and that under the 1943 Cairo Declaration, 'all territories stolen from China after 1914 would be returned by Japan, specifically including Formosa (Taiwan) and by implication the adjoining Senkakus'. The Japanese argue that through Japan's 'prescriptive (i.e. uncontested) ownership after the 1895 treaty' the PRC has no legal claim to the islands. It also points out that 'the Cairo Declaration referred to territories acquired after 1914 (whereas Japan obtained title in 1895)' and that the US included the islands as part of the Okinawa reversion to Japan in 1971. The PRC first claimed the Senkakus in December 1970, after a Nationalist gunboat planted a flag there in the previous September[12] and both Japan and Taiwan had granted oil exploration concessions in nearby waters. In 1971 Beijing made a formal claim to the Senkakus when the US transferred sovereignty over Okinawa to Japan.[13]

To emphasise the merits of its claim to the Senkakus, the PRC (in April 1978) marshalled some 200 fishing boats that violated the islands' territorial waters – an act of 'banditry', according to the Taiwan Government.[14] Because the Japanese and PRC Governments were near conclusion of another Sino-Japanese peace treaty (signed 12 August 1978), the PRC temporarily backed off.[15] Susumu Awanohara, in an article for the *Far Eastern Economic Review*, suggested that Beijing's action may have been, in part, a prod for Japan to get on with concluding its peace and friendship treaty negotiations with the PRC.[16] This 'fishing-boat diplomacy' was also a response to the imminent start-up of an exploration programme under the JDZ agreement (made 30 January 1974)[17] between Japan and South Korea, an action castigated by both the PRC and North Korea.[18]

The JDZ agreement provides that Japan and South Korea will co-manage exploration and development of marine petroleum resources in an 83,000 km² area in the east-central East China Sea. Under the JDZ arrangement Japan and South Korea concessionaires share equally any oil discovered. The governments receive no oil. The concessionaires share development costs, and Japan and South Korea receive royalties and taxes from profits, with Japan receiving royalty fees of 1 per cent and South Korea 12.5 per cent. As of mid-1987, however, no commercial petroleum finds had been made. Despite Beijing's protests when Japan and South Korea established the JDZ, the PRC in September of 1979 suggested that Japan and the PRC establish a similar joint-development programme for the Senkaku Islands area. At the time, George Lauriat and Melinda Liu noted, in the *Far Eastern Economic Review*, that the Japanese believed this suggestion was a ploy 'to create a condition of *de facto* joint ownership of the islands'. The Chinese Government's offer included an arrangement similar to one the two states have for offshore petroleum fields in China's Bo Hai Gulf area, whereby Japan provides financial and technical help and receives payment in petroleum.[19]

The problem of offshore boundaries and petroleum exploration and development concessions is linked to the US's relationship to the two Chinas. As the US's relations with Taiwan and mainland China have changed, the situation of petroleum enterprises working in the region has been altered.

Overtures by Washington DC, to Beijing for easing tensions between the US and China and a surprise visitation by the US's National Security Adviser Henry Kissinger to China in August 1971 caused at least two petroleum firms holding Taiwanese exploration concessions in waters near Taiwan to re-evaluate their status. When in the mid-1970s the US Government warned oil companies with Taiwanese concessions that it could not guarantee them protection, Conoco and Gulf Oil stopped work with gas wells that were already producing. These wells lay on the continental side of an undersea canyon situated between Taiwan and the mainland. Furthermore apprehension developed when the US Government intervened in the culmination of a contract between a US oil company and the Taiwan Government to lease a drilling rig for work only

60 km from the Senkakus.[20] At the same time, however, the US Export-Import Bank made a loan of US$11 million to Taiwan's Chinese Petroleum Corporation for its offshore programmes. This ambivalence may have been a reflection of the US's two-Chinas policy or a recognition (as became evident in the late 1970s) that 'Taiwan is not a top priority in Peking'.[21]

Some students of East Asia suggest that the PRC and Taiwan should co-operate to develop petroleum fields lying between them. Currently neither party is likely to risk such a venture, because it might be considered a tacit recognition of Taiwan's dependence or independence, depending on one's viewpoint. On the other hand, the two states are not so isolated from one another as both are prone publicly to claim. For years they have had indirect financial relationships. For example, an oil refinery in Thailand that processes PRC crude is financed by Taiwanese capital.[22]

South China Sea: sabre rattling from Taiwan to Natuna

As if China does not have troubled waters enough with its Yellow Sea and East China Sea neighbours, it has even more difficult problems along its southern shores and within the South China Sea. Within this region, the PRC, Taiwan, Vietnam, Malaysia, Indonesia, and the Philippines make conflicting offshore claims.[23]

South Vietnam had long been at odds with China over control of waters and islands (Paracels and Spratlys) in the South China Sea, as well as over a littoral boundary in the Gulf of Tonkin. Animosities came to a boil in late 1973 when South Vietnam proclaimed, as its territory, all of the Spratly Islands which sprawl over an area of 180,000 km.^{2} This action occurred despite the presence since 1946 of Taiwanese troops on one of the islands (Itu Aba or Taiping) and the previously professed control of seven of the islands by the Philippines.[24] Soon after South Vietnam made its Spratly decree, Beijing issued one of its own (11 January 1984), noting that it would not acquiesce to infringements on its traditional waters.[25]

On 16 January 1974, China put teeth into its pronouncement by invading the Paracel Islands, a group held by South Vietnam. A brief (January 16–20) military confrontation ensued with both sides losing one ship and the PRC occupying the islands and taking forty-eight prisoners, to the displeasure of the South China Sea's littoral states.[26] South Vietnam's successor state (Vietnam) has continued to argue the illegality of China's action, but Beijing has consolidated its position in the islands by constructing buildings, dredging a new harbour, and pursuing exploratory drilling programmes. Woodard and Davenport, in the *Journal of Northeast Asian Studies*, noted that the Paracels make an excellent shield for China's offshore activities farther to the north,[27] and Marwyn Samuels pointed out that Beijing's interest in the Paracels may have been stimulated by a leasing programme begun by Saigon off South Vietnam's southern shore and a similar effort by Hanoi in the Gulf of Tonkin off North Vietnam. More importantly, however, Beijing may have occupied the Paracels to pre-empt Hanoi's doing so once the South Vietnam Government fell.[28]

During 1973 prior to the Paracel Islands conflict, South Vietnam awarded Shell Oil (US) and a three-party consortium (Mobil Oil Vietnam Inc., Kaiyo Oil Company Ltd, and Société Nationale des Petroles Aquitaine)[29] a total of thirteen offshore concessions. Drillers made two potentially exploitable discoveries (about 100 and 200 km) off South Vietnam's south-eastern coast, but when the Saigon Government fell to North Vietnam in 1975, the wells were capped; four months later all concessions were nullified.[30] Although Beijing claimed the Spratlys in a decree issued soon after its occupation of the Paracels, the Hanoi Government in 1975 invaded and occupied the six islands of the archipelago held by South Vietnam.[31] The Philippines complicated the geopolitical situation by occupying seven of the eastern islands and granting (in 1976) exploration concessions in surrounding waters. Vietnam, Taiwan, and the PRC vehemently protested the Philippines' actions. The Philippines followed up on its oil leasing concessions (which resulted in several small oil discoveries in the Reed Bank and Palawan Island areas) with the construction of an airstrip on the Spratly Archipelago island of Pag-asa and a naval base on its island of Palawan. Taiwan has, also, made at least two gas condensate discoveries off its south-western coast.[32]

More recently a series of territorial disputes has arisen between Vietnam and Indonesia in waters north of Indonesia's Natuna Islands. Perhaps Hanoi has had its offshore ambitions whetted by its first commercial offshore production in the Bach Ho (White Tiger) field, situated 160 km south-east of Saigon (now Ho Chi Minh City), a site originally explored by the Mobil Oil-Kaiyo-Aquitaine Consortium. Like all of Vietnam's current offshore petroleum activities, the Bach Ho project is a venture of Vietsovpetro, a joint operator for Vietnam and the USSR. The reserve estimate for Bach Ho is a state secret but it may contain between 20 and 100 million bbl of crude.

Vietnam and China are also confronting one another in an especially sensitive area, the Gulf of Tonkin, which has shown good potential for petroleum. This region is the most politically explosive of all the PRC's continental shelf areas[33] and adjacent seas, because as Willy Østreng, of the Fridtjof Nansen Institute, has so aptly stated, 'no conflict is more irreconcilable than one between ideological brothers'.[34] The two states' antagonism, of course, is also a reflection of a distrust extending over the last thousand years, a distrust that continued even during the height of the Vietnam War.

On 18 January 1974 just after China invaded the South Vietnam-held Paracels, Hanoi, and Beijing agreed to begin talks for a division of the Gulf of Tonkin. In these negotiations Vietnam insisted that the Gulf of

Tonkin boundary should lie along the 108° 3' 13" meridian used in the 1887 Sino-French Convention on the Delimitation of the Frontier between China and Tonkin, which divided the ownership of islands controlled by the two parties. The Chinese feel this meridian comes too close to Hainan Island.

Like most states, China uses different principles (depending on their relative advantages) in arguing for the location of offshore boundaries. In the South China Sea it bases its claim on historical control and in the East China Sea and Yellow Sea, it adheres to historical control and to the natural prolongation of the continental shelf. In the Gulf of Tonkin it has not opted for either principle. Other principles – the median line and deepest water – might also be used here, but if either principle were applied in the north, the boundary would be pushed well to the east of the 108° meridian. Use of a median line in the south would put the boundary to the west of the 108° meridian.[35] The PRC's acceptance of a median line in the Gulf of Tonkin would compromise its Yellow and East China Seas boundary claims, locations where it rejects the median-line principle.

Although the two sides have not delimited a boundary, they have established (in 1974) a neutral zone in the Gulf of Tonkin's central portion (18°–20° N and 107°–108° E), within which neither party is actively to pursue petroleum exploration or development programmes (Figure 13.2).[36] Vietnam accuses China of violating this agreement.

Since 1979 the PRC has had seismic survey contractors working in areas east of the 108° meridian, a programme vigorously opposed by Vietnam,[37] and since 8 January 1981 drilling has been done both by foreign firms and the Chinese Ministry of Petroleum. Several small but productive wells have been developed.[38] Since the early 1980s incidents of harassment and firing on one another's fishing craft and gunboats have demonstrated the volatility of the region; these episodes reflect the larger, ideological and historical disputes between the two countries. Onshore military engagements along the China–Vietnam border, as recently as spring 1987, reflect the potential for further strife in these waters. The USSR's strong presence in the region exacerbates the problem. For example, Vietnam has negotiated a joint exploration agreement with the USSR, to be carried out by Vietsovpetro in areas west of the 108° meridian.[39] During winter the USSR sends mobile exploration rigs and drillships from its Sakhalin Island area to Vietnamese waters.[40]

The western embrace: assets and liabilities

In much the same way as Beijing has recognised the advantages of expanded political relationships with the non-communist world, so has it embraced the potential utility of western technology and operational expertise. This policy change is nowhere better seen than in the PRC's offshore petroleum exploration and development programmes. The PRC's change from a policy of self-reliance to one of seeking outside assistance has speeded the development of its offshore activities.

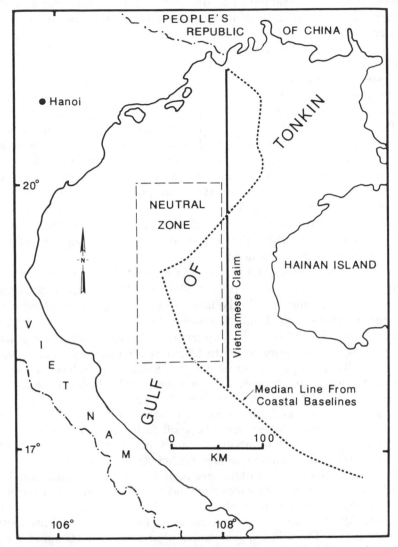

Figure 13.2 Vietnam and PRC Gulf of Tonkin boundary claims and neutral zone

Source: After K. Woodard and A. A. Davenport, 'The security dimension of China's offshore oil development', *Journal of Northeast Asian Studies*, vol. 1, no. 3 (1982), p. 5. With permission.

Beijing began purchasing oil exploration and production equipment in the early 1970s. Early efforts to make purchases from US suppliers ran into difficulties. China's governmental agencies wanted to buy sophisticated magnetometers (based on laser/maser technology). The US refused to sell the equipment because of its potential military applications.[41] In the late 1970s the PRC Government purchased offshore oil rigs.[42] It bought a slightly used semi-submersible, the *Borgny Dolphin*, from a

315

Norwegian firm, for example, and five new rigs from Japanese and US suppliers. Using these units as models, the Chinese soon produced their own deep-water rigs.[43] The Chinese also recognised the need for foreign-assisted offshore exploration and development, which began in earnest during the late 1970s and early 1980s.

The signing of the Sino-Japanese peace treaty on 12 August 1978 was a turning-point for outside participation in the PRC's offshore petroleum activities.[44] As was noted earlier, Japan has co-operated with China in the Bo Hai Gulf since 1980, supplying capital and technology and receiving oil in return. The five-year Sino-Japanese agreement proved so successful that it was extended by two years (to end in May 1987) and the original investment plan was increased from US$210 million to US$600 million. Today, some 10,000km^2 of the Bo Hai's shallows and beaches supply oil and gas for nearby cities and industries, such as fertiliser plants. China's energy officials expect that by 1990 the Bo Hai region (onshore and offshore) will provide perhaps 50 per cent of the country's annual petroleum needs.[45]

The PRC also participates in co-operative offshore programmes in waters along its eastern and southern shores. After extensive domestic and contracted seismic work in the late 1970s and early 1980s, along with some success in the East China Sea and Taiwan Strait areas, the PRC's Chinese National Offshore Oil Corporation (CNOOC) organised a sale for exploration concessions in South China Sea coastal waters. The sale was held in the spring and summer of 1982. Successful bidders, including Amoco, Mobil, Esso, Texaco, Phillips, and ARCO began exploratory drilling in 1983. CNOOC held a second round in 1985, but the oil companies were less enthusiastic because of the paucity of substantial first-round-concession discoveries. Many oil firms remained reluctant to bid, even though CNOOC abolished a 15 per cent royalty on fields producing less than 1 million tonnes (7,285,000 bbl) of oil per year.[46] Thirty-eight companies did obtain seismic data and twenty-three applied for drilling rights. With one exception, bidders in both rounds have not been rewarded by major discoveries.[47]

In 1983, ARCO made a major gas discovery (the Yinggehai field) south of Hainan Island. The field contains estimated reserves of 90,000 million m^3. In 1985 after two years' hard bargaining, ARCO and CNOOC signed a development contract, with production to begin in 1989. The contract called for ARCO to sell the gas to mainland China, with the price pegged to world oil prices. When ARCO signed the agreement, crude prices stood at about US$30/bbl. When the market declined in 1986, the agreement collapsed and renegotiations were begun. China's oil export earnings dropped sharply and reduced its available hard currency to pay for the project, which is expected to cost US$1,000 million mostly for pipeline connections to Hainan Island and the mainland cities of Zhangjiang, Zhuhai, Canton, and Hongkong. China also suffered a devaluation of its currency, a situation that created additional financial problems[48] for developing the Yinggehai gas field.

If China develops this gas field, it must deliver the gas at a price competitive with onshore coal and oil. It must also develop gas-consuming industries such as fertiliser and power plants. These development programmes will require time. Thus Beijing now wants to delay or scale-down the project, whereas ARCO is eager to proceed in order to recover its development costs. Paradoxically ARCO was not 'overjoyed' at the discovery of gas rather than oil, because gas usually provides less return on investments. This problem illustrates how the interests of national governments may not coincide with those of private industry, especially when that industry is foreign-based.[49] It also presents a quandary for China; on the one hand it wants, for the time being, to put development of the Yinggehai field on the 'back burner', but on the other, it wants to display to the world the only major find of its much touted deep-water petroleum exploration effort.

Norway's offshore neighbours

During the last two decades Norway has emerged as a major offshore petroleum producer.[50] It was only in 1962 that the Norwegian Government was first approached (by Phillips Petroleum) for permission to explore Norway's continental shelf. A year later Norway proclaimed sovereignty over its continental shelf resources;[51] the first finds were made in 1969, and by mid-1987 Norway ranked fifth for crude oil[52] and third for natural gas production[53] in the offshore. Drillers are making new discoveries in the North Sea nearly every year. Nevertheless, since 1980 Norway has pushed its exploration frontier out of the North Sea into the Norwegian and Barents Seas,[54] whose total area is about 1.5 million km^2 (more than four times the size of Norway's mainland).[55] This northward push is important economically and politically, both for Norway and for the world.

Norway shares oceanic boundaries with five states – the USSR, the UK, Iceland, Sweden, and Denmark, as well as with Denmark's two external semi-autonomous dependencies, the Faeroes and Greenland (Kalaallit Nunaat). Boundary issues in the North Sea were settled more than two decades ago. Not so in waters of the far north. Norway's northern petroleum frontier, in the Barents Sea adjacent to the USSR, is the focus of an important boundary dispute (Figure 13.3).

Svalbard

The discovery of coal in the early twentieth century on Svalbard (Spitzbergen) set the stage for one present-day dispute. Two decades later, several states interested in the coal deposits met to decide the sovereignty of Svalbard. From this meeting came the 1920 Svalbard Treaty. Under the treaty, Norway received full sovereignty over the cluster of islands and waters lying between 74° and 81°N latitude and 10° and 35°E longitude. The Svalbard Treaty, however, also provided that all

317

contracting parties should have access to carry out economic activities on the islands and in their territorial waters. These activities may be pursued on an equal basis with Norwegians and are not subject to Norwegian taxation. The Soviets (the only current claimant – other than Norway – to Svalbard's coal resources) operate collieries and claim that they also have a legal right to exploit continental shelf resources beyond the territorial sea limit.[56]

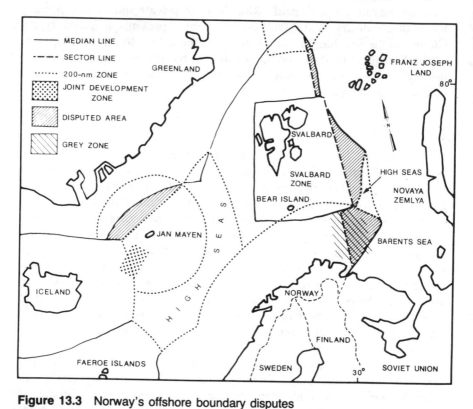

Figure 13.3 Norway's offshore boundary disputes

Source: After Bergen Bank, *Petroleum Activities in Norway 1987* (BB, Bergen, Norway, April 1987), p. 47. With permission.

The Norwegians contest this claim, noting that the treaty makes no mention of the continental shelf. Norway claims these waters are Norwegian, because they lie above an extension of the mainland's continental shelf and thus, under the 1958 GCCS, Norway controls the seabed and its resources, such as petroleum. Although the Soviets may be interested in the Svalbard Islands' oil prospects, they are more concerned for the overall continental shelf's strategic-military importance. In short, the Svalbard region is a gateway to and from the USSR's most important naval base, at Murmansk, on the Kola Peninsula. According to Willy Østreng, 75 per cent of the Soviets' strategic submarines are based here.[57]

A neighbourly stalemate

Since 1974, Norway and the USSR have disputed the ownership of other areas in the Barents Sea. The contested areas cover about 155,000 km^2 and disputed boundaries total more than 1,600km.[58] Waters under contention are of two types – a 'Grey Zone' and 'disputed areas' (see Figure 13.3).

The main focus of the boundary disputes lies north-east of Varanger Fjord. Moscow claims all waters east of the meridian (sector line) 32° 04' 35"E, with the exception of an eastward jog to 35°E in waters adjacent to the south-eastern part of the Svalbard zone. Oslo insists on a median-line boundary extending from Varanger Fjord 200nmi into the Barents Sea where it then follows an irregular pathway, finally joining with the Soviets' sector line to the north-east of Svalbard.

Early on, the two states' concerns focused on fisheries problems in waters north-east of Varanger Fjord. In 1978 after four years of negotiations, the two governments agreed to establish the so-called 'Grey Zone', which straddles the Soviets' sector line, causing an overlapping of the Grey Zone with disputed waters claimed by the Soviets. Norway has long considered these waters as part of its traditional fishing grounds. Under the Grey Zone Agreement a joint Norwegian–Soviet Fishery Commission sets quotas for the various fish stocks. Norway and the USSR police their own fishing boats and any foreign boats under contract to them, even when these craft are in waters claimed by the other party.[59] As occurs each year (on 1 July),[60] despite much controversy within Norway, the Grey Zone Agreement was again renewed in 1987.

Although Barents Sea boundary negotiators have been at work for fourteen years, they have made little progress, despite a December 1986 statement by Soviet Party Chairman Mikhail Gorbachev that the USSR is eager to resolve Barents Sea boundary issues.[61] The lack of progress stems not only from differing views on how the boundaries should be delimited, but also from (1) the sensitivity of the region relative to the USSR's access to its naval base at Murmansk; (2) Norway's NATO membership; and (3) an increasing recognition of the region's potential importance for petroleum. If petroleum should be discovered near or in the Grey Zone or disputed areas, negotiations could become even more difficult. The Norwegians, however, are used to bargaining under adversity, even under a stalemate; in 1983 the Norwegian Government discovered that the Soviets were negotiating from both sides of the political chesstable. It seems a Mr Arne Treholt, a political secretary for the Norwegian Ministry of Trade and a member of Norway's negotiating team, was working for the KGB![62]

The Norwegian Government, for environmental and political reasons, long delayed opening its offshore to petroleum exploration and development north of the 62°N parallel. Despite the political hazards of expanding oil exploration into waters adjacent to those being negotiated with the USSR, the Norwegians since 1984 have conducted seismic surveys in the south-western part of the Barents Sea. Officials of the petroleum firm ARCO Norway feel the seismic data indicate that giant reservoirs may occur in the Barents Sea, but it will require a minimum of 700 million bbl of oil to permit economic exploitation.[63] The Norwegian Government must agree with ARCO's assessment, because in an eleventh round of bidding (during 1987) Norwegian firms were granted six blocks in waters north and east of Hammerfest (see Figure 12.2). The Ministry of Petroleum and Energy (MPE) was to offer six or seven more blocks in the Barents Sea during the autumn of 1987, with bidding to close in January 1988 and awards to occur in April 1988.[64] Some proponents for developing petroleum resources in the region would like to push even farther eastward and allow foreign firms to participate, but so far, the government feels it wise only to admit Norwegian firms into this area.[65] In the twelfth round, however, foreign participation may be allowed.[66] The Storting, Norway's national parliament, has vigorously debated whether to extend exploration into this sens tive region. Unconstrained by public debate within the USSR or the sensitivities of the Norwegians, the Soviets (since the early 1980s) have sent drill ships into Barents Sea waters near the Grey Zone, and in one case, possibly inside it.

Perhaps the boundary issues will become less tense if, as the Norwegian and Soviet Governments hope, a newly formed consortium (BOCONOR A/S), composed of several Norwegian companies to service Soviet petroleum exploration activities in the Barents Sea, works effectively. BOCONOR (Barents Offshore Consortium of Norway) will help operate the fields and supply technical equipment and consumer goods.[67] Oslo hopes that co-operation with the Soviets will further stimulate industrial development in northern Norway,[68] as has occurred recently. In July 1987 two Norwegian firms (A/S Sydvaranger and Sor-Varanger Invest A/S) and a Finnish firm (Wartsila) established an offshore operators' fabrication and repair base (Kimek A.S.) near Kirkenes on Varanger Fjord.[69] Norway's Prime Minister, Gro Harlem Brundtland, during a Kremlin visit in mid-December 1986, suggested that perhaps, in future, private co-operative programmes should be extended to include state participation. Her rationale for this suggestion is that problems will likely arise because of different national regulations and legislation and the need for co-operative efforts in financial, environmental, and security matters. Although Brundtland did not suggest that Norway's government should participate directly with the USSR, her comments generated strong criticism by former Conservative Prime Minister, Kåre Willoch. He warned that intergovernmental co-operation could too easily encourage Soviet demands in other matters, and therefore, co-operative efforts should be limited to private enterprise.[70]

Some observers of Norway–USSR boundary issues suggest that the effort to establish 'absolute' boundaries should be abandoned and a joint-management regime established similar to Saudi Arabia and the Sudan's in the Red Sea[71] or Norway and Iceland's in waters south of Jan Mayen.

Prime Minister Bruntland's recent overtures to the USSR may be a step in this direction. On the other hand, as has been suggested by researchers with a detailed knowledge of the strategic importance of the Barents Sea, extensive development of offshore petroleum (independently or jointly) in the central Barents Sea region may conflict with the military needs of the USSR. Petroleum installations could (1) reduce the manoeuvring space for the Soviets' surface vesels; (2) increase 'background noise', making detection of foreign submarines more difficult, although this noise could obscure movements of their own submarines in and out of the area; and (3) cause the Soviets more apprehension for the region's overall security, if foreign firms participate in development programmes.[72]

Jan Mayen agreement

Rich fishing grounds surrounding Jan Mayen have long attracted seamen from both Norway and Iceland. Thus the two governments' initial jurisdictional negotiations centred on fishery disputes. These were settled on 28 May 1980. But after geologists began suggesting that petroleum resources might exist along the Jan Mayen Ridge that extends southward toward Iceland, the issue focused on claims to the continental shelf. To resolve the issue, the parties appointed a conciliation commission composed of Jens Evensen of Norway, Hans Andersen of Iceland, and Elliot Richardson of the US. The latter served as chairman. In May 1981 the commission suggested the establishment of a joint-development agreement. Norway and Iceland accepted the proposal in October 1981 and it took effect in June 1982.[73]

The joint-development zone lies between 70° 35' and 68°N latitude and 6° 30' and 10° 30'W longitude and includes an area of about 32,750km² on the Norwegian side and about 12,725km² on the Icelandic side of Iceland's 200-nmi EEZ boundary (see Figure 13.3). Under the final agreement, total participation by the two governments in the zone's joint venture petroleum exploration or development contracts would be at least 50 per cent. [74] Norway gets 25 per cent of the shelf resources on the Icelandic side of the 200-nmi EEZ boundary and Iceland receives a similar portion from the Norwegian side. Iceland not only has a potential areal-income advantage but also receives technical help from and has its seismic costs paid by Norway.[75] Although some critics believe Norway has been too generous with Iceland, others note that Iceland has supported Norway's claim to shelf areas surrounding Jan Mayen, a position that helps strengthen Norway's bargaining position with Denmark in establishing a boundary between Jan Mayen and Greenland.[76] But according to Østreng, some observers feel this precedent of compromise may weaken Norway's negotiating position with the USSR over the Barents Sea issue.[77]

From the Aegean Sea to the South Atlantic Ocean

The drama of geopolitical problems of offshore petroleum are not limited to the main actors and arenas of the contemporary international scene.

Figure 13.4 The Greek and Turkish Aegean Sea boundary region

Source: After G. Blake, 'The law of the inland sea', *Focus*, vol. 32, no. 2 (1981), p. 12. With permission.

Lesser known players and locations sometimes take centre stage. These confrontations and conflicts often have the potential for involving the major powers, directly or indirectly. Recent events in both the Aegean Sea and the South Atlantic illustrate the potential dangers when oil discoveries have been made or are anticipated in already troubled waters.

Greece vs Turkey in the Aegean Sea

For centuries Greeks and Turks have faced one another as arch-enemies across the Aegean Sea. The Greeks have not forgotten four centuries of Turkish rule (ending only in 1829). Their memory was freshened in 1974 with Turkey's occupation of northern Cyprus. The Turks, on the other hand, look upon a Greek presence (since just after the Second World War) on small islands within a few km of the Turkish mainland as a thorn in their side. This situation is especially difficult because Greece claims a 22-km territorial sea which effectively blocks Turkey's direct access to open seas in many areas, making the Aegean a 'Greek Lake'.[78]

The territorial dispute has been especially heated during the past fifteen years. During 1973 and 1974 Turkey unilaterally established a *de facto* boundary on the continental shelf by giving exploration concessions to the Turkish Petroleum Company (Figure 13.4). Greece in 1973 discovered petroleum in the northern Aegean. The potential for more oil discoveries in the region heightened tensions between the two countries, making the situation one of the most potentially dangerous in the Mediterranean. By 1976 the two states were on the brink of war. Although war did not materialise, the problem of disputed territorial waters remained, not only because of the potential for petroleum discoveries but also because of Turkey's sensitivity to strategically vital approaches to the Dardanelles.[79]

Consequently when in late March 1987 a Canadian firm – the North Aegean Petroleum Company (NAPC) – announced plans to fulfil a drilling contract made with Greece for work in disputed waters east of the island of Thásos, the Turkish Government sent (under naval escort) its scientific research vessel, the *Sismik I*, to explore for petroleum in waters claimed exclusively by Greece. Each side threatened military action if the other dared to begin exploratory work. The prospect of two NATO members coming to blows brought a vigorous response from the US Government and NATO's Secretary General, Lord Carrington, to mediate. It seems US Secretary of Defense, Caspar Weinberger, was successful (with a promise of increased arms supplies as the 'carrot') in getting the Turks to dampen their bellicose posturing, thus avoiding a violent clash between two unfriendly but political bedfellows, a dangerous and embarrassing situation for the NATO countries.[80]

How might this explosive problem be resolved? Gerald Blake, prior to this latest episode, suggested that the two states could fruitfully establish a joint, revenue-sharing petroleum enterprise,[81] but this would not be an easily achieved solution, given the two parties' long-standing antipathies and this latest confrontation. Rather than pursuing Blake's approach, the Greek Government seems bent on saving face by making NAPC a scapegoat.

It seems the Greek Government wants to give the impression that it had no prior knowledge of NAPC's plan to explore waters east of Thásos. Soon after NAPC's announcement, the government told the firm that it planned to expropriate a majority share of the company, because it had no governmental permission to drill and it is inappropriate for a foreign firm to work in a militarily sensitive area. NAPC insisted that it had received prior approval from the proper state agency (which owns part of NAPC) and it announced plans to go to the courts. By mid-1987 the government had backed off from its expropriation threat, but only after it had, in the previous spring, passed legislation providing for the government to acquire up to 51 per cent ownership in NAPC. Such action would give it a controlling interest in the firm and, therefore, a veto power in drilling-location decisions.[82]

The Malvinas/Falklands Dispute

During the spring and summer of 1982 world radio and television newscasts flashed political lightening and rumbled military thunder as a gathering storm developed over a remote corner of the South Atlantic Ocean. World attention was riveted on a little known group of islands called the Malvinas/Falklands (M/F), depending upon whether one's news release came from Argentina or the UK, the two contenders in the dispute. As each side attempted to impress the world public of its historical claim to this cluster of islands, some observers may have been led to ask why should anyone care about these specks of land – home to countless sheep and some 1,900 human inhabitants? They were soon to learn when, on 2 April 1982, Argentina invaded the islands and the UK picked up the gauntlet.

When the UK and Argentina went to war over the M/F, it was not merely a conflict for a few remote and relatively uninhabited islands. And it was not a dispute exclusively related to national self-determination by its peoples. Neither was it totally a ploy by the Argentine military to save its faltering junta, nor was it merely an opportunity to rescue Prime Minister Margaret Thatcher from her then current political weaknesses. These issues were important, but not the only ones. What we did not hear much about was the petroleum potential of the continental shelf surrounding the M/F, especially to the west toward the Tierra del Fuego region of Argentina.[83]

The M/F region is part of the Malvinas basin, an extension of the continental shelf of the mainland which has similar basins, as the Magallanes adjacent to Tierra del Fuego. Prior to the M/F conflict the Argentines had granted licences to Esso, Total, and Shell for exploration in the country's southern offshore from the city of Rio Gallegos in Santa Cruz on the north to Isla de Los Estados in Tierra del Fuego on the south (Figure 13.5). These firms, in addition to Argentina's state petroleum company, had made some small oil and gas strikes in this region. When the UK attempted to bring in a rig to explore the M/F shelf, the Argentine Government vehemently protested, claiming an infringement of its territorial waters. Significantly, seismic work done in the Malvinas basin indicates that the most promising area lies precisely along the median line between the M/F and the mainland.[84]

Argentina and the UK are still negotiating a settlement of the M/F

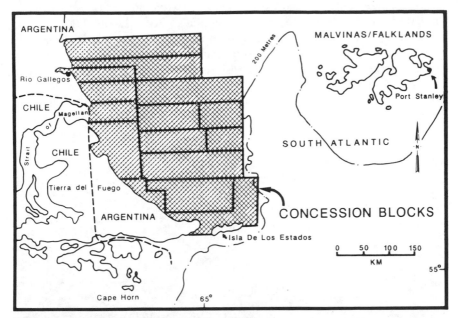

Figure 13.5 The Malvinas/Falklands region
Source: After F. C. F. Earney, *Ocean Mining: Geographic Perspectives*, Meddelelser fra Geografisk Institutt ved Norges Handelshøyskole og Universitetet i Bergen, no. 70 (Geografisk Institutt, Bergen, Norway, 1982), p. 27. With permission.

problem. The Argentines, despite the ouster of the military junta responsible for the M/F fiasco, still insist on negotiating the sovereignty of the islands; the UK, on the other hand, considers the sovereignty issue closed.[85] Some students of the problem believe the Argentines are interested not only in the petroleum (and hard mineral) potential of the M/F region but also in the enhanced political position that control of the area would give to Argentina's claim in Antarctica. Port Stanley would also provide Argentina with a good location for logistical support of its research efforts in Antarctica.[86] Research might be construed to mean exploration for hydrocarbons, given that these have been discovered in the Weddell Sea (within Argentina's Antarctica claim) and the nearby Bellingshausen Sea.[87] One irony of the M/F dispute is that the massive debt it caused Argentina has set back its offshore petroleum programmes.[88]

Although important, the potential for petroleum development in the M/F region should not be overstated. According to Richard Johnson, writing in the *Geographical Journal*, to be commercially exploitable the area must contain at least 1,000 million bbl of exploitable crude oil. Gas might be even less attractive, given the long distances to markets and high production costs in this difficult climatic area.[89]

Argentina's future co-operation in developing, transporting, or consuming M/F petroleum resources that might be discovered is doubtful. The Argentines since 1833 have waged a campaign to reclaim these islands (Figure 13.6), as exemplified by what Peter Beck terms its 'philatelic annexation'. Despite Argentina's protests, the UK and the Falklands have waged a philatelic campaign of their own. In 1933 they released a Falkland Islands centennial stamp and fifty years later, another issue commemorated the Falklands' sesquicentennial,[90] both of which (outside Argentina) sold handsomely. In sum, if there are winners in the Malvinas/Falklands conflict, it is the philatelists whose collections hold an unbroken record of this postage-stamp war.

Figure 13.6 Embarcation point in Buenos Aires, Argentina, for the ferry to Montevideo, Uruguay, August 1986
Source: Author.

Intranational geopolitics

Boundary and petroleum-resource-ownership disputes are not limited to the international arena. An especially bitter controversy over offshore petroleum rights was reconciled recently between Canada's national government and its province of Newfoundland.

St John's vs Ottawa

The offshore Newfoundland dispute arose after drillers discovered and made preliminary reserve estimates of 2,500 million bbl of oil in waters 300km south-east of St. John's.[91] This field – the Hibernia – is a major discovery, and initially geologists thought it would produce for 20–30 years, with a peak rate of perhaps 200,000bbl/d. By late 1985 commercial reserve estimates were put at 648 million bbl, and by 1986 the figure used was only 500 million.[92] Both Newfoundland and Canada's Governments hope future exploration in these waters and along the Labrador coast will provide major petroleum reserves. But the two governments' views have differed on how and at what speed these resources should be developed.

The recent controversy between Newfoundland and the national government reflects a long-standing conflict (starting in the 1960s) between St. John's and Ottawa over the ownership of the offshore. It also reflects a wider concern, that of federalism in resource ownership and management for all provinces in Canada. In 1967 the Supreme Court of Canada set a precedent by ruling that British Columbia's continental shelf extended only to the low-tide mark. But this decision did not discourage Newfoundland from disputing the national government's claim to both the ownership and revenues from resources on the province's OCS. From the beginning of the dispute, Newfoundland claimed that, when joining the Confederation (1949), it did not surrender control of its continental shelf.[93] In 1982 Newfoundland expressed a willingness to compromise on the ownership disagreement,[94] but the effort failed; both parties then went to the courts.

In February 1983 after long deliberation, the Supreme Court of Newfoundland Court of Appeal ruled that Newfoundland controlled seabed and subseabed resources only within its territorial sea limit of 22 km.[95] One year later the Supreme Court of Canada issued a decision that the national government owns all petroleum resources seaward from the territorial sea to the edge of the OCS.

Although the courts resolved the ownership question, problems of administrative functions and revenue sharing were not resolved. Previously the national government had full control of the OCS and most of the revenues it generated. Initially Ottawa offered Newfoundland revenues of 18–22 per cent. Subsequently, under Prime Minister John Turner, the national government sweetened its offer to 25 per cent. Then in April 1984 the Canadian Minister of Energy, Jean Chrétien, offered Newfoundland a revenue package similar to one made with Nova Scotia in 1982, attempting – some observers said – to play one province against another. Newfoundland was to receive all federal and provincial oil and gas taxes, and until its fiscal condition reached 110 per cent of the national average, no revenues had to be shared with other provinces. St. John's was still not ready for an agreement. While the revenue question remained unresolved, oil corporations let the parties know that exploration and development programmes off Newfoundland's shores would be deferred.[96]

In June 1984 during a national election campaign, a major breakthrough came. Under the leadership of Brian Mulroney, the Progressive Conservative Party (PCP) of Canada reached a tentative accord with the Newfoundland Government. The PCP agreed that, if the September election put them in power (which it did), it would make both offshore petroleum revenue and management compromises.[97] The Mulroney Administration met its pledge in early 1985. The Mulroney Accord allows Newfoundland to enact and administer offshore oil and gas resources revenue legislation just as it does for onshore resources. The national government retains the right to continue collecting petroleum corporate income and sales taxes.[98]

Distrust between St John's and Ottawa went deeper than the revenue dispute. Newfoundlanders wanted a voice in the speed and methods of offshore development, such as the use of fixed platforms rather than semi-submersibles.[99] People feared that first, an offshore oil industry would have a 'boom and bust' effect on the local economy and second, developers, if unconstrained, would export from the province any moneys generated, with little long-term improvement occurring in the region's industrial base. Some citizens in Labrador believe that neither Ottawa nor St. John's cares about their concerns for offshore petroleum activities, because the Mulroney Accord fails to protect the region's traditional industries (as fisheries), which could lose needed labour to the oil companies.

Protection of the marine environment and the creation of local jobs are also important concerns. The agreement provides for a joint development fund to expand onshore infrastructures for offshore petroleum and to speed offshore programmes. It also gives Newfoundland the final say on the methods and pace of development, which Ottawa may overrule 'only if the Province's choice, as determined by an independent objective process, delays unreasonably the attainment of self-sufficiency and security' in the country's petroleum resources. Offshore development decision-making occurs through a joint seven-member Canada–Newfoundland Offshore Petroleum Board composed of three members each from Newfoundland and the federal government, as well as a mutually acceptable chair. The Petroleum Board's main office is in Newfoundland. When possible, oil agency and company offices (including federal bodies and Petro-Canada) are located in the province.[100] In 1986 the board approved a development plan by Mobil Oil Canada Limited (and four partners) to develop the Hibernia field, scheduled to go on stream in 1992.[101] But Newfoundland's Premier, Brian Peckford, in early 1987 expressed his unhappiness with the government's slow progress in final development arrangements. Despite Peckford's fears, the Hibernia may yet produce oil by the early 1990s, if petroleum prices remain stable or rise.[102]

California's golden shore

California's coastal zone is a treasure of beauty and wildlife, a mecca for those seeking to commune with nature; it also holds a treasure of 'black gold' so needed in our petroleum-dependent society. Thus the stage has been set for a drama of environmental geopolitics.

The focus of the controversy concerns waters between Los Angeles in the south and San Luis Obispo in the north (Figure 13.7), although

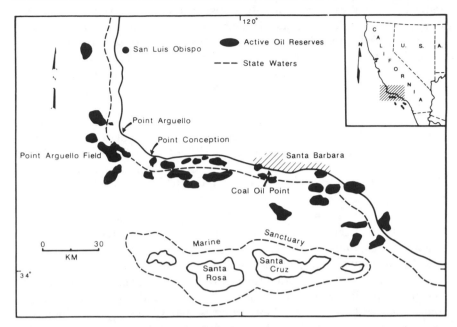

Figure 13.7 California's 'golden shore'

Source: After J. Redden, 'The world offshore: California', *Offshore*, vol. 46, no. 4 (1986), p. 11. With permission.

it spills over into areas extending from Mexico to Oregon. For nearly two decades, injunctions, lawsuits, demonstrations, and acrimony have characterised the relationships among private industry, residents, environmental groups, and federal, state, county, and local governmental agencies. Concern for California's relatively pristine coastal and marine environment was galvanised by the now infamous 1969 Santa Barbara oil spill. The outcry resulting from this spill echoed round the world to waters as distant as the North Sea, and these echoes continue to haunt the offshore of California.

Because oil discoveries have lagged in Alaska's offshore and the once highly touted potential of the US's eastern seaboard has not materialised, the DOI has placed much hope on developing California's OCS.[103] Oil discoveries by Phillips and Chevron in the early 1980s near Santa Barbara in the Santa Maria basin (especially in the Point Arguello field), demonstrated the potential of the region. In 1983 geologists put the Arguello field's estimated crude oil reserves at 100 million to 1,000 million bbl.[104] By 1987 the estimate was 300 million to 500 million bbl. Even this reduced estimate makes the Point Arguello field the largest known oil deposit in the US's entire OCS. Point Arguello oil was expected to come on stream by late 1987 or early 1988.[105] Not so in many other areas.

The Point Arguello discoveries encouraged the DOI in the early 1980s to plan a series of leasing sales within California's OCS. Although at first some communities showed a willingness to co-operate, strong resistance to petroleum exploration and development programmes soon developed, causing the DOI to delay several leasing sales. The frustration created by the delay of one sale prompted then Secretary of the Interior James Watt to assert that the issue was not development but patriotism and that opponents were 'enemies of liberty'.[106] Watt's hyperbole did nothing to bring about compromise.

The environmental battle for the California coast is joined at all levels – federal, state, county, and municipal. When the State of California moved to organise a sale in state waters extending from Point Conception in the south to Point Arguello in the north, Santa Barbara County sued the California Lands Commission to delay it. The suit accused the state of (1) failing to develop an adequate EIS, (2) neglecting to provide for oil spill control, and (3) hurrying to organise the sale because it feared crude oil within its waters would be syphoned off by federal OCS lessees.[107]

Federal sales have been challenged even by federal agencies, as when the Department of Defense opposed leasing in areas where the movement of submarines and the testing of missiles might be impeded.[108] When companies have sought to build landing and processing facilities, municipalities have denied their requests and several communities now require a public referendum for approval of onshore petroleum-related projects.[109]

Although the Congress in 1978 amended the OCSLA to allow more rapid offshore developments,[110] vigorous lobbying by environmental groups, private citizens, and the state government has caused the Congress (since 1982) to impose annually renewed moratoria on OCS leasing along California's central and northern coasts. The California Coastal Commission contends that leasing programmes should go slowly, if coherent coastal planning is to be achieved, and environmental groups argue that the current glut in the world's oil market makes hasty exploration and development unnecessary. The oil companies counter that even if new oil fields are discovered now, it will take another decade to bring them on stream.[111]

Charles DiBona, President and Chief Executive Officer of the American Petroleum Institute, emphasised the importance of not waiting, because the current surplus on the world market could quickly turn into a deficit, especially with consumption rising again. DiBona noted that during the decade 1984–94, the US must find at least 32,000 million bbl of domestic crude, merely to replace reserves consumed. He rebutted environmentalists' oil-spillage arguments by pointing out that as of 1983 more than 10,000 million bbl of oil had been produced in state and federal coastal waters, with only one episode (the 1969 Santa Barbara spill) occurring, whereby large amounts of oil reached the shore. DiBona further noted that natural oceanic oil seeps along the Santa Barbara Channel leak about 100bbl/d into coastal waters. If an estimated natural seepage of only 16bbl/d occurred during the period 1971–82, there would have been as much oil added to the sea as came from all spillage associated with offshore oil exploration and development during those same years.[112]

Producers, in fact (as noted in Chapter 11), have developed techniques for capturing petroleum released from natural seeps. ARCO in 1982 invested $8 million to install its collection system for the seep in waters near Coal Oil Point, near Santa Barbara (see Figures 11.16 and 11.17). This installation earned ARCO 'air emission credits for its exploratory drilling and future development work in the area'. Despite an initial support by California's State Lands Commission for ARCO's proposed Coal Oil Point (COP) project, the Commission's composition changed, resulting in a shift in attitude. Thus in the spring of 1987, the Lands Commission denied a permit for ARCO's COP project, because of the *visual blight* it might cause. The SLC also called for a major study of the 'cumulative effect of offshore oil and gas projects on the State's coastline'.[113] ARCO contends that the SLC's call for a general study is purely political. In its rebuttal to the Lands Commission's ruling on the COP project, ARCO stated that

> from the analysis of the issues of aesthetics in the staff report, it is possible to conclude that the history of the Coal Oil Point project has been one of years of dialogue, engineering design and environmental review to enable you to reach the decision that offshore production platforms are unattractive.[114]

According to ARCO, the SLC has no power to deny any development of leases whatsoever and had filed (July 1987) for damages of $796 million against Santa Barbara County and the State of California.[115] The company, in large part, bases its claims on two important court cases – *Union Oil Company of California et al.* v. *Rogers C. B. Morton et al.* and *First English Evangelical Lutheran Church* v. *Los Angeles, California*. In the *Union Oil* v. *Morton* case (decided 24 February 1975), the US Court of Appeals, Ninth Circuit Court, ruled that denying further development of an existing offshore leasehold is tantamount to cancellation of the lease and thus an illegal taking.[116] On 9 June 1987 the US Supreme Court ruled in the *Lutheran Church* v. *Los Angeles* case that 'landowners deprived of their property because of government regulations can sue for damages'.[117]

Occidental Petroleum also may test the Supreme Court's decision, if Los Angeles voters attempt to block its efforts to develop a project in the Pacific Palisades area, a programme that state and local governmental agencies have approved. Los Angeles voters may, in the near future, attempt to stymie the project in a ballot-box referendum.

In addition to individual company action, the Western Oil and Gas Association and the National Ocean Industries Association recently filed a suit in a Los Angeles federal court against San Diego, San Mateo, San Luis Obispo, Monterey, Santa Cruz, San Franciso, and Sonoma Counties, as well as the cities of San Diego, Oceanside, San Luis Obispo, Morro Bay, Monterey, Santa Cruz, and San Franciso. The suit charges that 'local ordinances restricting onshore facilities related to federal offshore oil development are unconstitutional'.[118]

What the outcome of all this litigation and animosity will be is hard to guess. Certainly the answer lies years ahead. It seems, however, that the oil companies – after nearly a decade of attempting to placate groups of environmental activists and lobby-sensitive state and local land-use agencies – are ready to do legal battle. What has worked for the environmental goose may work for the industrial gander. It is also possible that the two sides may sit down in a non-politicised setting to work out solutions to mutual problems. The petroleum and fishing industries have done just that through 'solutions ... shaped by those directly affected and not by some distant, administrative, political, or judicial decision maker'.[119] Unfortunately the 1988 presidential and congressional elections are at hand, and it is unlikely that the politicians will resist using an upcoming OCS Golden Shore leasing sale (scheduled for 1989) as grist for their campaign mills.

Conclusions

The geopolitics of offshore petroleum are many faceted, including issues concerning territorial disputes, sharing arrangements, and environmental litigation.

Because offshore oil and gas supplies will become more important in future, the potential for confrontations and conflicts will increase. Some of the most volatile offshore disputes presently occur in East and Southeast Asian waters, although several co-operative programmes there demonstrate the assets of accommodation, as in the Japan-South Korea JDZ and the Japan-PRC programme in the Bo Hai area. Assuredly the best geopolitical strategy is 'co-operation', not confrontation.

Norway and Iceland's joint-development agreement in specifically designated waters lying between Iceland and Norway's Jan Mayen Island also exemplifies the value of co-operation, as does Norway and the UK's sharing of the resources and management decisions for the gigantic Statfjord field and the lesser Murchison and Frigg fields that straddle their North Sea offshore boundary. Periodically the two states co-operate in making reserve-share adjustments. Similar arrangements exist in other areas, even in the contentious Persian Gulf. Two examples, among several, stand out. Saudi Arabia and Bahrain in 1985 signed an agreement establishing the Fasht Bu Saafa Hexagon (927km^2), whereby Saudi Arabia manages the petroleum resources contained and then shares half the revenues with Bahrain. Abu Dhabi and Qatar in 1969 also agreed to share revenues obtained from the Bunduq field situated near their boundary separating Das and Dayyinah Islands.[120]

As in international affairs, until national states and their subordinate political units reach agreements based on mutual trust and equity, coastal waters will not reach their full potential in meeting the needs of local citizens and regional and national economies. Where environmentalists and industry are locked in disputes, reasoned compromises must be reached to meet the desires and needs of individuals, interest groups and the larger community, whether local, regional, or national. Although the process of negotiation between the national government of Canada and the provincial government of Newfoundland took several years, an agreeable resource-sharing arrangement seems to have been accomplished. Let us hope a similar compromise is achieved for areas such as California's 'golden shore', where industry, government, and the public have so much at stake.

Notes

1 F. C. F. Earney, 'The geopolitics of minerals', *Focus*, vol. 31 no. 5 (1981), pp. 1–16; see also Committee on Energy and Natural Resources, US Senate, *The Geopolitics of Oil*, Publication no. 96, 96th Cong., 2nd sess. (USGPO, Washington, DC, 1980); F. C. F. Earney, 'Geopolitics of offshore petroleum', in H. E. Johansen, O. P. Matthews, and G. Rudzitis (eds), *Mineral Resources Development: Geopolitics, Economics and Policy* (Westview Press, Boulder, CO, 1987), pp. 51–82.
2 A. A. Meyerhoff and J-O. Willums, 'China's potential still a guessing game', *Offshore*, vol. 39, no. 1 (1979), p. 54.
3 S. S. Harrison, *China, Oil and Asia: Conflict Ahead?* (Columbia University Press, NY, 1977), p. 43.
4 P. C. Yuan, 'China's jurisdiction over its offshore petroleum resources', *Ocean Development and International Law Journal*, vol. 12, nos 3–4 (1983), p. 193.
5 'Xisha and Nansha Islands belong to China', *Beijing Review*, vol. 22, no. 21 (1979), pp. 23–6.
6 In reality, China may continue to support the 'one-Korea' concept as a means of assuring stable relations with North Korea and retaining its support in the unsettled Taiwan question, as well as remaining the champion of the LDCs. Once the Taiwan question is settled, China may 'prove more flexible about the "one-China" principle'. See H. Y. Lee, 'Korea's future: Peking's perspective', *Asian Survey*, vol. 17, no. 11 (1977), especially pp. 1,094–1,100.
7 The position of the median line itself can vary considerably, depending upon the positioning of the baseline from which it is measured.
8 Yuan, 'China's jurisdiction over its offshore petroleum resources', p. 203.
9 S. H. Lee, 'South Korea and the continental shelf issue: Agreements and disagreements between South Korea, Japan and China', *Korea and World Affairs*, vol. 10, no. 1 (1986), pp. 60–3.
10 Yuan, 'China's jurisdiction over its offshore petroleum resources', p. 200.
11 W. R. Feeney, 'Dispute settlement: the emerging law of the sea and East Asian maritime boundary conflicts', *Asian Profile*, vol. 8, no. 6 (1980), p. 582.
12 W. Glenn, 'Cool line on the Senkaku "Bandits"', *Far Eastern Economic Review*, vol. 100, no. 18 (1978), p. 33.
13 Feeney, 'Dispute settlement', p. 582.
14 Glenn, 'Cool line on the Senkaku "Bandits"', p. 34.
15 Feeney, 'Dispute settlement', p. 582.
16 S. Awanohara, 'An ill wind from the Senkakus', *Far Eastern Economic Review*, vol. 100, no. 17 (1978), p. 11.

17 R. Richardson, 'South Korea poised to drill', *Far Eastern Economic Review*, vol. 105, no. 37 (1979), p. 63.

18 J. Segal, 'Reality intrudes into oil plans', *Petroleum Economist*, vol. 46, no. 5 (1979), p. 186.

19 G. Lauriat and M. Liu, 'Pouring trouble on oily waters', *Far Eastern Economic Review*, vol. 105, no. 39 (1979), pp. 20–1; see also Yuan, 'China's jurisdiction over its offshore petroleum resources', p. 204; for a recent and useful look at the Senkaku problem, see W. Lee, 'Troubles under the water: Sino-Japanese conflict of sovereignty on the continental shelf in the East China Sea', *Ocean Development and International Law*, vol. 18, no. 5 (1987), pp. 585–611.

20 Harrison, *China, Oil and Asia*, p. 118.

21 'China sends a new message to Taiwan', *Business Week* (4 December 1978), p. 52.

22 Segal, 'Reality intrudes into oil plans', p. 187.

23 F. C. F. Earney, 'China's offshore petroleum frontiers: Confrontation? Conflict? Cooperation?', *Resources Policy*, vol. 7, no. 2 (1981), p. 122.

24 See M. J. Valencia, 'Oil and gas potential, overlapping claims, and political relations', in G. Kent and M. J. Valencia (eds), *Marine Policy in Southeast Asia* (University of California Press, Berkeley, CA, 1985), pp. 155–87.

25 M. S. Samuels, *Contest for the South China Sea* (Methuen, London, 1982), p. 100.

26 Earney, 'China's offshore petroleum frontiers', p. 120.

27 K. Woodard and A. A. Davenport, 'The security dimension of China's offshore oil development', *Journal of Northeast Asian Studies*, vol. 1, no. 3 (1982), pp. 10–11.

28 Samuels, *Contest for the South China Sea*, pp. 106–7.

29 Letter: S. R. Wyman, Foreign Negotiations Department, Exploration and Production Division, Mobil Oil Corporation, NY, NY, 11 August 1987.

30 ibid. As of mid-1987, no compensation had been paid to Mobil by the present Vietnamese Government.

31 'Xisha and Nansha Islands belong to China', pp. 24–6

32 L. LeBlanc, 'Nations scramble for unclaimed seabed', *Offshore*, vol. 37, no. 3 (1977), p. 43; for a detailed examination of the Paracel and Spratly disputes, see H. Chiu and C-H. Park, 'Legal status of the Paracel and Spratly Islands', *Ocean Development and International Law*, vol. 3, no. 1 (1975), pp. 1–28; M. H. Katchen, 'The Spratly Islands and the law of the sea: "dangerous ground" for Asian peace', *Asian Survey*, vol. 17, no. 12 (December 1977), pp. 1,167–81; for an analysis of all areas within the South China Sea, examine D. J. Dzurek, 'Boundary and resource disputes in the South China Sea', in E. M. Borgese and N. Ginsburg (eds), *Ocean Yearbook 5* (University of Chicago Press, Chicago, IL, 1985), pp. 254–84.

33 Woodard and Davenport, 'The security dimension of China's offshore oil development', p. 9.

34 W. Østreng, 'The politics of continental shelves: the South China Sea in a comparative perspective', *Co-operation and Conflict*, vol. 20, no. 4 (1985), p. 263.

35 Woodard and Davenport, 'The security dimension of China's offshore oil development', p. 7.

36 N. Chanda, 'China calls in the foreign rigs', *Far Eastern Economic Review*, vol. 105, no. 39 (1979), p. 21.

37 'Swell of oil disputes in the South China Sea', *Business Week* (15 October 1979), p. 44.

38 J. Wei, 'Oil exploitation in South China Sea: the eve of a massive battle', *Beijing Review*, vol. 27, no. 15 (1984), p. 22; see also J. Wei, 'Co-operation with foreign countries', *Beijing Review*, vol. 27, no. 16 (1984), pp. 25–7.

39 Woodard and Davenport, 'The security dimension of China's offshore oil development', p. 8.
40 'Vietnam steps up South China Sea effort', *Oil and Gas Journal*, vol. 85, no. 17 (1987), pp. 102–3.
41 Harrison, *China, Oil and Asia*, p. 77.
42 R. W. Hardy, *China's Oil Future: A Case of Modest Expectations* (Westview Press, Boulder, CO, 1978), pp. 22–4.
43 Earney, 'China's offshore petroleum frontiers', p. 120; see also B. Burnett, 'Drilling rigs slated for Chinese Waters', *Offshore*, vol. 38, no. 9 (1978), p. 11.
44 G. Lauriat, 'Japan Peace Treaty's effect on the oil industry', *Far Eastern Economic Review*, vol. 102, no. 40 (1979), pp. 60–1.
45 'Bohai Sea area to pump most oil', *Beijing Review*, vol. 30, no. 19 (1987), pp. 5–6.
46 I. Gorst, 'China: onshore key to increased output', *Petroleum Economist*, vol. 53, no. 12 (1986), p. 456.
47 D. F. Lomax, 'The investment implications of China's offshore oil development', *National Westminster Bank Quarterly Review* (February 1986), p. 51. In late 1987 the PRC reported the discovery of a 700 million bbl field off the coast of Guandong Province, the first major crude oil find in its offshore; see 'Major oil find reported in China', *OPEC Bulletin*, vol. 18, no. 8 (1987), p. 19.
48 E. Cheng, 'Stalling for more time', *Far Eastern Economic Review*, vol. 136, no. 22 (1987), p. 73.
49 ibid., pp. 73–4.
50 See F. C. F. Earney, 'Norway's offshore petroleum industry', *Resources Policy*, vol. 8, no. 2 (1982), pp. 133–42.
51 F. Laursen, 'Security aspects of Danish and Norwegian law of the sea policies', *Ocean Development and International Law*, vol. 18, no. 2 (1987), p. 215.
52 'Worldwide offshore daily average oil production (000b/d)', *Offshore*, vol. 47, no. 5 (1987), p. 52.
53 'Offshore gas production (MMcfd)', *Offshore*, vol. 47, no. 5 (1987), p. 51.
54 'Oil prospecting moves north', *Norinform*, no. 31 (8 October 1985), pp. 2–3.
55 W. Østreng, 'Norway's law of the sea policy in the 1970s', *Ocean Development and International Law*, vol. 11, nos 1/2 (1982), pp. 77–8.
56 For an entertaining discussion of the relationships of Norwegians and Russians on Svalbard, see R. Moseley, 'A case of coexistence', *Chicago Tribune Magazine* (16 August 1987), section 10, pp. 14–15, 18–23. Norway recently undertook a full review of Svalbard's status. The government is especially concerned that only one-third of Svalbard's residents are Norwegian. The government is also under pressure to find alternate economic activities for the populace, because the collieries have become prohibitively uneconomic. See 'Svalbard's future under review', *Norinform*, no. 14 (5 May 1987), p. 8.
57 W. Østreng, 'Delimitation arrangements in Arctic seas', *Marine Policy*, vol. 10, no. 2 (1986) p. 134. For a detailed examination of the strategic significance of the USSR's northern naval forces and Norway's northern waters as a gateway to and from the Kola Peninsula, see W. Østreng, *The Soviet Union in Arctic Waters*, Law of the Sea Institute Occasional Paper no. 36 (LSI, University of Hawaii, Honolulu, HA, 1987).
58 Bergen Bank, *Petroleum Activities in Norway 1987* (BB, Bergen, Norway, April 1987), p. 46.
59 R. R. Churchill, 'Maritime boundary problems in the Barents Sea', in G. Blake (ed.), *Maritime Boundaries and Ocean Resources* (Croom Helm, London, 1987), pp. 149–50.

60 Østreng, 'Delimitation arrangements in Arctic Seas', p. 136.
61 'Continued Grey Zone in the Barents Sea', *Norinform*, no. 1 (13 January 1987), p. 7.
62 'Twenty-year sentence for espionage', *Norinform*, no. 20 (25 June 1985), pp. 1–2.
63 R. Vielvoye, 'Barents Sea frontier', *Oil and Gas Journal*, vol. 85, no. 12 (1987), p. 29; see also 'Will Barents Sea be new arena for oil activities?', *Norinform*, no. 7 (24 February 1987), p. 9.
64 'Twenty new North Sea blocks in 12th concession round', *Norinform*, no. 18 (2 June 1987), p. 3.
65 The Norwegian Government has decided to extend seismic exploration as far east as 32°E; see H. O. Bergesen, A. Moe, and W. Østreng, *Soviet Oil and Security Interests in the Barents Sea* (St. Martin's Press, NY, 1987), p. 46.
66 'Eight operator tasks awarded in 11th round of concessions', *Norinform*, no. 12 (31 March 1987), p. 2.
67 Earney, 'Geopolitics of offshore petroleum', p. 67.
68 'Soviet continental shelf opens for Norwegian firms', *Norinform*, no. 25 (27 August 1985), p. 1; 'Industrial co-operation between Finnmark and USSR?', *Norinform*, no. 16 (19 May 1987), p. 9.
69 'Wartsila, Norwegian partners establish Barents Sea base', *Offshore*, vol. 47, no. 8 (1987), p. 71.
70 'Soviet/Norwegian co-operation on oil resources in north?', *Norinform*, no. 43 (16 December 1986), p. 2.
71 Churchill, 'Maritime boundary problems in the Barents Sea', p. 158.
72 Bergesen, Moe, and Østreng, *Soviet Oil and Security Interests in the Barents Sea*, pp. 96–7.
73 Laursen, 'Security aspects of Danish and Norwegian law of the sea policies', pp. 222–3.
74 ibid., p. 223.
75 'The Jan Mayen agreement – a model for the Barents Sea?', *Norinform*, no. 17 (18 May 1983), p. 2; 'Oil – Norway to assist Iceland', *Norinform*, no. 8 (1 March 1983), p. 4; for a useful and more detailed examination of the joint development arrangements see Østreng, 'Delimitation arrangements in Arctic Seas', pp. 137–40.
76 Laursen, 'Security aspects of Danish and Norwegian law of the sea policies', pp. 223–4.
77 Østreng, 'Delimitation arrangements in Arctic Seas', p. 142.
78 G. Blake, 'The Law of the Inland Sea', *Focus*, vol. 32, no. 2 (1981), p. 12.
79 ibid., pp. 13–14.
80 J. Bierman, 'Storm clouds over the Aegean', *Maclean's*, vol. 100, no. 14 (1987), pp. 26–7.
81 Blake, 'The law of the inland sea', p. 14.
82 'Greece not seeking N. Aegean group's assets', *Oil and Gas Journal*, vol. 85, no. 20 (1987), p. 24.
83 F. C. F. Earney, *Ocean Mining: Geographic Perspectives*, Meddeleser fra Geografisk Institutt ved Norges Handelshøysokle og Universitetet i Bergen, no. 70 (Geografisk Institutt, Bergen Norway, 1982), p. 25.
84 R. Johnson, 'The geography of the Falkland Islands: The Islands' resources', *Geographical Journal*, vol. 149, pt I, no. 1 (1983), p. 7.
85 For a useful discussion of the principles involved, see M. J. Levitin, 'The law of force and the force of law: Grenada, the Falklands, and humanitarian intervention', *Harvard International Law Journal*, vol. 27, no. 2 (1986), pp. 621–57.
86 M. Van Sant Hall, 'Argentine policy motivations in the Falklands War and the aftermath', *Naval War College Review*, vol. 36, no. 6/seq. 300 (1983),

p. 22; see also Lord [E. E. A.] Shackleton, 'The Falkland Islands and their history', *Geographical Journal*, vol. 149, pt I, no. 1 (1983), p. 4; M. A. Morris, 'Southern cone maritime security after the 1984 Argentine–Chilean Treaty of Peace and Friendship', *Ocean Development and International Law*, vol. 18, no. 2 (1987), p. 243; J. S. Roucek, 'The geopolitics of the Antarctic: the land is free for scientific work but its wealth of minerals has excited Imperialist claims', *American Journal of Economics and Sociology*, vol. 45, no. 1 (1986), pp. 69–77.

87 Johnson, 'Geography of the Falkland Islands', p. 7.
88 'Drilling: Latin America', *Offshore*, vol. 44, no. 5 (1984), p. 105.
89 Johnson, 'Geography of the Falkland Islands', p. 7.
90 P. J. Beck, 'Argentina's philatelic annexation of the Falklands', *History Today*, vol. 33, no. 2, (1983), pp. 39–44.
91 J. Redden, 'The world offshore: Newfoundland', *Offshore*, vol. 44, no. 9 (1984), p. 11.
92 M. Nichols, 'A gathering frontier storm', *Maclean's*, vol. 99, no. 27 (1986), p. 33.
93 'A framework for agreement: a paper presented by Newfoundland in the Canada–Newfoundland offshore mineral resources negotiations, 12 November 1981, Montreal' (St John's Newfoundland, 1981), p. 3.
94 'Canada–Newfoundland offshore negotiations: a proposal for settlement presented by Newfoundland, 25 January 1982, Montreal' (St John's, Newfoundland, 1982), p. 2.
95 'In the Supreme Court of Newfoundland Court of Appeal, no. 23, Decided: 17 February 1983' (St John's, Newfoundland, 1983), pp. 18, 68–9.
96 J. Lewington, 'Agreements may increase East Coast drilling', *Globe and Mail* (Toronto) (6 April 1984), p. B4.
97 Letter to the Honourable Brian Peckford, Premier of Newfoundland and Labrador, 14 June 1984, pp. 4–5, as provided by personal correspondence with J. G. Fitzgerald, Executive Director, Petroleum Directorate, Government of Newfoundland and Labrador, St. John's, 7 September 1984.
98 'Canadian Arctic production prospects brighten', *Oil and Gas Journal*, vol. 83, no. 1 (1985), p. 60.
99 W. Steif, 'Newfoundland's oil economy', *Multinational Monitor*, vol. 6, no. 14 (1985), pp. 6–7.
100 Letter to the Honourable Brian Peckford, pp. 1–6.
101 'Drilling programs hit roadblock off Nova Scotia', *Oil and Gas Journal*, vol. 85, no. 1 (1987), p. 20.
102 Nichols, 'A gathering frontier storm', p. 32.
103 Low prices for petroleum in 1987 encouraged a reduction of exploration in Alaskan waters, which require especially high investments. See L. A. LeBlanc, 'Target: giant fields: costly and barren Alaska, Atlantic frontiers leave fewer options in US search', *Offshore*, vol. 47, no. 1 (1987), pp. 36–7.
104 R. Baum, 'Controversy bubbles over offshore oil development', *Chemical and Engineering News*, vol. 61, no. 21 (1983), p. 21.
105 B. Williams, 'Development off California hobbled', *Oil and Gas Journal*, vol. 85, no. 28 (1987), p. 26.
106 Baum, 'Controversy bubbles over offshore oil development', pp. 22–3. For an excellent discussion of the California-federal government controversy, see P. S. Ruffra, 'Watt v. California: Supreme Court sinks consistency review of offshore oil leases', *Columbia Journal of Environmental Law*, vol. 10, no. 1 (1985), pp. 131–47.
107 Baum, 'Controversy bubbles over offshore oil development', p. 24.
108 J. Redden, 'The world offshore: California', *Offshore*, vol. 44, no. 7 (1984), p. 11.

109 'Oil's up: Santa Barbara', *The Economist*, vol. 301, no. 7,476 (1986), p. 33.
110 C. J. DiBona, 'Offshore oil and gas drilling: a necessity for US energy security', *Public Utilities Fortnightly*, vol. 114, no. 12 (1984), p. 37.
111 'The offshore oil dilemma', *Sunset*, vol. 177, no. 5 (1986), p. 276.
112 C. J. DiBona, 'Offshore oil and gas drilling', pp. 36–8.
113 B. Williams, 'Environmental delays continue to plague projects off California', *Oil and Gas Journal*, vol. 85, no. 26 (1987), p. 15.
114 ibid., pp. 16–17.
115 'California municipalities sued over anti-oil rules', *Oil and Gas Journal*, vol. 85, no. 33 (1987), p. 20.
116 US Court of Appeals, Ninth Circuit, *Union Oil Company of California et al., Plaintiffs and Appellants, v. The Honorable Rogers C. B. Morton, Secretary of the Department of the Interior of the United States of America et al., Defendants-Appellees*, no. 73–1692 (24 February 1975), *Federal Reporter*, vol. 512, F.2d, pp. 743–52.
117 US Supreme Court, *First English Evangelical Lutheran Church of Glendale Appellant v. County of Los Angeles, California*, Decision no. 85–1199, decided 9 June 1987; see *United States Law Week*, vol. 55, no. 48 (1987), pp. 4,781–92.
118 'California municipalities sued over Anti-oil rules', p.20.
119 G. W. Cormick and A. Knaster, 'Oil and fishing industries negotiate: mediation and scientific issues', *Environment*, vol. 28, no. 10 (1986), p. 30.
120 J. R. V. Prescott, *The Maritime Political Boundaries of the World* (Methuen, London, 1985), pp. 169–70.

Chapter fourteen

Epilogue

The time has come to return to the question posed at the beginning of the book: can the oceans provide us with golden mineral riches or only with an illusion – fools' gold? Ofcourse, neither alternative provides an adequate answer, certainly not one applicable to all marine minerals.

As identified early on, seabed mining has several advantages

1 an accessible supply of numerous metallic and non-metallic mineral resources;
2 an available and relatively inexpensive transport mode – the water;
3 a fairly well distributed network of terminals for receiving, processing, and distributing marine mineral materials;
4 a probably somewhat less rigorous environmental regulatory regime than that which will exist onshore;
5 a potential for providing some states with a greater mineral self-sufficiency.

Although not applicable for all marine minerals, liabilities include

1 long distances from mine sites to markets;
2 presently over-supplied mineral markets;
3 environmentally and technologically difficult working conditions;
4 large capital investments needed to develop more sophisticated mining and processing methods;
5 a yet unsettled management regime for deep seabed mining.

Given the scientific community and mineral resource producers' knowledge of nearshore minerals and a likely growth in demand for some of the placer metals (gold, platinum, and tin), the fertiliser and chemical materials (phosphorite, sulphur, coral, and shells) and the aggregates and industrial sands, a greater use of the continental shelves will occur in coming decades. As new technologies evolve, offshore petroleum producers will penetrate into harsher and deeper oceanic regions, as exemplified by the arctic waters of the US, the USSR, and Canada, perhaps even those of Antarctica. How rapidly this penetration will occur depends, in part, on first, governmental leasing and taxing policies that encourage or discourage industry's interest, and second, advances in deep-water and ice-environment technologies.

Local, regional, and national governments and regulatory agencies, as well as the general populace, must recognise that the localised geopolitical problems of offshore petroleum important to them are also an integral part of the interests of the international economic community (OPEC, COMECON, EEC, and IEA) and of military alliances (NATO and WTO – the Warsaw Treaty Organisation). We must adjust to the reality that oil is no longer merely an economic commodity but has become an international political commodity[1] and an Achilles' heel for many regions. For example, if the Barents Sea area (Norwegian and Soviet) becomes a major source of gas for Western Europe, its supply reliability would hinge on relationships and actions of the superpowers (US and USSR) in the region.[2] Petroleum is also an important focus for political posturing and debate within states (as in Norway and the UK) and between states. In future, increased frictions may occur between states sharing proved or potentially petroleum-rich continental shelf areas, as between Italy and Yugoslavia; Nigeria and Cameroon, Colombia and Venezuela, and Canada's Northwest Territories and Denmark's Greenland.

As the offshore becomes more important as a resource region, both for petroleum and other minerals, there may be intensified confrontations among private industry, national, regional, and local governmental agencies and the general public over the pace and method of development programmes. And governmental inter-agency rivalries will likely emerge in many offshore mineral-producing countries, as has occurred between the US's DOC and DOI. Such competition and diverse agendas may demand an increased integration of planning and decision making through regional offshore resource management bodies, as demonstrated by recently developed joint efforts of coastal states and federal agencies in the US.

Although the world community during the past two decades has moved forward in establishing a programme of international management of deep seabed mineral resources, years of negotiations lie ahead to resolve detailed procedures and regulations for extracting these minerals. Assuming that the necessary sixty states ratify the LOS Convention, a major question remains of whether parallel ocean mining regimes – one under the LOS Convention and the other under a reciprocating states agreement – can mutually accommodate. Despite the current refusal of the US and several other countries with deep seabed mining capabilities to sign or ratify the Convention, the majority of world states may decide, in future, that it can be modified to bring those states under its umbrella. For example, PREPCOM (in August 1987) showed significant flexibility in accommodating the US's concerns, as in delaying its acceptance of the USSR's mine site until US consortia (in concert with Canadian, European, and Japanese corporations) could resolve their claim overlaps, even though the USSR is a treaty signatory and the US is not.[3]

In late 1987 a panel of distinguished specialists on law-of-the-sea problems issud a statement containing a call for the US and other world states to renegotiate the LOS Convention, until it is acceptable to all.

Such a consensus should be obtainable, given the general agreement of nearly all states on matters other than deep seabed mining. The panel stressed, however, that rapidly changing world mineral market conditions have reduced the urgency for finalising a deep seabed mining regime and made it unlikely that a full-scale International Seabed Authority 'could become self-supporting from revenues produced by seabed minerals development'. In addition, changed technological capabilities (and those likely in future) make obsolete the present Convention's provisions for the number, size, and production tonnages for first generation mine sites. Finally the panelists emphasised that recently discovered onshore mineral sources and improved mining and processing technologies 'will result in a long postponement (if not abandonment) of, and slower projected rates of growth for, commercial deep seabed mining operations'.[4]

Even if deep seabed mining should begin several decades after the turn of the century, time will demonstrate to the LDC community that mining in the 'Area' alone will never provide the sweeping improvement they seek in economic development and the sharing of earth's resource wealth. The wave of sanguine anticipation of the late 1960s and throughout the 1970s for a new international economic order based on seabed mining will not lift the Third World from its poverty, a reality most Third World leaders have always recognised. On the other hand, the LOS negotiation process and its resulting Convention have helped demonstrate that international management of ocean resources – whether spatial, biological, or mineralogical – is a potentially workable concept. Elisabeth Mann Borgese described this concept, embodied in the Convention, as an opportunity to use the oceans as a laboratory for building a new world order, an 'order we may hope will prove more rational, more humane and more responsive to the real needs of the world than the old order that is disintegrating in hunger and violence'.[5]

Notes

1 Ø. Noreng, 'The international petroleum game and Norway's dilemma', *Co-operation and Conflict*, vol. 17, no. 2 (1982), p. 85.
2 M. Saeter, 'Natural gas: new dimensions of Norwegian foreign policy', *Co-operation and Conflict*, vol. 17, no. 2 (1982), p. 145.
3 Council on Ocean Law, *The United States and the 1982 Law of the Sea Convention: A Synopsis of the Status of the Treaty in 1987* (COL, Washington, DC, Oct. 1987), p. 3.
4 Panel on the Law of Ocean Uses, *Deep Seabed Mining and the 1982 Convention on the Law of the Sea* (Council on Ocean Law, Washington, DC, 25 Sept. 1987), pp. 3–4.
5 E. M. Borgese, 'The law of the sea', *Scientific American*, vol. 248, no. 3 (1983), p. 42.

Appendix

United Nations, Article 76, 'Definition of the continental shelf', *The Law of the Sea: Official Text of the United Nations Convention on the Law of the Sea* (UN, NY, 1983), pp. 27–8.

1 The continental shelf of a coastal State comprises the seabed and subsoil of the submarine areas that extend beyond its territorial sea throughout the natural prolongation of its land territory to the outer edge of the continental margin, or to a distance of 200 nautical miles from the baselines from which the breadth of the territorial sea is measured where the outer edge of the continental margin does not extend up to that distance.

2 The continental shelf of a coastal State shall not extend beyond the limits provided for in paragraphs 4 to 6.

3 The continental margin comprises the submerged prolongation of the land mass of the coastal State and consists of the seabed and subsoil of the shelf, the slope and the rise. It does not include the deep ocean floor with its oceanic ridges or the subsoil thereof.

4 (a) For the purposes of this Convention, the coastal State shall establish the outer edge of the continental margin wherever the margin extends beyond 200 nautical miles from the baselines from which the breadth of the territorial sea is measured, by either:

  (i) a line delineated in accordance with paragraph 7 by reference to the outermost fixed points at each of which the thickness of sedimentary rocks is at least 1 per cent of the shortest distance from such point to the foot of the continental slope; or

  (ii) a line delineated in accordance with paragraph 7 by reference to fixed points not more than 60 nautical miles from the foot of the continental slope.

(b) In the absence of evidence to the contrary, the foot of the continental slope shall be determined as the point of maximum change in the gradient at its base.

5 The fixed points comprising the line of the outer limits of the continental shelf on the seabed, drawn in accordance with paragraph 4 (a) (i) and (ii), either shall not exceed 350 nautical miles from the

baselines from which the breadth of the territorial sea is measured or shall not exceed 100 nautical miles from the 2,500 metre isobath, which is a line connecting the depth of 2,500 metres.

6 Notwithstanding the provisions of paragraph 5, on submarine ridges, the outer limit of the continental shelf shall not exceed 350 nautical miles from the baselines from which the breadth of the territorial sea is measured. This paragraph does not apply to submarine elevations that are natural components of the continental margin, such as its plateaux, rises, caps, banks, and spurs.

7 The coastal State shall delineate the outer limits of its continental shelf, where that shelf extends beyond 200 nautical miles from the baselines from which the breadth of the territorial sea is measured, by straight lines not exceeding 60 nautical miles in length, connecting fixed points, defined by co-ordinates of latitude and longitude.

8 Information on the limits of the continental shelf beyond 200 nautical miles from the baselines from which the breadth of the territorial sea is measured shall be submitted by the coastal State to the Commission on the Limits of the Continental Shelf set up under Annex II on the basis of equitable geographical representation. The Commission shall make recommendations to coastal States on matters related to the establishment of the outer limits of their continental shelf. The limits of the shelf established by a coastal State on the basis of these recommendations shall be final and binding.

9 The coastal State shall deposit with the Secretary-General of the United Nations charts and relevant information, including geodetic data, permanently describing the outer limits of its continental shelf. The Secretary-General shall give due publicity thereto.

10 The provisions of this article are without prejudice to the question of delimitation of the continental shelf between States with opposite or adjacent coasts.

Bibliography

Aamo, B. S., 'Norwegian oil policy: basic objectives', in M. Saeter and I. Smart (eds), *The Political Implications of North Sea Oil and Gas* (Universitetsforlaget, Oslo, 1975), pp. 81–92.

Alcoforado do Couteo, A. C., Interview, Engineer, Companhia Nacional de Alcalis, Rio de Janeiro, Brazil, 18 Aug. 1986.

Alexander, L. M. and L. C. Hanson, 'Regionalizing the U.S. EEZ', *Oceanus*, vol. 27, no. 4 (1984/85), pp. 7–12.

Alexander, L. M. and R. D. Hodgson, 'The impact of the 200-mile economic zone on the law of the sea', *San Diego Law Review*, vol. 12, no. 3 (1975), pp. 569–99.

Allison, A. P., 'The Soviet Union and UNCLOS III: pragmatism and policy evolution', *Ocean Development and International Law*, vol. 16, no. 2 (1986), pp. 109–36.

Amann, H., 'Development of ocean mining in the Red Sea', *Marine Mining*, vol. 5, no. 2 (1985), pp. 103–16.

'The American dimension and Norwegian oil policy', *Norwegian Oil Review*, vol. 12, no. 6 (1986), pp. 10–11.

'American Mining Declaration of Policy', *Mining Congress Journal*, vol. 73, no. 10 (1987), pp. 11–22.

Ampian, S. G., 'Barite', in *Mineral Facts and Problems 1985*, USBM Bulletin 675 (USGPO, Washington, DC, 1985), pp. 65–74.

Anand, R. P., *Origin and Development of the Law of the Sea: History of International Law Revisited* (Martinus Nijhoff Publishers, The Hague, 1983).

Andrews, B. V., J. E. Flipse, and F. C. Brown, *The Economic Viability of a Four-Metal Pioneer Deep Ocean Mining Venture*, TAMU–SG–84–201 (Sea Grant College Program, Texas A&M University, College Station, TX, Oct. 1983).

Andrews, J. E. and G. H. W. Friedrich, 'Distribution patterns of manganese nodule deposits in the Northeast Equatorial Pacific', *Marine Mining*, vol. 2, nos 1/2 (1979), pp. 1–43.

Antinori, C. M., 'The Bering Sea: a maritime delimitation dispute between the United States and the Soviet Union', *Ocean Development and International Law*, vol. 18, no. 1 (1987), pp. 1–47.

Aoyagi, T., 'Magnesium supply and demand report', in *Magnesium in the Auto Industry: Prospects for the Future, Proceedings 44th Annual World Magnesium Conference, Tokyo, Japan, 17–20 May 1987* (International Magnesium Association and Japan Light Metal Association, McLean, VA, 1987), pp. 32–3.

'Aragonite: white gold in the Bahamas', brochure supplied by Marcona Ocean Industries Ltd, n.d.

'Aragonite: white gold in the Bahamas', *Carib*, vol. 1 (1978).

Archer, A. A., 'Sand and gravel demands on the North Sea – present and future', in E. D. Goldberg (ed.), *North Sea Science* (MIT Press, Cambridge, MA, 1973), pp. 337–49.

ARCO Oil and Gas Company, Atlantic Richfield Company, *Santa Barbara Channel Seep Containment Project* (ARCO, Los Angeles, CA, 1985).

Arden, J., Letter, for the Secretary of Energy, New Zealand Ministry of Energy, Wellington, NZ, 30 April 1987.

Argarwal, J. C. *et al.*, 'Comparative economics of recovery of metals from ocean nodules', *Marine Mining*, vol. 2, nos 1/2 (1979), pp. 119–30.

Arita, M. *et al.*, 'Exploration and exploitation of offshore sand in Japan', ESCAP, UN, *CCOP Technical Bulletin*, vol. 17 (Dec. 1985), pp. 81–99.

Arrow, D. W., 'Seabeds, sovereignty and objective regions', *Fordham International Law Journal*, vol. 7, no. 2 (1983–84), pp. 169–243.

'Arusha understanding', *Oceans Policy News* (May 1986), pp. 5–7.

Atkinson, F. and S. Hall, *Oil and the British Economy* (Croom Helm, Beckenham, 1983).

Auburn, F. M., 'The international seabed area', *International and Comparative Law Quarterly*, 4th series, vol. 20, pt 2 (1971), pp. 173–94.

Ausland, J. C., *Norway, Oil and Foreign Policy* (Westview Press, Boulder, CO, 1979).

Awanohara, S., 'An ill wind from the Senkakus', *Far Eastern Economic Review*, vol. 100, no. 17 (1978), pp. 10–12.

Bäcker, H., Letter, Preussag AG, Hannover, FRG, 7 July 1987.

Bailey, R. J., Letter, OCS Co-ordinator, Department of Land Conservation and Development, State of Oregon, Portland, OR, 10 Aug. 1987.

—— 'Marine minerals in the Exclusive Economic Zone: implications for coastal states and territories', paper presented at the Western Legislative Conference, Pacific States/Territories Ocean Resource Group, San Francisco, CA, 28 Feb. 1987.

Baldwin, P. L. and M. F. Baldwin, *Onshore Planning for Offshore Oil: Lessons from Scotland* (Conservation Foundation, Washington, DC, 1975).

Ballard, R. D. 'The exploits of *Alvin* and *Angus*: exploring the East Pacific Rise', *Oceanus*, vol. 27, no. 3 (1984), pp. 7–14.

Balzer, S., *Survey of Foreign Offshore Development Activities for Minerals other than Oil and Gas* (Oil and Gas Lands Administration, Energy, Mines and Resources Canada, Ottawa, 1986).

Band, G. C., 'U.K. North Sea production prospects to the year 2000', *Journal of Petroleum Technology*, vol. 39, no. 1 (1987), pp. 64–70.

'Barite and clay minerals', *Minerals and Materials: A Bimonthly Survey* (June/July 1986), p. 32.

Bartlett, P. M., 'Republic of South Africa coastal and marine minerals potential', *Marine Mining*, vol. 6, no. 4 (1987), pp. 361–9.

Baum, R., 'Controversy bubbles over offshore oil development', *Chemical and Engineering News*, vol. 61, no. 21 (1983), pp. 21–5.

'Beach sand bonanza in Madagascar', *Mining Journal*, vol. 308, no. 7,914 (1987), p. 309.

Beall, R., Letter, Public Relations Co-ordinator, Lockheed Missiles and Space Co., Sunnyvale, CA, 21 April 1987.

Beck, P. J., 'Argentina's philatelic annexation of the Falklands', *History Today*, vol. 33, no. 2 (1983), pp. 39–44.

Beiersdorf, H., H-R. Kudrass and U. von Stackelberg, *Placer Deposits of Ilmenite and Zircon on the Zambezi Shelf*, Geologisches Jahrbuch, Reihe D, Heft 36, Bundesanstalt für Geowissenschaften und Rohstoffe und den Geologischen Landesämtern in der Bundesrepublik Deutschland (Alfred-Bentz-Haus, Hannover, FRG, 1980).

Benedict, M., T. H. Pigford, and H. W. Levi, *Nuclear Chemical Engineering*, 2nd edn (McGraw-Hill, NY, 1981).

Bergen Bank, *Petroleum Activities in Norway 1987* (BB, Bergen, Norway, April 1987).

Bergesen, H. O., A. Moe, and W. Østreng, *Soviet Oil and Security Interests in the Barents Sea* (St Martin's Press, NY, 1987).

Bergsager, E., 'First drilling in Norwegian Sea off Norway yields encouraging results', *Oil and Gas Journal*, vol. 79, no. 23 (1981), pp. 115, 118, 120, 124.

Bernier, L., 'Ocean mining activity shifting to Exclusive Economic Zones', *Engineering and Mining Journal*, vol. 185, no. 7 (1984), pp. 57–60.

Best, F. R. and M. J. Driscoll, 'Prospects for the recovery of Uranium from seawater', *Nuclear Technology*, vol. 73, no. 1 (1986), pp. 55–68.

Best, F. R. and S. Yamamoto, 'A review on Uranium recovery processes and the international meeting on the recovery of uranium from seawater', *Scientific Bulletin*, Office of Naval Research, vol. 9, no. 1 (1984), pp. 93–100.

'Beyond the Norwegian 62nd Parallel: ever further north', *Oil and Enterprise: The Oil and Gas Review*, no. 25 (Aug. 1985), pp. 38–9.

Bierman, J., 'Storm clouds over the Aegean', *Maclean's*, vol. 100, no. 14 (1987), pp. 26–7.

Bignell, R. D., 'Genesis of the Red Sea: metalliferous sediments', *Marine Mining*, vol. 1, no. 3 (1978), pp. 209–35.

Blake, G., 'The law of the inland sea', *Focus*, vol. 32, no. 2 (1981), pp. 12–15.

Bland, D., *UK Oil Taxation* (Oyez-Longman, London, 1984).

Bleicher, S. A., 'Reflections on the failure of NOAA's ocean management office', *Coastal Zone Management Journal*, vol. 11, no. 4 (1984), pp. 353–67.

Bleiwas, D. I., A. E. Sabin, and G. R. Peterson, *Tin Availability – Market Economy Countries: A Minerals Availability Appraisal*, USBM Information Circular 9086 (DOI, Washington, DC, 1986).

'Bohai Sea area to pump most oil', *Beijing Review*, vol. 30, no. 19 (1987), pp. 5–6.

Bokuniewicz, H., 'Sand mining in New York Harbor: a chronology', paper presented at Offshore Sand and Gravel Mining Workshop, 18–20 March 1986, Stony Brook, NY.

Bonatti, E. and Y. R. Nayudu, 'The origin of manganese nodules on the ocean floor', *American Journal of Science*, vol. 263, no. 1 (1965), pp. 17–39.

Borgese, E. M., Letters, Chairman, International Ocean Institute, Halifax, Nova Scotia, Canada, 31 Dec. 1987, and Valletta, Malta, 26 Aug. 1987.

—— 'The law of the sea', *Scientific American*, vol. 248, no. 3 (1983), pp. 42–9.

—— 'The Preparatory Commission for the International Sea-Bed Authority and for the International Tribunal for the Law of the Sea: third session', in E. M. Borgese and N. Ginsburg (eds), *Ocean Yearbook 6* (University of Chicago Press, Chicago, IL, 1986), pp. 1–14.

Bowen, R. and T. Hennessey, 'U.S. EEZ relations with Canada and Mexico', *Oceanus*, vol. 27, no. 4 (1984/85), pp. 41–7.

Boy de la Tour, X. and D. Champlon, 'Economic aspects of hydrocarbons production in arctic areas', *Oil and Enterprise: The Oil and Gas Review*, no. 29 (Dec. 1985), pp. 47–52.

Brabyn, H., 'Blue planet', *UNESCO Courier* (Feb. 1986), pp. 4–10.

'Brazil continues to hit oil in 13-year-old Campos play', *Offshore*, vol. 47, no. 11 (1987), pp. 48, 50–1.

'Brazil slates deepwater development', *Oil and Gas Journal*, vol. 84, no. 22 (1986), pp. 32–3.

'Britain', *OPEC Bulletin*, vol. 14, no. 8 (1983), pp. 30–40.

British Coal, *Coal in Scotland* (BC, London, 1986).

British Petroleum Company plc, *BP Statistical Review of World Energy* (BP, London, June 1987).

Broadus, J. M., 'Seabed materials', *Science*, vol. 235, no. 4,791 (1987), pp. 853–60.

——— Testimony in *The Ocean and the Future*, 'Hearings', before the Subcommittee on Oceanography of the Committee on Merchant Marine and Fisheries, House of Representatives, 99th Cong., 1st. sess., 24 Oct. 1985 (USGPO, Washington, DC, 1986), pp. 132–67.

Broadus, J. M. and P. Hoagland III, 'Conflict resolution in the assignment of area entitlements for seabed mining', *San Diego Law Review*, vol. 21, no. 3 (1984), pp. 541–76.

——— 'Rivalry and coordination in marine hard minerals regulation', in *Exclusive Economic Zone Papers: Oceans '84 Conference Proceedings* (NOAA, DOC, Rockville, MD, 1984), pp. 56–61.

Brooke, R. L. 'The current status of deep seabed mining', *Virginia Journal of International Law*, vol. 24, no. 2 (1984), pp. 361–417.

Brown, E. D., 'The United Nations Convention on the Law of the Sea 1982: the British Government's dilemma', *Current Legal Problems*, vol. 37 (1984), pp. 259–93.

Bryan, C. E., Letter, Director, Fisheries Resource Programs, Texas Parks and Wildlife Department, Austin, TX, 12 Feb. 1987.

Bryce, D. H., Letter, Project Co-Director, Fletcher Challenge Ltd, Auckland, NZ, 3 April 1987.

Bullis, L. H. and J. E. Mielke, *Strategic and Critical Materials* (Westview Press, Boulder, CO, 1985).

Burnett, B., 'Drilling rigs slated for Chinese waters', *Offshore*, vol. 38, no. 9 (1978), p. 11.

Burnett, W. C., 'Phosphorites in the U.S. Exclusive Economic Zone', in M. Lockwood and G. Hill (eds), *Proceedings: Exclusive Economic Zone Symposium Exploring the New Ocean Frontier, Washington, D.C., 2–3 October 1985* (DOC, Rockville, MD, 1986), pp. 135–40.

Burroughs, T., 'Ocean mining: boom or bust?', *Technology Review*, vol. 87, no. 3 (1984), pp. 54–60.

'California municipalities sued over antioil rules', *Oil and Gas Journal*, vol. 85, no. 33 (1987), pp. 20–1.

'Campos: 60% of output and new strikes keep coming', *Petrobrás News*, no. 106 (Sept. 1987), pp. 4–5.

'Canada–Newfoundland offshore negotiations: a proposal for settlement, presented by Newfoundland, 25 January 1982, Montreal' (St John's, Newfoundland, 1982).

'Canadian Arctic production prospects brighten', *Oil and Gas Journal*, vol. 83, no. 1 (1985), pp. 57–60.

Cape Breton Development Corporation, *Coal: The Energy Opportunity* (CBDC, Sydney, Nova Scotia, Canada, 1982).

Carlin, J. F. Jr, 'Tin', in *Mineral Facts and Problems*, USBM Bulletin 675 (USGPO, Washington, DC, 1985), pp. 847–58.

——— 'Tin', in *Minerals Yearbook, 1984, Vol. I: Metals and Minerals* (USGPO, Washington, DC, 1985), pp. 901–12.

Carter, J., 'The changing structure of energy industries in the United Kingdom', in J. M. Hollander, H. Brooks, and D. Sternlight (eds), *Annual Review of Energy*, vol. 11 (Annual Reviews, Palo Alto, CA, 1986), pp. 451–69.

Center for the Study of Marine Policy in Association with the Coastal States Organization, Coastal States are Ocean States: Proceedings of a Conference Held at the Mayflower Hotel, *Washington, D.C., 1–3 April 1987* (CSMP, University of Delaware, Newark, DL, 1987).

Central Bureau of Statistics, *Norges Offisielle Statistikk B 690: Statistical Yearbook of Norway 1987* (CBS, Oslo, 1987).

——— *Norges Offisielle Statistikk B 388: Statistical Yearbook of Norway 1983* (CBS, Oslo, 1983).

Chanda, N., 'China calls in the foreign rigs', *Far Eastern Economic Review*, vol. 105, no. 39 (1979), p. 21.

Chapman, K., 'North Sea hydrocarbons too precious to burn', *Geographical Magazine*, vol. 50, no. 5 (1978), pp. 492, 494, 498.

—— *North Sea Oil and Gas: A Geographical Perspective* (David & Charles, London, 1976).

—— 'There's too much oil!', *Geographical Magazine*, vol. 58, no. 8 (1986), pp. 378–9.

Charlier, R. H., 'Other ocean resources', in E. M. Borgese and N. Ginsburg (eds), *Ocean Yearbook 1* (University of Chicago Press, Chicago, IL, 1978), pp. 160–210.

Cheng, E., 'Stalling for more time', *Far Eastern Economic Review*, vol. 136, no. 22 (1987), pp. 73–4.

'China sends a new message to Taiwan', *Business Week* (4 Dec. 1978), p. 52.

Chirinos Garcia, M., Interview, Director General, Ministry of Energy and Mines, Lima, Peru, 7 Aug. 1986.

Chiu, H. and C-H. Park, 'Legal status of the Paracel and Spratly Islands', *Ocean Development and International Law*, vol. 3, no. 1 (1975), pp. 1–28.

Chowdhury, J. and W. P. Stadig, 'Solar energy vies for a desalting niche', *Chemical Engineering* (US), vol. 91, no. 1 (1984), pp. 42–3.

Chung, J. S., 'Advances in manganese nodule mining technology', *Marine Technology Society Journal*, vol. 19, no. 4 (1985), pp. 39–44.

Churchill, R. R., 'Maritime boundary problems in the Barents Sea', in G. Blake (ed.), *Marine Boundaries and Ocean Resources* (Croom Helm, London, 1987), pp. 147–61.

Clark, A. L., P. Humphrey, C. J. Johanson, and D. K. Pak, *Resource Assessment: Cobalt-Rich Manganese Crust Potential – Exclusive Economic Zones: US Trust and Affiliated Territories in the Pacific*, OCS Study MMS85–0006 (Minerals Management Service, DOI, Washington, DC, 1985).

Clarke, G. M. (ed.) *Industrial Minerals Directory – First Edition 1987* (Metal Bulletin Books, London, 1986).

—— Industrial Minerals Directory (Metal Bulletin Books, London, 1984).

Clemons, J. A., 'Recent developments in the law of the sea 1983–1984', *San Diego Law Review*, vol. 22, no. 4 (1985), pp. 801–38.

Clow, B. B., Letters, Executive Director, International Magnesium Association, McLean, VA, 23 June and 13 Jan. 1987.

Coakley, G. J., *Namibia: Mineral Perspectives* (USBM, DOI, Washington, DC, 1983).

Coal Mining Research Centre, *Japan's Coal Mining Industry Today* (CMRC, Tokyo, Japan, 1986).

'"Coastal States Are Ocean States" Conference', *Oceans Policy News* (April 1987), pp. 5–6.

'Cobalt: Zaïre and Zambia reaffirm price ceiling', *Minerals and Materials: A Bimonthly Survey* (April/May 1987), p. 15.

Committee on Energy and Natural Resources, US Senate, *The Geopolitics of Oil*, publication no. 96, 96th Cong., 2nd sess. (USGPO, Washington, DC, 1980).

Comptroller General of the United States, *Impediments to US Involvement in Deep Ocean Mining Can Be Overcome: Report to the Congress of the United States*, EMD–82–31 (General Accounting Office, Washington, DC, 3 Feb. 1982).

—— *Uncertainties Surround Future of US Ocean Mining: Report to the Congress of the United States*, GAO/NSIAD–83–41 (General Accounting Office, Washington, DC, 6 Sept. 1983).

'Continued Grey Zone in the Barents Sea', *Norinform*, no. 1 (13 Jan. 1987), p. 7.

'Convention on the continental shelf', Article 1, adopted 29 April 1958, UN Document A/Conf. 13/L.53.

Cooke, L. W., *Estimates of Undiscovered, Economically Recoverable Oil and Gas Resources for the Outer Continental Shelf as of July 1984* (MMS, DOI, Washington, DC, 1985).

Cooper, J., 'Delimitation of the maritime boundary in the Gulf of Maine Area', *Ocean Development and International Law*, vol. 16, no. 1 (1986), pp. 59–90.

Cooper, J. C. B., 'The oil industry of Scotland', *Scottish Bankers Magazine*, vol. 76, no. 303 (1984), pp. 85–92.

Corell, R. W., 'Research more focused for autonomous underwater vehicles', *Sea Technology*, vol. 28, no. 8 (1987), pp. 41–5.

Cormick, G. W. and A. Knaster, 'Oil and fishing industries negotiate: mediation and scientific issues', *Environment*, vol. 28, no. 10 (1986), pp. 6–15, 30.

'Corporate interests and activities in seabed mining', mimeographed data supplied by M. B. Fisk, Law of the Sea Officer, UN, NY, 8 Jan. 1987.

Corti, G. and F. Frazer, *The Nation's Oil: A Story of Control* (Graham & Troutman, London, 1983).

Council on Ocean Law, *The United States and the 1982 Law of the Sea Convention: A Synopsis of the Status of the Treaty in 1987* (COL, Washington, DC, Oct. 1987).

Couper, A. (ed.), *The Times Atlas of the Oceans* (published by Van Nostrand Reinhold for Times Books, NY, 1983).

Cowley, C., Letters, Public Relations Manager, CDM (Proprietary) Ltd, Windhoek, South West Africa/Namibia, 5 May and 12 Feb. 1987.

Craynon, J., Letter, Division of International Minerals, US Bureau of Mines, DOI, Washington, DC, 27 Feb. 1987.

Cronan, D. S., 'Marine mineral resources: reaping the mineral harvest of the deep', *Geology Today*, vol. 1, no. 1 (1985), pp. 15–19.

—— *Underwater Minerals* (Academic Press, London, 1980).

Crorkan, P. H., Interview, Manager of Operations, Lota Mine, Empresa Nacional del Carbon SA, Lota, Chile, 9 Aug. 1986.

—— Letter, Manager of Operations, Lota Mine, Empresa Nacional del Carbon SA, Lota, Chile, 1 April 1987.

Crowe, A. L. *Shell Management Annual Report, September 1982–April 1983*, Management Data Series no. 70 (Coastal Fisheries Branch, Texas Parks and Wildlife Department, Austin, TX, 1984).

Cruickshank, M., 'Marine mineral resources survey (interview)', *Sea Technology*, vol. 27, no. 8 (1986), pp. 28–31.

—— 'Marine sand and gravel mining and processing technologies', paper presented at Offshore Sand & Gravel Mining Workshop, 18–20 March 1986, Stony Brook, NY.

Cruickshank, M., J. P. Flanagan, B. Holt, and J. W. Padan, *Marine Mining on the Outer Continental Shelf*, Environmental Effects Overview, OCS Report 87–0035 (Minerals Management Service, DOI, Washington, DC, 1987).

'Cut in Norwegian oil exports', *Norinform*, no. 30 (16 Sept. 1986), p. 1.

Daintith, T. C. and I. W. Gault, 'Oil and gas national regimes', in C. M. Mason (ed.), *The Effective Management of Resources* (Frances Pinter, London, 1979).

Dam, K. W., *Oil Resources: Who Gets What How?* (University of Chicago Press, Chicago, IL, 1976).

D'Amato, A., 'An alternative to the Law of the Sea Convention', *American Journal of International Law*, vol. 77, no. 2 (1983), pp. 281–5.

Davis, J. B., Letter, Co-ordinator, Public Affairs Department, EXXON Company, USA, Houston, TX, 8 Oct. 1987.

Davis, L. L. and V. V. Tepordei, 'Sand and gravel', in *Mineral Facts and Problems 1985*, USBM Bulletin 675 (USGPO, Washington, DC, 1985), pp. 689–703.

'Declaration of Policy of the American Mining Congress', *Mining Congress Journal*, vol. 64, no. 11 (1978), pp. 79–91.

'Declaration on national licenses', *Oceans Policy News* (May 1986), p. 4.

Deep Ocean Working Group, Departmental Coordinating Committee on Ocean Mining, *Deep Ocean Mining Study: Final Report*, Division Document no. 1983–2 (Department of Energy, Mines and Resources, Ottawa, Canada, 1983).

'Deep Seabed Hard Mineral Resources Act', Public Law 96–283, 96th Cong., 28 June 1980, 94 Stat, pp. 553–86.

'Deep Seabed Mining License Amendment', *Oceans Policy News* (Nov. 1985), p. 7.

Dehais, J. A., P. L. Guyette, and W. A. Wallace, 'Onshore pressures make offshore mining viable', *Rock Products*, vol. 84, no. 6 (1981), pp. 72–6.

Dehais, J. A. and W. A. Wallace, 'Economic aspects of offshore sand and gravel mining', paper presented at Offshore Sand and Gravel Mining Workshop, 18–20 March 1986, Stony Brook, NY.

Delegation of Australia, Special Commission 2, Preparatory Commission for the International Sea-Bed Authority and for the International Tribunal for the Law of the Sea, *The Enterprise: Economic Viability of Deep Sea-Bed Mining of Polymetallic Nodules*, LOS/PCN/SCN.2/WP. 10 (UN, NY, 14 Jan. 1986).

Demarffy, A., Letter, Senior Officer, UN, NY, 14 Jan. 1986.

Dempsey, P., 'Norwegian initiative', *World Oil*, vol. 204, no. 5 (1987), p. 23.

Department of Energy, *Brown Book 1983* (DE, London, 1983).

Department of International Economic and Social Affairs, UN, *Methodologies for Assessing the Impact of Deep Sea-Bed Minerals on the World Economy*, ST/ESA/168 (UN, NY, 1986).

Department of Planning and Economic Development, State of Hawaii, and Minerals Management Service, DOI, *Mining Development Scenario for Cobalt-Rich Manganese Crusts in the Exclusive Economic Zones of the Hawaiian Archipelago and Johnston Island*, Ocean Resources Branch Contribution no. 38 (DPED and MMS, Honolulu, HA, Jan. 1987).

Desalting Plant Inventories, Inventory no. 7 (International Desalination Association, Topsfield, MA, 1980), p. 1.

'Developments concerning the international area (as of 9 July 1986)', mimeographed data supplied by M. B. Fisk, Law of the Sea Officer, UN, NY, 8 Jan. 1987.

DeVereux, M. P. and C. N. Morris, *North Sea Oil Taxation: The Development of the North Sea Tax System*, Report Series no. 6 (Institute for Fiscal Studies, London, 1983).

De Vorsey, L., 'Historical geography and the Canada–United States seaward boundary on Georges bank', in G. Blake (ed.), *Maritime Boundaries and Ocean Resources* (Croom Helm, London, 1987), pp. 182–207.

DeYoung, J. H. Jr *et al.*, *International Strategic Minerals Inventory Summary Report – Nickel*, USGS Circular 930–D (DOI, Washington, DC, 1985).

DiBona, C. J., 'Offshore oil and gas drilling: a necessity for U.S. energy security', *Public Utilities Fortnightly*, vol. 114, no. 12 (1984), pp. 36–8.

Dillon, W. P., 'Mineral resources of the Atlantic Exclusive Economic Zone', in *Exclusive Economic Zone Papers: Exclusive Economic Zone Symposium '84* (Marine Technology Society and the Institute of Electrical and Electronics Engineers Council on Oceanic Engineering, n. 1., 1984), pp. 72–8.

Dillon, W. P. and F. T. Manheim, 'Resource potential of the Western North Atlantic Basin', in P. R. Vogt and B. E. Tucholke (eds), *The Geology of North America, The Western North Atlantic Region: Vol. M* (Geological Society of North America, Boulder, CO, 1986), pp. 661–76.

Dimok, B., *An Assessment of Alluvial Sampling Systems for Offshore Placer Exploration* (Oil and Gas Lands Administration, Energy, Mines and Resources, Canada, Ottawa, 1986).

Director General, [name illegible], Letter, Geological Survey Department, Colombo, Sri Lanka, 8 April 1987.

Donges, J. B. (ed.), *The Economics of Deep-Sea Mining* (Springer-Verlag, Berlin, FRG, 1985).

'Door to OPEC Ajar', *Norinform*, no. 17 (20 May 1986), p. 1.

Doran, C. F., *Myth Oil, and Politics: Introduction to the Political Economy of Petroleum* (The Free Press, NY, 1977).

Douglass, G. Jr (comp.), *The Louisiana Shell Industry*, revised (Louisiana Shell Producers Association, n. 1., 1986).

'Drilling: Latin America', *Offshore*, vol. 44, no. 5 (1984), p. 105.

'Drilling programs hit roadblock off Nova Scotia', *Oil and Gas Journal*, vol. 85, no. 1 (1987), p. 20.

Driscoll, M. J., Letter, Professor, Department of Nuclear Engineering, Massachusetts Institute of Technology, Cambridge, MA, 13 July 1987.

Duane, D. B., 'Sedimentation and ocean engineering: placer mineral resources', in D. J. Stanley and D. J. P. Swift (eds), *Marine Sediment Transport and Environmental Management* (John Wiley, NY, 1976), pp. 535–56.

Duggan, J., 'Now is the time for the United Kingdom to support OPEC', *OPEC Bulletin*, vol. 18, no. 1 (1987), pp. 4–7.

Dzurek, D. J., 'Boundary and resource disputes in the South China Sea', in E. M. Borgese and N. Ginsburg (eds), *Ocean Yearbook 5* (University of Chicago Press, Chicago, IL, 1985), pp. 254–84.

Earney, F. C. F., 'China's offshore petroleum frontiers: confrontation? conflict? co-operation?', *Resources Policy*, vol. 7, no. 1 (1981), pp. 118–28.

—— 'Geopolitics of offshore petroleum', in H. E. Johansen, O. P. Matthews, and G. Rudzitis (eds), *Mineral Resources Development: Geopolitics, Economics, and Policy* (Westview Press, Boulder, CO, 1987), pp. 51–82.

—— 'Law of the Sea, resource use, and international understanding', *Journal of Geography*, vol. 84, no. 3 (1985), pp. 105–10.

—— 'Mining, planning, and the urban environment', *CRC Critical Reviews in Environmental Control*, vol. 7, no. 1 (1977), pp. 1–89.

—— 'New ores for old furnaces: pelletized iron', *Association of American Geographers, Annals*, vol. 59, no. 3 (1969), pp. 512–34.

—— 'Norway's offshore petroleum industry', *Resources Policy*, vol. 8, no. 2 (1982), pp. 133–42.

—— *Ocean Mining: Geographic Perspectives*, Meddelelser fra Geografisk Institutt ved Norges Handleshøyskole og Universitetet i Bergen, no. 70 (Geografisk Institutt, Bergen, Norway, 1982).

—— 'Ocean space and seabed mining', *Journal of Geography*, vol. 74, no. 9 (1975), pp. 539–47.

—— *Petroleum and Hard Minerals from the Sea* (Edward Arnold, London, 1980).

—— 'Seashells and cement in Iceland', *Marine Mining*, vol. 5, no. 3 (1986), pp. 307–19.

—— 'The geopolitics of minerals', *Focus*, vol. 31, no. 5 (1981), pp. 1–16.

—— 'The United States Exclusive Economic Zone: mineral resources', in G. Blake (ed.), *Maritime Boundaries and Ocean Resources* (Croom Helm, London, 1987), pp. 162–81.

Eckert, R. D., *The Enclosure of Ocean Resources: Economics and The Law of the Sea* (Hoover Institution Press, Stanford, CA, 1979).

Economic and Social Commission for Asia and the Pacific, UN, *Committee for Co-ordination of Joint Prospecting for Mineral Resources in Asian Offshore Areas, Proceedings of the Twenty-First Session, Bandung, Indonesia, 26 November–7 December 1984, Part I, Report of the Committee* (UN, Bangkok, Thailand, 1985).

Edmond, J. M., 'The geochemistry of ridge crest hot springs', *Oceanus*, vol. 27, no. 3 (1984), pp. 15–19.

Ehrlich, H. L., 'The role of microbes in manganese nodule genesis and degradation', in D. R. Horn (ed.), *Papers from a conference on Ferromanganese Deposits on the Ocean Floor* (National Science Foundation, Washington, DC, 1972), pp. 63–70.

'Eight Operator tasks awarded in 11th round of concessions', *Norinform*, no. 12 (31 March 1987), p. 2.

Elkan, W. and R. E. D. Bishop, 'North Sea oil: responses to employment opportunities', *Energy Economics*, vol. 7, no. 2 (1985), pp. 127–33.

Empresa Nacional del Carbon SA, *Proyecto Aumento Productividad Mina Lota, 1986–2000*, 3rd edn (ENACAR, Lota Alto, Chile, May 1986).

Endo, K., Letter, Manager, Energy and Mineral Resources Research Office, Mitsui Coal Mining Co. Ltd, Tokyo, 8 Aug. 1987.

Evans, J. R., G. S. Dabai, and C. R. Levine, 'Mining and marketing sand and gravel: outer continental shelf Southern California', *California Geology*, vol. 35, no. 12 (1982), pp. 259–76.

Falconer, R. K. H., Letter, GeoResearch Associates, Wellington, New Zealand, 25 Feb. 1987.

Fan, Y. F., Letter, Director, Department of Mines, Ministry of Economic Affairs, Taipei, Taiwan, 31 March 1987.

Farris, J., Letter, Office of Public Information, Texas Department of Water Resources, Austin, TX, 25 Oct. 1978.

'Federal ocean budgets for 1988, part I', *Sea Technology*, vol. 28, no. 2 (1987), pp. 30–4.

Federal Register, vol. 53, no. 128 (5 July 1988), pp. 25,242–60; vol. 52, no. 58 (26 March 1987), pp 9,758–66; vol. 51, no. 143 (25 July 1986), pp. 26,794–824; vol. 51, no. 68 (9 April 1986), p. 12,163; vol. 50, no. 10 (15 Jan. 1985), p. 2,264; vol. 49, no. 237 (7 Dec. 1984), p. 47,871; vol. 48, no. 60 (28 March 1983), p. 12,840; vol. 47, no. 236 (8 Dec. 1982), p. 55,313; vol. 46, no. 56 (24 March 1981), p. 18,448.

Fee, D. A. and J. O'Dea, *Technology for Developing Marginal Offshore Oilfields* (Elsevier Applied Science Publishers, London, 1986).

Feeney, W. R., 'Dispute settlement: the emerging law of the sea and East Asian maritime boundary conflicts', *Asian Profile*, vol. 8, no. 6 (1980), pp. 573–95.

Felix, D., 'Some problems in making nodule abundance estimates from sea floor photographs', *Marine Mining*, vol. 2, no. 3 (1980), pp. 293–302.

Fellerer, R., Letters, Managing Director, Arbeitsgemeinschaft meertechnisch gewinnbare Rohstoffe, Hannover, FRG, 28 Jan. and 8 Jan. 1987.

—— 'Prospecting and evaluating remote sensing techniques', Report no. 8, Seminar on the Exploitation of the Deep Seabed, presented under the auspices of the EEC for the benefit of the ACP experts to the United Nations Conference on the Law of the Sea, Brussels, Belgium, 22–25 Feb. 1977.

Ferraro Costa, E., 'Peru and the law of the sea convention', *Marine Policy*, vol. 11, no. 1 (Jan. 1987), pp. 45–57.

Figg, C., Letter, Oil and Gas Division, Department of Energy, London, 24 Nov. 1987.

Fischer, D. W., *North Sea Oil – An Environment Interface* (Universitetsforlaget, Bergen, Norway, 1981).

Fisk, M. B., Letters, Law of the Sea Officer, UN, NY, 8 Dec. 1987 and 8 Jan. 1988.

Fitzgerald, J. G., Letter, Executive Director, Petroleum Directorate, Government of Newfoundland and Labrador, St John's, 7 Sept. 1984.

Flåm, S. and G. Stensland, 'Exploration and taxation: some normative issues', *Energy Economics*, vol. 7, no. 4 (1985), pp. 237–40.

Flanagan, J., Letter, National Oceanic and Atmospheric Administration, DOC, Washington, DC, 11 Oct. 1988.

Flipse, J. E., Letter, Associate Deputy Chancellor, College of Engineering, Texas A&M University System, College Station, TX, 5 Aug. 1987.
—— An Economic Analysis of a Pioneer Deep Ocean Mining Venture, TAMU-SG-82–201, COE Report no. 262 (Sea Grant College Program, Texas A&M University, College Station, TX, Aug. 1982).
'Four thousand jobs lost in the North Sea', Norinform, no. 26 (19 Aug. 1986), p. 2.
Fowler, J. H., 'Aggregate resources in Nova Scotia', Open File Report 465 (Nova Scotia Department of Mines and Energy, Halifax, NS, Canada, 1982).
'A framework for agreement: a paper presented by Newfoundland in the Canada–Newfoundland offshore mineral resources negotiations, 12 November 1981, Montreal' (St John's, Newfoundland, 1981).
Fraser, M. J., Letter, Vice President of Technology, Cyprus Minerals Company, Englewood, CO, 30 Jan. 1987.
Frazer, J. Z., 'Manganese nodule reserves: an updated estimate', Marine Mining, vol. 1, nos 1/2 (1977), pp. 103–23.
Frazer, J. Z. and M. B. Fisk, 'Geological factors related to characteristics of sea-floor manganese nodule deposits', Deep-Sea Research, vol. 28a (1980), pp. 1,533–51.
Freeport-McMoRan Inc., Annual Report 1985 (New Orleans, LA, 1985).
Freeport-McMoRan Inc., Form 10–K, Annual Report Pursuant to Section 13 on 15(d) of the Securities Exchange Act of 1934 (New Orleans, LA, 14 April 1986).
Freeport-McMoRan Resource Partners, Limited Partnership: Prospectus (Dean Witter Reynolds Inc. et al., 20 June 1986).
Freer, R., Letter, Technical Executive, Sand and Gravel Association Ltd, London, UK, 1 April 1987.
Freire, W., 'Taking the plunge into deepwater petroleum exploration', OPEC Bulletin, vol. 17, no. 10 (Dec. 1986–Jan. 1987), pp. 12–17.
Galenson, W., A Welfare State Strikes Oil: The Norwegian Experience (University Press of America, Lanham, MD, 1986).
Gamble, J. K. Jr and M. Frankowska, 'The significance of signature to the 1982 Montego Bay Convention on the Law of the Sea', Ocean Development and International Law Journal, vol. 14, no. 2 (1984), pp. 121–60.
'Geevor tin mines expansion proceeds at a slower pace', Engineering and Mining Journal, vol. 186, no. 10 (1985), p. 22.
Glasby, G. P., 'The role of submarine volcanism in controlling the genesis of marine manganese nodules', Oceanography and Marine Biology: An Annual Review, vol. 11 (1973), pp. 27–44.
Glassner, M. I., Letter, Professor and Chairman, Department of Geography, Southern Connecticut State University, New Haven, CT, 25 Nov. 1987.
—— 'Land-locked states and 1982 Law of the Sea Convention', Marine Policy Reports, vol. 9, no. 1 (1986), pp. 8–14.
Glenn, W., 'Cool line on the Senkaku "Bandits"', Far Eastern Economic Review, vol. 100, no. 18 (1978), pp. 33–4.
Goff, R., 'Ice islands may aid Beaufort development', in Harsh Environment and Deepwater Handbook (PennWell Publishing, Tulsa, OK, 1985), pp. 121–5.
Goldberg, E. D. and G. O. S. Arrhenius, 'Chemistry of Pacific Ocean pelagic sediments', Geochimica et Cosmochimica Acta, vol. 13, nos 2–3 (1958), pp. 153–212.
Good, J. W., 'Prospects for nearshore placer mining in the Pacific Northwest', unpublished paper, Marine Resource Management Program, College of Oceanography, Oregon State University, Corvallis, OR (1987).
Gorst, I., 'China: onshore key to increased output', Petroleum Economist, vol. 53, no. 12 (1986), pp. 451–7.
Grassle, J. F., 'Hydrothermal vent animals: distribution and biology', Science, vol. 229, no. 4,715 (1985), pp. 713–17.

'Greece not seeking N. Aegean group's assets', *Oil and Gas Journal*, vol. 85, no. 20 (1987), p. 24.

Green, C. J. B., 'London Metal Exchange', *Mining Annual Review 1986* (Mining Journal Ltd, London, 1986), pp. 41–3.

Greenslate, J., 'Microorganisms in the construction of manganese nodules', *Nature*, vol. 249, no. 5,453 (1974), pp. 181–3.

Greenwald, R. J., Letter, Ocean Mining Associates, Gloucester Point, VA, 7 Dec. 1986.

Grogan, R. L., Letter, Director, Office of Management and Budget Division of Governmental Co-ordination, State of Alaska, Juneau, AK, 1 June 1987, to R. Stone, MMS, Long Beach, CA, provided by R. J. Bailey, OCS Co-ordinator, Department of Land Conservation and Development, State of Oregon, Portland, OR, 10 Aug. 1987 (mimeographed).

Gross, G. A., 'Mineral deposits on the deep seabed', *Marine Mining*, vol. 6, no. 2 (1987), pp. 109–19.

Grote, P. B. and J. Q. Burns, 'System design considerations in deep ocean mining lift systems', *Marine Mining*, vol. 2, no. 4 (1981), pp. 357–83.

Grover, D. H., 'Mining sand in the EEZ', *Sea Technology*, vol. 26, no. 2 (1985), pp. 40–2.

Halbach, P. and R. Fellerer. 'The metallic minerals of the Pacific seafloor', *GeoJournal*, vol. 4, no. 5 (1980), pp. 407–22.

Halbach, P. and F. T. Manheim, 'Potential of cobalt and other metals in ferromanganese crusts on seamounts of the Central Pacific Basin', *Marine Mining*, vol. 4, no. 4 (1984), pp. 319–36.

Halbach, P., M. Segl, D. Puteanus, and A. Mangini, 'Relationships between co-fluxes and growth rates in ferromanganese deposits from Central Pacific seamount areas', *Nature*, vol. 304, no. 5,928 (1983), pp. 716–19.

Halbouty, M. T., 'World petroleum reserves and resources with special reference to developing countries', in *Petroleum Exploration Strategies in Developing Countries: Proceedings of a United Nations Meeting Held in The Hague 16–20 March 1981* (UN Natural Resources Division, Department of Technical Co-operation for Development, published in co-operation with the UN by Graham & Troutman, London, 1982), pp. 3–16.

Halcon, N. C., Letter, Mineral Economics and Information Division, Bureau of Mines and Geo-Sciences, Ministry of Natural Resources, Manila, Philippines, 26 March 1987.

Hale, P. B., *A Re-Appraisal of Offshore Non-Fuel Mineral Development Potential*, Ocean Mining Division Document no. 1,984–2 (Energy, Mines and Resources Canada, Ottawa, 1984).

—— 'Canada's offshore non-fuel mineral resources – opportunities for development', *Marine Mining*, vol. 6, no. 2 (1987), pp. 89–108.

Hann, D., 'Political and bureaucratic pressures on U.K. oil taxation policy', *Scottish Journal of Political Economy*, vol. 32, no. 3 (1985), pp. 278–95.

—— 'The process of government and UK oil participation policy', *Energy Policy*, vol. 14, no. 3 (1986), pp. 253–61.

Hansen, J. C., 'Regional policy in an oil economy: the case of Norway', *Geoforum*, vol. 14, no. 4 (1983), pp. 353–61.

'Harald Norvik new Statoil chief', *Norinform*, no. 1 (12 Jan. 1988), p. 1.

Hardy, R. W., *China's Oil Future: A Case of Modest Expectations* (Westview Press, Boulder, CO, 1978).

Harrington, C. E., Letter, Chief Geologist, Nautical Charting Division, National Oceanic and Atmospheric Administration, DOC, Washington, DC, 3 Nov. 1987.

Harris, P. M., Letter, Institute of Geological Sciences, London, UK, 7 Nov. 1978.

Harrison, S. S., *China, Oil and Asia: Conflict Ahead?* (Columbia University Press, NY, 1977).

Haynes, B. W., 'Characterization of ocean floor minerals', in J. R. Pederson (comp. and ed.), *Research 1985: A Summary of Significant Results in Mineral Technology and Economics* (USGPO, Washington, DC, 1985), pp. 49–50.

Haynes, B. W., M. J. Magyar, and E. G. Godoy, 'Extractive metallurgy of ferromanganese crusts from the Necker Ridge Area, Hawaiian Exclusive Economic Zone', *Marine Mining*, vol. 6, no. 1 (1987), pp. 23–36.

Haynes, B. W., S. L. Law, and D. C. Barron, 'An elemental description of Pacific manganese nodules', *Marine Mining*, vol. 5, no. 3 (1986), pp. 239–76.

Haynes, B. W., *et al.*, *Laboratory Processing and Characterization of Waste Materials from Manganese Nodules*, USBM Report of Investigations 8938 (USBM, Washington, DC, 1985).

Heap, P., Letters, Senior Press Officer, British Coal, London, 16 and 30 March 1987.

Heath, G. R., 'Manganese nodules: unanswered questions', *Oceanus*, vol. 25, no. 3 (1982), pp. 37–41.

Hein, J. R. *et al.*, 'Geological and geochemical data for seamounts and associated ferromanganese crusts in and near the Hawaiian, Johnston Island and Palmyra Island Exclusive Economic Zones', USGS Open File Report 85–292 (USGS, Washington, DC, 1985).

Hein, J. R., F. T. Manheim, and W. C. Schwab, *Cobalt-Rich Ferromanganese Crusts from the Central Pacific*, OTC 5234, Offshore Technology Conference, May 1986, pp. 119–26.

Hein, J. R., L. A. Morgenson, D. A. Glague, and R. A. Koski, 'Cobalt-rich ferromanganese crusts from the Exclusive Economic Zone of the United States and nodules from the Oceanic Pacific', in D. Scholl, A. Grantz, and J. Vedder (eds), *Geology and Resource Potential of the Continental Margins of Western North America and Adjacent Ocean Basins*, American Association of Petroleum Geologists, Memoir 1986 (in press).

Heye, D. and V. Marchig, 'Relationship between the growth rate of manganese nodules from the Central Pacific and their chemical constitution', *Marine Geology*, vol. 23, nos 1–2 (1977), pp. M19–M25.

Hillman, T. C., *Manganese Nodule Resources of Three Areas in the Northeast Pacific Ocean: With Proposed Mining-Beneficiating Systems and Costs*, USBM Information Circular 8933 (USGPO, Washington, DC, 1983).

Hillman, T. C. and B. B. Gosling, *Mining Deep Ocean Manganese Nodules: Description and Economic Analysis of a Potential Venture*, USBM Information Circular 9015 (USGPO, Washington, DC, 1985).

Hinz, K., 'Assessment of the hydrocarbon potential of the continental margins' in F. Bender (ed.), *New Paths to Mineral Exploration: Proceedings of the Third International Symposium, Held in Hannover, Federal Republic of Germany at the Federal Institute of Geosciences and Mineral Resources, 27–29 October 1982* (E. Schweizerbart'sche, Verlagsbuchhandlung, Stuttgart, FRG, 1983), pp. 55–78.

Hirota, J., 'Potential effects of deep-sea minerals mining on macrozooplankton in the North Equatorial Pacific', *Marine Mining*, vol. 3, nos 1/2 (1981), pp. 19–57.

Hoagland, P. III, 'Performance requirements in ocean mineral development', *Marine Policy Reports*, vol. 9, no. 3 (1987), pp. 5–10.

—— 'Seabed mining patent activity: some first steps toward an understanding of strategic behavior', *Journal of Resource Management and Technology*, vol. 14, no. 3 (1986), pp. 211–22.

Hodgson, D. L., 'Mining the beach for diamonds at CDM', *Engineering and Mining Journal*, vol. 178, no. 6 (1977), pp. 145–51.

Holden, W. M., 'Miners under the sea – right now', *Oceans*, vol. 8, no. 1 (1975), pp. 55–7.

Hollick, A. L., 'Managing the oceans', *Wilson Quarterly*, vol. 8, no. 3 (1984), pp. 70–87.

Holly, J. H., 'Zinc', in *Mineral Facts and Problems 1985*, USBM Bulletin 675 (USGPO, Washington, DC, 1985), pp. 923–39.

Holmstrom, D., 'CIDS: island in the ice', *EXXON USA*, vol. 24, no. 3 (1985), pp. 10–15.

Holt-Jensen, A., Letter, Geografisk Institutt, Universitetet i Bergen, Bergen, Norway, 9 Nov. 1987.

—— 'Norway and the sea: the shifting importance of marine resources through Norwegian history', *GeoJournal*, vol. 10, no. 4 (1985), pp. 393–9.

Horn, D. R., B. M. Horn, and M. N. Delach, *Ferromanganese Deposits of the North Pacific Basin*, Technical Report no. 1, NSF GX–33616 (National Science Foundation, Washington, DC, 1972).

Horsfield, B. and P. B. Stone, *The Great Ocean Business* (Coward, McCann, & Geoghegan, NY, 1972).

Hotta, H., 'Recovery of uranium from seawater', *Oceanus*, vol. 30, no. 1 (1987), pp. 44–7.

Hough, G.V., 'World survey – natural gas: steady progress is maintained', *Petroleum Economist*, vol. 54, no. 8 (1987), pp. 294–6.

Hundal, H. S., R. B. Trattner, and P. N. Cheremisinoff, 'Freshwater from seawater, part 4: from screening to filtration – treating the water supply', *Water and Sewage Works*, vol. 126, no. 12 (1979), p. 25.

'Hydrocarbons correlate with sea heights', *World Oil*, vol. 204, no. 4 (1987), pp. 39–40.

'Implementation of resolution II', *Oceans Policy News* (May 1987), p. 4.

'In search of: minerals', *Marine Resource Bulletin*, vol. 18, no. 3 (1986), p. 5.

'In search of: oyster shell', *Marine Resource Bulletin*, vol. 18, no. 3 (1986), p. 4.

'In the Supreme Court of Newfoundland Court of Appeal, no. 23, Decided: 17 Febuary 1983' (St John's, Newfoundland, 1983).

'Industrial co-operation between Finmark and USSR? *Norinform*, no. 16 (19 May 1987), p. 9.

Inouye, D. K., 'Resource development in the EEZ: lead or follow, the time has come', *Sea Technology*, vol. 28, no. 6 (1987), pp. 36–7.

'Interior Department's proposed regulations for marine minerals mining on the OCS', *Oceans Policy News* (May 1987), pp. 12–13.

'International', *Oceans Policy News* (Sept. 1987), p. 1.

'International Energy Agency (IEA) ministerial communiqué', *OECD Observer*, vol. 136 (Sept. 1985), pp. 33–7.

'International issues and actions: cobalt', *Minerals and Materials: A Bimonthly Survey* (Dec. 1985/Jan. 1986), p. 11.

'International issues and actions: nickel', *Minerals and Materials: A Bimonthly Survey* (Feb./March 1986), p. 5.

'International issues and actions: nickel', *Minerals and Materials: A Bimonthly Survey* (Oct./Nov. 1985), pp. 5–6.

'Italian position on the Law of the Sea Convention', *Italy and the Law of the Sea Newsletter*, no. 15 (Jan. 1986), pp. 2–5.

Ives, G., 'Ice platform concept proven for arctic offshore drilling', *Petroleum Engineer International*, vol. 46, no. 6 (1974), p. 10.

Iwasaki, T., H. Okamura, A. Takata, and H. Tsurusaki, 'The status and prospects of sea bed mining in Japan', in J. S. Chung *et al.*, *Proceedings of the Fifth International Offshore Mechanics and Arctic Engineering (OMAE) Symposium, 13–18 April, 1986, New York, Vol. II*, (American Society of Mechanical Engineers, New York, 1986), pp. 524–30.

Jacobson, J. L. 'Law of the Sea – What now?', *Naval War College Review*, vol. 37, no. 2/seq. 302 (1984), pp. 82–99.

Jagota, S. P., 'The United Nations Convention on the Law of the Sea, 1982', in E. M. Borgese and N. Ginsburg (eds), *Ocean Yearbook 5* (University of Chicago Press, Chicago, IL, 1985), pp. 10–28.

'The Jan Mayen Agreement – A model for the Barents Sea?', *Norinform*, no. 17 (18 May 1983), p. 2.

Jannasch, H. W., 'Chemosynthesis: the nutritional basis for life at deep-sea vents', *Oceanus*, vol. 27, no. 3 (1984), pp. 73–8.

Jannasch, H. W. and M. J. Mottl, 'Geomicrobiology of deep-sea hydrothermal vents', *Science*, vol. 229, no. 4,715 (1985), pp. 717–25.

'Japan's energy evolution', *Mining Journal*, vol. 308, no. 7,906 (1987), pp. 141–3.

Jenkin, M., *British Industry and the North Sea: State Intervention in a Developing Industrial Sector* (Macmillan, London, 1981).

Johnson, C. J., *Economic and Business Investment Climates for Manganese Nodule Processing in Six Pacific Countries*, prepared for DOC Institutional Grant no. NO81AA–D–00070 (University of Hawaii Sea Grant Program, Honolulu, HA, 1985).

Johnson, C. J. and J. M. Otto, 'Manganese nodule project economics', *Resources Policy*, vol. 12, no. 1 (1986), pp. 17–28.

Johnson, C. J. et al., *Resource Assessment of Cobalt-Rich Ferromanganese Crusts in the Hawaiian Archipelago*, Report of the East-West Center and the Research Corporation of the University of Hawaii for the Minerals Management Service, DOI, Co-operative Agreement no. 14–12–0001–30177, 'Environmental studies in support of the proposed leasing for cobalt-rich crusts' (Resource Systems Institute, East-West Center, Honolulu, HA, May 1985).

Johnson, R., 'The geography of the Falkland Islands: the island's resources', *Geographical Journal*, vol. 149, pt. I, no. 1 (1983), pp. 4–7.

Jolly, J. L. W., 'Copper', in *Mineral Facts and Problems 1985*, USBM Bulletin 675 (USGPO, Washington, DC, 1985), 197–219.

Jones, M. E., *Deepwater Oil Production and Manned Underwater Structures* (Graham & Troutman, London, 1981).

Jones, T. S., 'Manganese', in *Mineral Facts and Problems 1985*, USBM Bulletin 675 (USGPO, Washington, DC, 1985), pp. 483–98.

Jones, W. B., 'Risk assessment: corporate ventures in deep seabed mining outside the framework of the UN Convention on the Law of the Sea', *Ocean Development and International Law*, vol. 16, no. 4 (1986), pp. 341–51.

Jordan, A., Letter, Manager, Public Relations, Dravo Basic Materials Co. Inc., Kenner, LA, 14 Oct. 1987.

Joyner, C. C., 'Normative evolution and policy process in the law of the sea', *Ocean Development and International Law*, vol. 15, no. 1 (1985), pp. 61–76.

Juda, L. 'The Exclusive Economic Zone: compatibility of national claims and the UN Convention on the Law of the Sea', *Ocean Development and International Law*, vol. 16, no. 1 (1986), pp. 1–58.

Jumars, P. A., 'Limits in predicting and detecting benthic community responses to manganese nodule mining', *Marine Mining*, vol. 3, nos 1/2 (1981), pp. 213–29.

Juneau, C. L. Jr, *Shell Dredging in Louisiana 1914–1984* (Louisiana Department of Wildlife and Fisheries, New Iberia, LA, 1984).

Kalyniuk, G. W., Letter, Support Manager, ESSO Resources Canada, Edmonton, Alberta, 13 March 1979.

Katchen, M. H., 'The Spratly Islands and the law of the sea: "dangerous ground" for Asian peace', *Asian Survey*, vol. 17, no. 12 (1977), pp. 1,167–81.

Kaufman, D. W. (ed.), *Sodium Chloride: The Production and Properties of Salt* (Hafner Publishing, NY, 1971).

Kemp, A. G., C. P. Hallwood, and P. W. Wood, 'The benefits of North Sea oil', *Energy Policy*, vol. 11, no. 1 (1983), pp. 119–30.

Kemp, A. G. and D. Rose, 'Impact of budget proposals', *Petroleum Economist*, vol. 54, no. 5 (1987), pp. 169–70.

Kennedy, D., 'Ocean uranium: limitless energy', *Technology Review*, vol. 87, no. 7 (1984), pp. 73–4.

Kimball, L., 'Is there a mini-treaty? Will there be one?', *Neptune*, no. 19 (March 1982), pp. 1, 7–8.

Kimbell, C. L. and W. L. Zajac, 'Minerals in the world economy', preprint from *USBM Minerals Yearbook, Vol. I: Metals and Minerals* (USGPO, Washington, DC, 1986).

Kinney, G. L., Letter, Division of International Minerals, USBM, DOI, Washington, DC, 20 Feb. 1987.

—— 'The mineral industry of Thailand', in *Minerals Yearbook, 1984, Vol. III, Area Reports: International* (USGPO, Washington, DC, 1986), pp. 797–807.

Kirk, W. S., 'Cobalt', in *Mineral Facts and Problems 1985*, USBM Bulletin 675 (USGPO, Washington, DC, 1985), pp. 171–83.

Knecht, R. W., 'Deep seabed ocean mining', *Oceanus*, vol. 25, no. 3 (1982), pp. 3–11.

Koski, R. A., W. R. Normark, J. L. Morton, and J. R. Delaney, 'Metal sulfide deposits on the Juan de Fuca Ridge', *Oceanus*, vol. 25, no. 3 (1982), pp. 42–8.

Kramer, D. A., Letters, Division of Nonferrous Metals, USBM, DOI, Washington, DC, 3 and 9 March 1987.

—— 'Magnesium', in *Mineral Facts and Problems 1985*, USBM Bulletin 675 (USGPO, Washington, DC, 1985), pp. 471–82.

—— 'Magnesium compounds', preprint from *USBM Minerals Yearbook 1985, Vol. I, Metals and Minerals* (USGPO, Washington, DC, 1986).

Ku, T. L. and W. S. Broeker, 'Radiochemical studies on manganese nodules of deep-sea origin', *Deep Sea Research*, vol. 16 (1969), pp. 625–37.

Kuczynski, I., *British Offshore Oil and Gas Policy* (Garland Publishing, NY, 1982).

Latorre, J. Cortez, Interview, Assistant Manager, Production Division, Lota Mine, Empresa Nacional del Carbon SA, Lota Alto, Chile, 9 Aug. 1986.

Lauriat, G., 'Japan Peace Treaty's effect on the oil industry', *Far Eastern Economic Review*, vol. 102, no. 40 (1979), pp. 60–1.

Lauriat, G. and M. Liu, 'Pouring trouble on oily waters', *Far Eastern Economic Review*, vol. 105, no. 39 (1979), pp. 19–21.

Laursen, F., 'Security aspects of Danish and Norwegian law of the sea policies', *Ocean Development and International Law*, vol. 18, no. 2 (1987), pp. 199–233.

Lavelle, J. W. and E. Ozturgut, 'Dispersion of deep-sea mining particulates and their effect on light in ocean surface layers', *Marine Mining*, vol. 3, nos 1/2 (1981), pp. 185–212.

Lavelle, J. W., E. Ozturgut, S. A. Swift, and B. H. Erickson, 'Dispersal and resedimentation of the benthic plume from deep-sea mining operations: a model with calibration', *Marine Mining*, vol. 3, nos 1/2 (1981), pp. 59–93.

'Law of the Sea (LOS) Convention', *Oceans Policy News* (Aug. 1987), p. 1.

'Law of the Sea Ratifications', *Special Report: The Preparatory Commission, February 27–March 23 1989* (Council on Ocean Law, Washington, DC, 1989) p. 1.

LeBlanc, L. A., 'Advanced technology', *Offshore*, vol. 47, no. 4 (1987), p. 21.

—— 'Hydrocarbon deposits affect sea level', *Offshore*, vol. 47, no. 4 (1987), p. 21.

—— 'Nations scramble for unclaimed seabed', *Offshore*, vol. 37, no. 3 (1977), pp. 41–6.

—— 'Operators probe for least-cost production', in *Harsh Environment and Deepwater Handbook* (PennWell Publishing, Tulsa, OK, 1985), pp. 49–52, 54–7.

—— 'Target: giant fields: costly and barren Alaska, Atlantic frontiers leave fewer options in U.S. search', *Offshore*, vol. 47, no. 1 (1987), pp. 36–7.

Lee, H. Y., 'Korea's future: Peking's perspective', *Asian Survey*, vol. 17, no. 11 (1977), pp. 1,088–1,102.

Lee, S-H., 'South Korea and the continental shelf issue: agreements and disagreements between South Korea, Japan and China', *Korea and World Affairs*, vol. 10, no. 1 (1986), pp. 55–71.

Lee, W-C., 'Troubles under the water: Sino-Japanese conflict of sovereignty on the continental shelf in the East China Sea', *Ocean Development and International Law*, vol. 18, no. 5 (1987), pp. 585–611.

Lenoble, J-P., Letter, President, Afernod, Ifremer, Paris, France, 30 Jan. 1987.

Levitin, M. J., 'The law of force and the force of law: Grenada, the Falklands, and humanitarian intervention', *Harvard International Law Journal*, vol. 27, no. 2 (1986), pp. 621–57.

Lewington, J., 'Agreements may increase East Coast drilling', *Globe and Mail* (Toronto), 6 April 1984, p. B4.

Lewis, T.M. and I. H. McNicoll, *North Sea Oil and Scotland's Economic Prospects* (Croom Helm, London, 1978).

Lind, T. and G. A. Mackay, *Norwegian Oil Policies* (C. Hurst, London, 1980).

Lindelof, L.A., Figure 1 of United States Patent no. 3,731,975, patented 8 May 1973.

Lockwood, M., Letter, Oceanographer, National Oceanic and Atmospheric Administration, DOC, Rockville, MD, 30 July 1987.

Loebenstein, J. R., 'Platinum-group metals', in *Mineral Facts and Problems 1985*, USBM Bulletin 675 (USGPO, Washington, DC, 1985), pp. 595–616.

Lomax, D.F., 'The investment implications of China's offshore oil development', *National Westminster Bank Quarterly Review* (Feb. 1986), pp. 50–69.

Lowe, A. V., 'The development of the concept of the contiguous zone', *British Yearbook of International Law*, vol. 52 (1981), pp. 109–69.

Lucas, J. M., 'Gold', in *Mineral Facts and Problems 1985*, USBM Bulletin 675 (USGPO, Washington, DC, 1985), pp. 323–38.

Lundgren, B., Letter, President, PetroScan AB, Göteborg, Sweden, 5 Nov. 1987.

Luoma, R. T., 'A comparative study of national legislation concerning the deep sea mining of manganese nodules', *Journal of Maritime Law and Commerce*, vol. 14, no. 2 (1983), pp. 243–68.

Lyday, P. A., 'Bromine', in *Mineral Facts and Problems 1985*, USBM Bulletin 675 (USGPO, Washington, DC, 1985), pp. 103–10.

—— 'Bromine', *Mining Engineering*, vol. 37, no. 5 (1985), pp. 463–5.

Lynd, L. E., 'Titanium', in *Mineral Facts and Problems 1985*, USBM Bulletin 675 (USGPO, Washington, DC, 1985), pp. 859–79.

Lynk, E. L. and M. G. Webb, 'An unintended consequence of the taxation system for U.K. North Sea Oil', *Scottish Journal of Political Economy*, vol. 33, no. 1 (1986), pp. 58–73.

Mabro, R., 'Significance of abolishing British National Oil Corporation', *Financial Times* (21 March 1985), as quoted in *OPEC Bulletin*, vol. 16, no. 3 (1985), pp. 33, 79.

Mabro, R., *et al.*, *The Market for North Sea Crude* (Oxford University Press, Oxford, 1986).

McDonald, A., 'Mines in a lawless sea', *Geographical Magazine*, vol. 54, no. 9 (1982), pp. 501–3.

Macdonald, E. H., *Alluvial Mining: The Geology, Technology and Economics of Placers* (Chapman & Hall, London, 1983).

McGregor, B. A. and M. Lockwood, *Mapping and Research in the Exclusive Economic Zone* (DOC, Washington, DC, 1985).

MacIssac, D., Letters, Information Officer, Cape Breton Development Corporation, Sydney, Nova Scotia, Canada, 9 April and 9 March 1987.

MacKay, D. I. and G. A. Mackay, *The Political Economy of North Sea Oil* (Westview Press, Boulder, CO, 1975).

McKelvey, V. E., *Subsea Mineral Resources*, Chapter A of USGS Bulletin 1689 (USGPO, Washington, DC, 1986).
—— 'The U.S. phosphate industry: revised prospects and potential', *Marine Technology Society Journal*, vol. 19, no. 4 (1985), pp. 65–7.
McKelvey, V. E., J. I. Tracey Jr, G. E. Stoertz, and J. G. Vedder, *Subsea Mineral Resources and Problems Related to their Development*, USGS Circular 619 (USGS, Washinton, DC 1969).
McKelvey, V. E., N. A. Wright, and R. W. Bowen, *Analysis of the World Distribution of Metal-Rich Subsea Manganese Nodules*, USGS Circular 886 (USGS, Arlington, VA, 1983).
McManus, R. 'Legal status and 1983–1984 developments', in M. B. Hatem (ed.), *Marine Polymetallic Sulfides – A National Overview and Future Needs: Workshop Proceedings, 19–20 January 1983*, Maryland Sea Grant Publication no. UM–SG–TS–83–04 (University of Maryland, College Park, MD, 1983), pp. 81–90.
McMullin, R. H., Memorandum, Acting Regional Director, Minerals Management Service, Alaska Outer Continental Shelf Region, DOI, Anchorage, AK, to T. Giordano, Program Director, Office of Strategic and International Minerals, MMS, DOI, Long Beach, CA, 10 Sept. 1987.
McMurray, G., Letter, Marine Minerals Co-ordinator, Department of Geology and Mineral Industries, State of Oregon, Portland, OR, 2 July 1987.
—— 'The Gorda Ridge technical task force: a cooperative federal-state approach to offshore mining issues', *Marine Mining*, vol. 5, no. 4 (1986), pp. 467–75.
MacQuarrie, W., Letter, Mineral Development Co-ordinator, Department of Energy and Forestry, Prince Edward Island, Charlottetown, PEI, Canada, 9 Feb. 1987.
MacRae, J. R., Letter, Wetland Resources Co-ordinator, Environmental Assessment Branch, Resource Protection Division, Texas Parks and Wildlife Department, Austin, TX, 19 May 1987.
'Magnesium: uses in cars to spur structural consumption', *Modern Metals*, vol. 40, no. 12 (1985), p. 80.
Magnuson, R. G., 'The coastal state challenge', address presented to the North Carolina Coastal States Ocean Policy Conference, Raleigh, NC, 31 Oct. 1985.
'Major oil find reported in China', *OPEC Bulletin*, vol. 18, no. 8 (1987), p. 19.
'Making the best of a bad job', *Tin International*, vol. 59, no. 8 (1986), p. 267.
El Mallakh, R., Ø. Noreng, and B. W. Poulson, *Petroleum and Economic Development: The Cases of Mexico and Norway* (D. C. Heath, Lexington, MA, 1984).
Malone, J. L., 'Freedom and opportunity: foundation for a dynamic oceans policy', *Department of State Bulletin*, vol. 84, no. 2,093 (1984), pp. 76–9.
—— 'Who needs the sea treaty?', *Foreign Policy*, no. 54 (Spring 1984), pp. 44–63.
Mangone, G. J. (ed.), *American Strategic Minerals* (Crane Russak, NY, 1984).
Manheim, F. T., Letter, Geologist, Office of Energy and Marine Geology, Atlantic-Gulf branch, USGS, DOI, Woods Hole, MA, 1 Nov. 1987.
—— Book Review of G. P. Glasby, *Marine Manganese Deposits*, in *Geochimica et Cosmochimica Acta*, vol. 42, no. 5 (1978), pp. 541–2.
—— 'Marine cobalt resources', *Science*, vol. 232, no. 4,750 (1986), pp. 600–8.
—— *Mineral Resources off the Northeastern Coast of the United States*, USGS Circular 669 (USGS, Washington, DC, 1972).
Manners, G., 'North Sea oil: benefits, costs and uncertainties', *Geoforum*, vol. 15, no. 1 (1984), pp. 57–64.
Manners, I. R., *North Sea Oil and Environmental Planning: The United Kingom Experience* (University of Texas Press, Austin, TX, 1982).
Margler, L. W., *Project Summary: Environmental Implications of Changes in the Brominated Chemicals Industry* (Environmental Protection Agency, Cincinnati, OH, Sept. 1982).

Maris, A. T. and J. R. Paulling, 'Analysis and design of the cold-water pipe (CWP) for the OTEC system with application to OTEC-1', *Marine Technology*, vol. 17, no. 3 (1980), pp. 281–9.

Marr, A., *Foreign Nationals in North-East Scotland*, North Sea Study Occasional Paper no. 7 (Department of Political Economy, University of Aberdeen, Scotland, 1975).

Marshall, H. R. Jr, 'Disputed areas influence OCS leasing policy', *Offshore*, vol. 45, no. 5 (1985), pp. 99–102.

Marsteller, T. F. Jr and R. L. Tucker, 'Problems of the technology transfer provisions in the Law of the Sea Treaty', *IDEA – The Journal of Law and Technology*, vol. 24, no. 4 (1983), pp. 167–80.

Martino, O., Letter, Division of International Minerals, USBM, DOI, Washington, DC, 3 March 1987.

May, E. B., 'A survey of the oyster and oyster shell resources of Alabama', *Alabama Marine Resources Bulletin*, no. 4 (Feb. 1971), pp. 1–53.

Mayo, E., 'Tin', *Mining Annual Review 1986* (Mining Journal Ltd, London, 1986), pp. 34–40.

Mead, W. J., A. Moseidjord, D. D. Muraoka and P. E. Sorensen, *Offshore Lands: Oil and Gas Leasing and Conservation on the Outer Continental Shelf* (Pacific Institute for Public Policy, San Francisco, CA, 1985).

Melancon, J. L. and S. G. Bokun, *Evaluation of Reef Shell Embankment: Final Report*, Louisiana Highway Research Report no. FHWA/LA–81/129 (Louisiana Department of Transportation and Development, Baton Rouge, LA, 1980).

Mercando, L., Telephone Interview, Director of Process Metallurgy, Kennocott Corporation, Salt Lake City, UT, 30 Jan. 1987.

Mero, J. L., 'Ocean mining: an historical perspective', *Marine Mining*, vol. 1, no. 3 (1978), pp. 243–55.

—— *The Mineral Resources of the Sea* (Elsevier, Amsterdam, 1965).

—— *The Mining and Processing of Deep-Sea Manganese Nodules* (Institute of Marine Mineral Resources, Berkeley, CA, 1959).

Messalum, L., Letter, Vice President, Ocean Management Inc., NY, NY, 9 Jan. 1987.

Metal Mining Agency of Japan, *Recovery of Uranium from Seawater* (MMAJ, Tokyo, Aug. 1986).

Mew, M., 'Phosphate rock', *Mining Annual Review 1986* (Mining Journal Ltd, London, 1986), pp. 101–3.

Meyerhoff, A. A. and J-O. Willums, 'China's potential still a guessing game', *Offshore*, vol. 39, no. 1 (1979), p. 54.

Mikami, H. M., Letter, Consultant, Pleasanton, CA, 17 June 1987.

—— 'Refractory magnesia', paper presented at the Conference for Raw Materials for Refractories, Tuscaloosa, AL, 8–9 Feb. 1982.

Miles, A. J., 'The marine sand and gravel industry of the United Kingdom', *World Dredging and Marine Construction*, vol. 21, no. 8 (1985), pp. 28–32.

Miller, C. K. and J. H. Fowler, 'Development potential for offshore placer and aggregate resources of Nova Scotia, Canada', *Marine Mining*, vol. 6, no. 2 (1987), pp. 121–39.

Miller, M. M., Letter, Department of Nuclear Engineering and Energy Laboratory, Massachusetts Institute of Technology, Cambridge, MA, 29 Jan. 1987.

Mineral Policy Sector, Energy, Mines and Resources Canada, *The Canadian Minerals and Metals Sector: A Framework for Discussion and Consultation* (EMRC, Ottawa, Feb. 1985).

Minerals Management Service, DOI, *Briefing Book: OCS Marine Mining* (DOI, Washington, DC, July 1987).

—— *Draft Environmental Impact Statement: Proposed Marine Mineral Lease Sale in the Hawaiian Archipelago and Johnston Island Exclusive Economic Zones*,

Ocean Resources Branch Contribution no. 40 (Department of Planning and Economic Development, Honolulu, HA, Jan, 1987).

—— *Proposed Notice of Sale: Oil and Gas Lease Sale No. 97, Outer Continental Shelf Beaufort Sea (January 1988)* (DOI, Washington, DC, 1987).

Ministry of Industry and Crafts, *Petroleum Exploration North of 62° N*, Report no. 91 to the Storting (1975–6) (MIC, Oslo, 1976).

Mitchell, R. and P. Benson, 'Control of marine biofouling in heat exchanger systems', *Marine Technology Society Journal*, vol. 15, no. 4 (1981), pp. 11–20.

Moffitt, D., 'Ocean mining: a framework for North Carolina and other coastal states', *Legal Tides*, vol. 1, no. 4 (1986), p. 7.

Moore, J. R., Letters, Professor, Department of Marine Studies, University of Texas, Austin, TX, 12 March 1987 and 21 Dec. 1986.

—— 'Alternative sources of strategic minerals from the seabed', in G. J. Mangone (ed.), *American Strategic Minerals* (Crane Russak, NY, 1984), pp. 85–108.

Morris, M. A., 'Southern cone maritime security after the 1984 Argentine–Chilean Treaty of Peace and Friendship', *Ocean Development and International Law*, vol. 18, no. 2 (1987), pp. 235–54.

Morse, D. E., 'Salt' and 'Sulfur', in *Mineral Facts and Problems 1985*, USBM Bulletin 675 (USGPO, Washington, DC, 1985), pp. 679–88 and 783–97, respectively.

—— 'Salt', in *Minerals Yearbook 1984, Vol. I, Metals and Minerals* (USGPO, Washington, DC 1985), pp. 763–74.

Moseley, R., 'A case of coexistence', *Chicago Tribune Magazine* (16 Aug. 1987), section 10, pp. 14–15, 18–23.

Mullins, H. T. and R. F. Rasch, 'Sea-floor phosphorites along the Central California Margin', *Economic Geology*, vol. 80, no. 3 (1985), pp. 696–715.

Murphy, C., 'LOS Preparatory Commission begins work', *Soundings*, vol. 8, no. 2 (1983), p. 1.

Murray, J., 'On the distribution of volcanic debris over the floor of the ocean and its character, source and some of the products of its disintegration and decomposition', *Proceedings of the Royal Society of Edinburgh*, vol. 9 (Royal Society, Edinburgh, 1876), pp. 247–61.

Murray, J. and A. F. Renard, 'Report on deep-sea deposits based on specimens collected during the voyage of H.M.S. *Challenger*, 1872–1876', in C. W. Thomson (ed.), *Report on the Scientific Results of the Voyage of H.M.S. 'Challenger'*, vol. 5 (Eyre & Spottiswoode, London, 1981), pp. 1–525.

Murray, J. and R. Irvine, 'On the manganese oxides and manganese nodules in marine deposits', *Transactions: Royal Society of Edinburgh*, vol. 37 (Royal Society, Edinburgh, 1894), pp. 721–42.

Myklebust, S., B. Stenseth, and P. E. Søybe (eds), *North Sea Oil and Gas Yearbook: Comparative Statistical Analysis, 1986* (Central Bureau of Statistics of Norway and Universitetsforlaget AS, Stavanger, 1987).

'NRD used to chart Norwegian hydrocarbons', *Offshore*, vol. 46, no. 4 (1986), p. 68.

National Academy of Sciences, *Oil in the Sea: Inputs, Fates, and Effects* (National Academy Press, Washington, DC, 1985).

National Advisory Committee on Oceans and Atmosphere, *A Report to the President and the Congress, Fifteenth Annual Report* (National Technical Information Center, DOC, Springfield, VA, 30 June 1986).

—— *An Assessment of the Roles and Missions of the National Oceanic and Atmospheric Administration* (Texas A&M University, College Station, TX, 1987).

'National Advisory Committee on Oceans and Atmosphere (NACOA)', *Oceans Policy News* (May 1986), p. 13.

National Coal Board, *Coal in Northumberland and Durham* (NCB, London, 1983).

'National mining licenses', *Oceans Policy News* (March-April 1985), p. 4.

'National Oceans/Marine Policy Commission', *Oceans Policy News* (March 1987), pp. 7–8.

'National Oceans Policy Commission', *Oceans Policy News* (Aug. 1987), p. 5.

'National Oceans Policy Commission', *Oceans Policy News* (June 1987), pp. 5–6.

National Science Foundation, *Deep Sea Searches: The Story of the Seabed Assessment Program* (NSF, Washington, DC, 1975).

'Natural resources of the subsoil and sea bed of the continental shelf', Presidential Proclamation no. 2,667, *Federal Register*, vol. 10 (28 Sept. 1945), p. 12,303.

Nelsen, B., Letter, Fulbright Scholar, Norsk Utenrikspolitik Institutt, Oslo, Norway, 6 Nov. 1987.

Nelson, C. W., *Magnesium Supply and Demand Report* (International Magnesium Association, McLean, VA, 1985).

Neudecker, S., 'Coral mining in Sri Lanka', *Sea Frontiers*, vol. 22, no. 4 (1976), pp. 215–22.

'New analytical technique boosts odds in exploration', *Ocean Industry*, vol. 20, no. 5 (1985), pp. 23–4.

'New offshore fields boost Brazilian gas reserves', *Offshore*, vol. 46, no. 11 (1986), pp. 66, 69.

Nichols, M., 'A gathering frontier storm', *Maclean's*, vol. 99, no. 27 (1986), pp. 32–3.

'Nickel', *Mining Annual Review, 1986* (Mining Journal Ltd, London, 1986), pp. 66–7.

Nordquist, M. H., 'Foreword', *San Diego Law Review*, vol. 22, no. 4 (1985), pp. 725–31.

Noreng, Ø., 'Energy policy and prospects in Norway', in J. M. Hollander, H. Brooks, and D. Sternlight (eds), *Annual Review of Energy*, vol. 11 (Annual Reviews, Palo Alto, CA, 1986), pp. 393–415.

—— 'The international petroleum game and Norway's dilemma', *Co-operation and Conflict*, vol. 17, no. 2 (1982), pp. 85–93.

—— *The Oil Industry and Government Strategy in the North Sea* (Croom Helm, London, 1980).

'North Sea strike affects Norway's treasury', *Norinform*, no. 13 (22 April 1987), p. 7.

'Norway curbs oil production', *Norinform*, no. 2 (20 Jan. 1987), p. 1.

'Norway edges towards staggered field developments', *OPEC Bulletin*, vol. 18, no. 1 (1987), pp. 57–8.

'Norway makes no changes in oil taxation', *Norinform*, no. 14 (5 May 1987), p. 6.

'Norway trims oil output', *OPEC Bulletin*, vol. 17, no. 8 (1987), p. 18.

'Norway will not cut oil production yet', *Norinform*, no. 34 (29 Oct. 1985), p. 3.

'Norway would like early meeting with OPEC', *Norinform*, no. 26 (1 Sept, 1987), p. 2.

'Norway's tax plan sparks opposition', *Oil and Gas Journal*, vol. 84, no. 30 (1986), pp. 42–3.

Norwegian Petroleum Directorate, *Petroleum Outlook 1986* (Stavanger, 1985).

The Ocean and the Future, 'Hearings', before the Subcommittee on Oceanography of the Committee on Merchant Marine and Fisheries, House of Representatives, 99th Cong., 1st sess., 24 Oct. 1985 (USGPO, Washington, DC, 1986).

'Ocean Mineral Resources Development Act', H. R. 2048, House of Representatives, 99th Cong., 1st sess., 16 April 1985.

Odell, P. R., 'The economic background to North Sea oil and gas development', in M. Saeter and I. Smart (eds), *Political Implications of North Sea Oil and Gas* (Universitetsforlaget, Oslo, 1975), pp. 51–63.

Office for Ocean Affairs and the Law of the Sea, UN, *Law of the Sea Bulletin*, no. 9 (April 1987), pp. 3–17.

Office of Ocean and Coastal Resource Management, NOAA, DOC, *Deep Seabed Mining: Draft Environmental Impact Statement* (DOC, Washington, DC, May 1984).

Office of Saline Water, DOI, *The A-B-Seas of Desalting* (USGPO, Washington, DC, 1968).

Office of Technology Assessment, US Congress, *Marine Minerals: Exploring Our New Ocean Frontier* (USGPO, Washington, DC, 1987).

—— *Oil and Gas Technologies for the Arctic and Deepwater* (USGPO, Washington, DC, 1985).

Office of the Special Representative of the Secretary-General for the Law of the Sea, UN, *Law of the Sea Bulletin*, no. 2 (March 1985), pp. ii–iv.

Office of Water Research and Technology, DOI, *Electrodialysis Technology* (USGPO, Washington, DC, 1979).

—— *The A-B-C of Desalting* (USGPO, Washington, DC, 1977).

'Offshore drilling for Thai Sn', *Mining Magazine*, vol. 148, no. 1 (1983), p. 12.

'Offshore gas production (MMcfd)', *Offshore*, vol. 47, no. 5 (1987), p. 51.

'The offshore oil dilemma', *Sunset*, vol. 177, no. 5 (1986), p. 276.

Oglesbee, R. M. and L. G. Kuhlman, 'Weather, depths impact subsea well projects', in *Harsh Environment and Deepwater Handbook* (PennWell Publishing, Tulsa, OK, 1985), pp. 59–63.

Ohuchi, T., Letter, Mining Engineer, Production Management Department, Mitsui Coal Mining Co. Ltd, Tokyo, Japan, 12 June 1987.

'Oil – Norway to assist Iceland', *Norinform*, no. 8 (1 March 1983), p. 4.

'Oil price fall threatens state revenue', *Norinform*, no. 3 (28 Jan. 1986), p. 1.

'Oil prospecting moves north', *Norinform*, no. 31 (8 Oct. 1985), pp. 2–3.

'Oil revenues may be decimated', *Norinform*, no. 25 (15 July 1986), p. 2.

'Oil taxation', *Finance Act 1987*, chap. 16, pt V (HMSO, London, 15 May 1987), pp. 41–9.

'Oil's up: Santa Barbara', *The Economist*, vol. 301, no. 7,476 (1986), pp. 33–4.

Osborne, D. G. and K. Atkinson, 'Tin off the north coast of Cornwall and offshore testing of beneficiation equipment', *Marine Mining*, vol. 2, nos 1/2 (1979), pp. 45–57.

Østreng, W., 'Delimitation arrangements in Arctic seas', *Marine Policy*, vol. 10, no. 2 (1986), pp. 132–54.

—— 'Norway's law of the sea policy in the 1970s', *Ocean Development and International Law*, vol. 11, nos 1/2 (1982), pp. 77–8.

—— 'Norwegian petroleum policy and ocean management: the need to consider foreign interests', *Cooperation and Conflict*, vol. 17, no. 2 (1982), pp. 117–38.

—— 'The politics of continental shelves: the south China Sea in a comparative perspective', *Cooperation and Conflict*, vol. 20, no. 4 (1985), pp. 253–77.

—— *The Soviet Union in Arctic Waters*, Law of the Sea Institute Occasional Paper no. 36 (LSI, University of Hawaii, Honolulu, HA, 1987).

Owen, R. M. and J. R. Moore, 'Sediment dispersal patterns as clues to placer-like platinum accumulation in and near Chagvan Bay, Alaska', paper presented at the Eighth Annual Offshore Conference, Houston, TX, 3–6 May 1976.

Oyama, T., Letter, General Manager, Technical Department, Technology Research Association of Manganese Nodules Mining System, Tokyo, Japan, 25 June 1987.

Ozretich, R. J., 'Increased oxygen demand and microbial biomass', *Marine Mining*, vol. 3, nos 1/2 (1981), pp. 109–18.

Ozturgut, E. *et al.*, *Deep Ocean Mining of Manganese Nodules in the North Pacific: Pre-Mining Environmental Conditions and Anticipated Mining Effects*,

NOAA Technical Memorandum, ERL MESA–33 (DOC, Washington, DC, 1978).

Padan, J., 'Development of metalliferous oxides from manganese nodules', *Marine Technology Society Journal*, vol. 19, no. 4 (1985), pp. 31–8.

Palmore, R. D., Telephone Interview, Manager, Geology and Survey Department, Dravo Basic Materials Co. Inc., Mobile, AL, 31 March 1987 and Letter, 6 March 1987.

Panel on the Law of Ocean Uses, *Deep Seabed Mining and the 1982 Convention on the Law of the Sea* (Council on Ocean Law, Washington, DC, 25 Sept. 1987).

Pardo, A., 'Ocean space and mankind', *Third World Quarterly*, vol. 6, no. 3 (1984), pp. 559–73.

Pasho, D. W., 'Canada and ocean mining', *Marine Technology Society Journal*, vol. 19, no. 4 (1985), pp. 26–30.

—— *Continuous Line Bucket Consortia: Appendix*, Internal Report RMB 1976–8 (Resource Management and Conservation Branch, Energy, Mines and Resources Canada, EMR, Ottawa, n.d.).

—— *The United Kingdom Offshore Aggregate Industry: A Review of Management Practices and Issues* (Ocean Mining Division, Canada Oil and Gas Lands Administration, Ottawa, 1986).

Pasho, D. W. and D. E. C. King, *Processing Sector Cost Estimates for the Ocean Management Inc. Manganese Nodule Processing Facility*, Mineral Policy Sector Internal Report MRI 82/7 (Department of Energy, Mines and Resources Canada, Ottawa, Aug. 1980).

Paul, R. G., Telephone Interview, Minerals Management Service, DOI, Long Beach, CA, 24 July 1987.

—— 'Development of metalliferous oxides from cobalt-rich manganese crusts', *Marine Technology Society Journal*, vol. 19, no. 4 (1985), pp. 45–9.

Peck, D. L., 'The U.S. Geological Survey Program and Plans in the EEZ', in *Symposium Proceedings: A National Program for the Assessment and Development of the Mineral Resources of the United States Exclusive Economic Zone, 15–17 November 1983*, USGS Survey Circular 929 (USGS, Alexandria, VA, 1984).

Peck, D. L. and G. W. Hill, 'The United States Geological Survey's Program in the Exclusive Economic Zone (EEZ)', in *Proceedings: The Exclusive Economic Zone Symposium Exploring the New Ocean Frontier, Washington, DC., 2–3 October, 1985* (DOC, Washington, DC, 1986), pp. 21–30.

Peckford, B. (Letter to Peckford), Premier of Newfoundland and Labrador, 14 June 1984, pp. 4–5, as provided by personal correspondence with J. G. Fitzgerald, Executive Director, Petroleum Directorate, Government of Newfoundland and Labrador, St John's, 7 Sept. 1984.

Pederson, J. R. (comp. and ed.), 'Minerals availability: titanium', in *Bureau of Mines Research 1985: A Summary of Significant Results in Mineral Technology and Economics* (USGPO, Washington, DC, 1985), pp. 75–6.

Peters, R. and M. Williamson, 'Design for a deep-ocean rock core drill', *Marine Mining*, vol. 5, no. 3 (1986), pp. 321–9.

Peters, W. C., *Exploration and Mining Geology* (John Wiley, NY, 1978).

PetroScan AB, 'Company information', mimeographed data provided to the author, 29 Sept. 1986, Göteborg, Sweden.

—— *The NRD-Method for Location of Offshore Hydrocarbon Deposits* (PetroScan, Göteborg, Sweden, 1985).

—— *The NRD-Method: What You May Not Realize about the Waves of the Sea* (PetroScan, Göteborg, Sweden, 1987).

'Phillips gives up field stake', *Petroleum Economist*, vol. 53, no. 4 (1987), p. 143.

Pina-Jordan, J. G., 'Measurement and modelling of uranium and strategic element sorption by amidoxime resins in natural seawater', unpublished MS Thesis, Texas A&M University, College Station, TX, 1985.

Pinto, M. C. W., 'Emerging concepts of the law of the sea: some social and cultural impacts', in J. G. Richardson (ed.), *Managing the Ocean: Resources, Research, Law* (Lomond Publications, Mt Airy, MD, 1985), pp. 297–309.

Piper, D. Z., J. R. Basler, and J. L. Bischoff, 'Oxidation state of marine manganese nodules', *Geochimica et Cosmochimica Acta*, vol. 48, no. 11 (1984), pp. 2,347–55.

Piper, D. Z. and M. E. Williamson, 'Composition of Pacific Ocean ferromanganese nodules', *Marine Geology*, vol. 23, no. 4 (1977), pp. 285–303.

Playford, P. E., Letter, Director, Geological Survey of Western Australia, Perth, WA, 11 May and 16 March 1987.

Post, A. M., *Deepsea Mining and the Law of the Sea* (Martinus Nijhoff, The Hague, 1983).

Potter, M. J., 'Feldspar', in *Mineral Facts and Problems 1985*, USBM Bulletin 675 (USGPO, Washington, DC, 1985), pp. 253–63.

Power, L. D., L. D. Finn, R. W. Beck, and J. M. Hulett, 'Tests recognize guyed tower potential', *Offshore*, vol. 38, no. 5 (1978), pp. 218, 221–2, 224, 226, 229.

Prescott, J. R. V., *The Maritime Political Boundaries of the World* (Methuen, London, 1985).

Pressler, J. W., 'Gem stones', in *Mineral Facts and Problems 1985*, USBM Bulletin 675 (USGPO, Washington, DC, 1985), pp. 305–15.

Prime, G., Letters, Geologist, Nova Scotia Department of Mines and Energy, Halifax, NS, 7 March and 7 Jan. 1987.

'Production progresses in 2 U.K. fields', *Offshore*, vol. 38, no. 5 (1978), pp. 230, 232.

'Proposed revision of OMI deep seabed mining license', *Oceans Policy News* (June 1986), p. 5.

Quarry Materials Act, No. 45, An Act Respecting the Acquisition of Rights to Quarry Materials within the Province (St John's, Newfoundland, 11 June 1976).

Quarry Materials (Amendment) Act, Chapter 33, An Act to Amend the Quarry Materials Act, 1976, (St John's, Newfoundland, 5 April 1977).

Quinlan, M., 'Companies press for urgent action', *Petroleum Economist*, vol. 53, no. 6 (1986), pp. 205–10.

Rabb, W., 'Physical and chemical features of Pacific deep sea manganese nodules and their implication to the genesis of nodules', in D. R. Horn (ed.), *Papers from a Conference on Ferromanganese Deposits on the Ocean Floor* (National Science Foundation, Washington, DC, 1972), pp. 31–49.

von Rad, U., Interview, Bundesanstalt für Geowissenschaften und Rohstoffe, Hannover, FRG, 26 Sept. 1986.

—— 'Outline of SONNE Cruise SO-17 on the Chatham Rise phosphorite deposits east of New Zealand', in U. von Rad and H-R. Kudrass (comps), *Phosphate Deposits on the Chatham Rise, New Zealand*, Geologisches Jahrbuch, Reihe D, Heft 65, Bundesanstalt für Geowissenschaften und Rohstoffe und den Geologischen Landesämtern in der Bundesrepublik Deutschland (Alfred-Bentz-Haus, Hannover, FRG, 1984), pp. 5–24.

Ramakrishna, K., R. E. Bowen, and J. H. Archer, 'Outer limits of continental shelf: a legal analysis of Chilean and Ecuadorian Island claims and US response', *Marine Policy*, vol. 11, no. 1 (1987), pp. 58–68.

Ratiner, L. S., 'The law of the sea: a crossroads for American foreign policy', *Foreign Affairs*, vol. 60, no. 5 (1982), pp. 1,006–21.

'Record Norwegian oil production but reduced income', *Norinform*, no. 1 (12 Jan. 1988), p. 6.

Redden, J., 'Brazil sees giant possibilities', *Offshore*, vol. 47, no. 2 (1987), pp. 29–30.

—— 'The world offshore: California', *Offshore*, vol. 46, no. 4 (1986), p. 11.
—— 'The world offshore: California', *Offshore*, vol. 44, no. 7 (1984), p. 11.
—— 'The world offshore: Newfoundland', *Offshore*, vol. 44, no. 9 (1984), pp. 11–13.
Reed, S. A., *Desalting Seawater and Brackish Water: 1981 Cost Update*, Contract no. W-7405-eng-26, OWRT no. 14–34–0001–1440 (Oak Ridge National Laboratory, Oak Ridge, TN, 1982), pp. 32–41.
Regan, R., 'Nickel keeps looking for a better price', *Iron Age Metals Producer*, vol. 229, no. 9 (1986), pp. 62–4.
'Registration of Pioneer Investors in the International Sea-Bed Area in Accordance with Resolution II of the Third United Nations Conference on the Law of the Sea', *Law of the Sea Bulletin*, Special Issue II (April 1988), pp. 6, 25, 82, 116.
'Regulation and taxation of United Kingdom oil and gas exploration and production' (Department of Energy, London, 1987), p. 3 (mimeographed), supplied by Letter: C. Figg, Oil and Gas Division, Department of Energy, London, UK, 24 Nov. 1987.
Reimnitz, E., P. Barnes, and L. Phillips, 'Evidence of 60 metre deep arctic pressure-ridge keels', in *Harsh Environment and Deepwater Handbook* (PennWell Publishing, Tulsa, OK, 1985), pp. 41–4, 46–7.
'Reverse osmosis desalination', *Mechanical Engineering*, vol. 103, no. 6 (1981), p. 72.
Richardson, R., 'South Korea poised to drill', *Far Eastern Economic Review*, vol. 105, no. 37 (1979), p. 63.
'Riding out the storm: United Kingdom', *World Oil*, vol. 205, no. 2 (1987), pp. 48, 50–1.
Riggs, S. R., 'Future frontier for phosphate in the Exclusive Economic Zone ' Continental shelf of southeastern United States', in M. Lockwood and G. Hill, *Proceedings: Exclusive Economic Zone Symposium: Exploring the New Ocean Frontier, Washington, DC., 2–3 October 1985* (DOC, Rockville, MD, 1986), pp. 97–107.
—— 'Geologic framework of phosphate research in Onslow Bay, North Carolina continental shelf', *Economic Geology*, vol. 80, no. 3 (1985), pp. 716–38.
Riley, J. G., 'How Imperial built first arctic island', *Petroleum Engineer International*, vol. 46, no. 1 (1974), pp. 25–8.
—— 'The construction of artificial islands in the Beaufort Sea', *Journal of Petroleum Technology*, vol. 28, no. 4 (1976), pp. 365–71.
Riva, J. P. Jr, *World Petroleum Resources and Reserves* (Westview Press, Boulder, CO, 1983).
Robinson, C., 'Oil depletion policy in the United Kingdom', *Three Banks Review*, no. 135 (Sept. 1982), pp. 3–16.
—— 'The Errors of North Sea Policy', *Lloyds Bank Review* (July 1981), pp. 14–33.
Robson, G. G., *Platinum: 1987* (Johnson Matthey, London, May 1987).
—— *Platinum: 1986 Interim Review* (Johnson Matthey, London, 1986).
da Rocha Lima, L., Interview, Director, Companhia Nacional de Alcalis, Rio de Janeiro, Brazil, 18 Aug. 1986.
Rockwell, M. C. and K. A. MacDonald, 'Processing technology for the recovery of placer minerals', *Marine Mining*, vol. 6, no. 2 (1987), pp. 161–75.
Roels, O.A., 'From the deep sea: food, energy, and fresh water', *Mechanical Engineering*, vol. 102, no. 6 (1980), pp. 36–43.
Rogich, D. G., Internal Report on the Eleventh Joint Meeting of the Marine Mining Panel of the United States and Japan Program on National Resources, at Tokyo, Fukuoka, Nagasaki, and Tsukuba, 12–26 Oct. 1986, US Bureau of Mines, (Washington, DC, 1986, mimeographed).

Rona, P. A., 'Hydrothermal mineralization at slow-spreading centers: Red Sea, Atlantic Ocean, and Indian Ocean', *Marine Mining*, vol. 5, no. 2 (1985), pp. 117–45.

—— 'Hydrothermal mineralization at slow spreading centers: the Atlantic model', *Marine Mining*, vol. 6, no. 1 (1987), pp. 1–7.

—— 'Potential mineral and energy resources at submerged plate boundaries', *Marine Technology Society Journal*, vol. 19, no. 4 (1985), pp. 18–25.

—— *et al.*, 'Black smokers, massive sulphides and vent biota at the Mid-Atlantic Ridge', *Nature*, vol. 321, no. 6,065 (1986), pp. 33–7.

Rose, F., 'Fish – and perhaps oil – lie at the bottom of US and Canada dispute', *Wall Street Journal* (13 April 1984), p. 30.

Ross, A. C., 'The United Kingdom's experience with North Sea oil and gas', *IDS Bulletin*, vol. 17, no. 4 (1986), pp. 42–7.

Rothe, P., 'Marine geology: mineral resources of the sea', in J. G. Richardson (ed.), *Managing the Ocean: Resources, Research, Law* (Lomond Publications, Mt Airy, MD, 1985), pp. 17–28 and by the same title in *Impact of Science on Society*, vol. 33, nos 3/4 (1983), pp. 357–66.

Roucek, J. S., 'The geopolitics of the Antarctic: the land is free for scientific work but its wealth of minerals has excited imperialist claims', *American Journal of Economics and Sociology*, vol. 45, no. 1 (1986), pp. 69–77.

Rowland, C. and D. Hann, *The Economics of North Sea Taxation* (St Martin's Press, NY, 1987).

Rowland, R. W., M. R. Goud, and B. A. McGregor, *The US Exclusive Economic Zone – A Summary of its Geology, Exploration, and Resource Potential*, USGS Circular 912 (DOI, Alexandria, VA, 1983).

Rowland, T.J., 'Non-energy marine mineral resources of the world oceans', *Marine Technology Society Journal*, vol. 19, no. 4 (1985), pp. 6–17.

Ruffra, P. S., 'Watt v. California: Supreme Court sinks consistency review of offshore oil leases', *Columbia Journal of Environmental Law*, vol. 10, no. 1 (1985), pp. 131–47.

de Ruiter, P. A. C., 'The Gabon and Congo Basin salt deposits', *Economic Geology*, vol. 74, no. 2 (1979), pp. 419–31.

Ryall, P. J. C., 'Remote drilling technology, *Marine Mining*, vol. 6, no. 2 (1987), pp. 149–60.

Ryan, P. R., 'The Exclusive Economic Zone', *Oceanus*, vol. 27, no. 4 (1984/85), pp. 3–4.

Saeter, M., 'Natural gas: new dimensions of Norwegian foreign policy', *Cooperation and Conflict*, vol. 17, no. 2 (1982), pp. 139–50.

Samson, J., 'Compilation of information on polymetallic sulfide deposits and occurrences off the West Coast of Canada', Draft Working Paper (Canada Oil and Gas Lands Administration, Energy, Mines and Resources Canada, Ottawa, Oct. 1985).

Samuels, M. S., *Contest for the South China Sea* (Methuen, London, 1982).

Sanderson, P., 'North Sea oil and the UK economy: boon or bane?', *Barclays Review*, vol. 59, no. 4 (1984), pp. 84–9.

Schade, J. P., 'Current and future uses of nickel: defending existing markets and searching for new ones', *Minerals and Materials: A Bimonthly Survey* (April/May 1986), pp. 6–9.

Scott, M. R. *et al.*, 'Rapidly accumulating manganese deposits from the median valley of the Mid-Atlantic Ridge', *Geophysical Research Letters*, vol. 1, no. 8 (1974), pp. 355–8.

'Sea law – "A rendezvous with history"', *UN Chronicle*, vol. 19, no. 6 (1982), pp. 3–15.

'Sea law: their reasons why', *UN Chronicle* vol. 19, no. 6 (1982), pp. 16–22.

'Seabed mining: Soviet Bloc forms seabed mining venture', *Minerals and Materials: A Bimonthly Survey* (April/May 1987), p. 17.

See, D. S., 'Solar salt', in D. W. Kaufman (ed.), *Sodium Chloride: The Production and Properties of Salt* (Hafner Publishing, NY, 1971), pp. 96–108.

Segal, J., 'Reality intrudes into oil plans', *Petroleum Economist*, vol. 46, no. 5 (1979), pp. 186–7.

Sekino, M. *et al.*, 'Reverse osmosis modules for water desalination', *Chemical Engineering Progress*, vol. 81, no. 12 (1985), pp. 52–6.

Shackleton, Lord [E. A. A.], 'The Falkland Islands and their history', *Geographical Journal*, vol. 149, pt I, no. 1 (1983), pp. 1–4.

Sharpless, J., Letters, Secretary for Environmental Affairs, State of California, Sacramento, CA, 19 June 1987 and 30 Sept. 1986, to the Minerals Management Service, DOI, provided by R. J. Bailey, OCS Co-ordinator, Department of Land Conservation and Development, State of Oregon, Portland, OR, 10 Aug. 1987 (mimeographed).

Shingleton, B., 'UNCLOS III and the struggle for law: the elusive customary law of seabed mining', *Ocean Development and International Law Journal*, vol. 13, no. 1 (1983), pp. 33–63.

Shoda, Y., Letter, Director, Technical Development Department, Metal Mining Agency of Japan, Tokyo, Japan, 3 July 1987.

Sibley, S. F., 'Nickel', in *Mineral Facts and Problems 1985*, USBM Bulletin 675 (USGPO, Washington, DC, 1985), pp. 535–51.

Siconolfi, M. and N. Behrmann, 'Platinum prices reach 2½-month high amid Japanese and European demand', *Wall Street Journal* (31 July 1987), p. 18.

Silverstein, D., 'Proprietary protection for deepsea mining technology transfer: new approach to seabeds controversy', *Journal of the Patent Office Society*, vol. 60, no. 3 (1978), pp. 143, 145–6, 169.

Singleton, R. H., 'The mineral industry of Ireland', *Minerals Yearbook 1984, Vol. III, Area Reports International* (USGPO, Washington, DC, 1986), pp. 435–45.

Skånland, H., 'Economic perspectives', *Norges Bank Economic Bulletin*, vol. 58, no. 1 (1987), pp. 3–24.

Smith, D. B. and A. Crosby, 'The regional and stratigraphic context of Zechstein 3 and 4 potash deposits in the British sector of the southern North Sea and adjoining land areas', *Economic Geology*, vol. 74, no. 2 (1979), pp. 397–408.

Smith, H. D., A. Hogg, and A. M. Hutcheson, 'Scotland and offshore oil: the developing impact', *Scottish Geographical Magazine*, vol. 92, no. 2 (1976), pp. 74–91.

Smith, J. B., 'Managing nonenergy marine mineral development – Genesis of a program', paper presented at Oceans 85 Conference, 12–14 Nov. 1985, San Diego, CA.

Smith, J. B., B. R. Holt, and R. G. Paul, 'The minerals management service's nonenergy leasing program for the Outer Continental Shelf/Exclusive Economic Zone', *Minerals and Materials: A Bimonthly Survey* (April/May 1985), pp. 35–43.

Smoak, J. F., 'Diamond-industrial', in *Mineral Facts and Problems 1985*, USBM Bulletin 675 (USGPO, Washington, DC, 1985), pp. 233–47.

Sollen, R., 'Air pollution trapped on ocean floor', *Oceanus*, vol. 26, no. 3 (1983), pp. 26–7.

Somero, G. N., 'Physiology and biochemistry of the hydrothermal vent animals', *Oceanus*, vol. 27, no. 3 (1984), pp. 67–72.

'Soviet continental shelf opens for Norwegian firms', *Norinform*, no. 25 (27 Aug. 1985), p. 1.

'Soviet/Norwegian Cooperation on oil resources in north?', *Norinform*, no. 43 (16 Dec. 1986), p. 2.

al-Sowayegh, A., *Arab Petropolitics* (St Martin's Press, NY, 1984).

Speigler, K. S., *Salt-Water Purification*, 2nd edn (Plenum Press, NY, 1977).

Spies, R. B., 'Natural submarine petroleum seeps', *Oceanus*, vol. 26, no. 3 (1983), pp. 24–6, 28–9.

Spooner, J., 'Tin – Trial of errors', *Mining Magazine*, vol. 154, no. 6 (1986), pp. 475–7.

Spreull, W. J. and J. M. L. Uren, *Marine Aggregates: Offshore Dredging for Sand and Gravel* (St Albans Sand & Gravel Co. Ltd, UK, 1986).

Sridhar, R., W. E. Jones, and J. S. Warner, 'Extraction of copper, nickel and cobalt from sea nodules', *Journal of Metals*, vol. 28, no. 4 (1976), pp. 32–7.

von Stackelberg, U., 'Influence of hiatuses and volcanic ash rains on the origin of manganese nodules of the Equatorial North Pacific (*Valdivia* Cruises VA-13/2 and VA-18)', *Marine Mining*, vol. 3, nos 3/4 (1982), pp. 297–314.

State of Oregon, *Senate Bill 606*, 'A Bill for an Act relating to hard mineral deposits in submersible and submerged lands', 64th Oregon Legislative Assembly – 1987 Regular Session, Salem, OR, 19 May 1987.

State of Oregon, *Senate Bill 630*, 'A Bill for an Act relating to ocean resources planning', 64th Oregon Legislative Assembly – 1987 Regular Session, Salem, OR, 24 June 1987.

'State oil power in Norway', *Petroleum Economist*, vol. 53, no. 2 (1986), p. 38.

'Statfjord C on stream', *Norinform*, no. 22 (9 July 1985), pp. 2–3.

'Statoil buoyant', *Norinform*, no. 28 (17 Sept. 1985), p. 5.

'Statoil buys up Dansk Esso A/S', *Norinform*, no. 15 (6 May 1986), p. 8.

'Statoil overspend stirs political strife', *Norinform*, no. 31 (6 Oct. 1987), pp. 1–2.

'Statoil plans gigantic gas storage facilities on continent', *Norinform*, no. 33 (7 Oct. 1986), p. 4.

'Statoil profits drop', *Norinform*, no. 10 (17 March 1987), p. 3.

'Status of the United Nations Convention on the Law of the Sea', *Law of the Sea Bulletin*, no. 8 (Nov. 1986), pp. 1–6.

Steif , W., 'Newfoundland's oil economy', *Multinational Monitor*, vol. 6, no. 14 (1985), pp. 6–7.

Stephen-Hassard, Q. D. *et al.*, *The Feasibility and Potential Impact of Manganese Nodule Processing in Hawaii* (Department of Planning and Economic Development, State of Hawaii, Honolulu, Feb. 1978).

Stern, J., 'After Sleipner: a policy for UK gas supplies', *Energy Policy*, vol. 14, no. 1 (1986), pp. 9–14.

—— 'Norwegian Troll gas: the consequences for Britain, continental Europe and energy security', *World Today*, vol. 43, no. 1 (1987), pp. 1–4.

Stokes, P., 'Decision soon on tin mine', *Daily Telegraph* (8 Oct. 1986), p. 10.

Stowasser, W. F., Letter, Division of Industrial Minerals, USBM, DOI, Washington, DC, 7 April 1987.

—— 'Phosphate rock', in *Mineral Facts and Problems 1985*, USBM Bulletin 675 (USGPO, Washington, DC, 1985), pp. 579–93.

Stowe, K. S., *Ocean Science* (John Wiley, NY, 1979).

Streicher, D., Letter, Executive Assistant, Marcona Ocean Industries Ltd, Apopka, FL, 30 July 1987.

Stubblefield, W. L., 'Phosphate minerals on the sea floor: geologic economic and social aspects of mining', *Sea Grant Research Advances*, Research Note 10 (NOAA, DOC, Aug. 1987).

Surace-Smith, K., 'United States activity outside of the Law of the Sea Convention: deep seabed mining and transit passage', *Columbia Law Review*, vol. 84, no. 4 (1984), pp. 1,032–58.

'Svalbard's future under review', *Norinform*, no. 14 (5 May 1987), p. 8.

El-Swaify, S. A. and F. W. Chromec, 'The agricultural potential of manganese nodule waste material', in P. B. Humphrey (ed.), *Marine Mining: A New Beginning*, Conference Proceedings, *18–21 July 1982, Hilo, Hawaii* (Department of Planning and Economic Development, State of Hawaii, Honolulu, 1985).

'Swell of oil disputes in the South China Sea', *Business Week* (15 Oct. 1979), p. 44.

Taitt, B. M., 'The Exclusive Economic Zone: a Caribbean perspective, Part I – Evolution of a concept', *West Indian Law Journal*, vol. 7, no. 1 (1983), pp. 36–55.

Takeshima, T., Letter, Director, Japan Sand and Gravel Association, Tokyo, 28 May 1987.

Tanzosh, F. J., 'OTEC is charting two courses', *Chemical Engineering*, vol. 91, no. 22 (1984), pp. 29, 31.

Tarcisio de Almeida, J., Letters, Chief Geologist, Economic Minerals Section, Departmento Nacional de Produção Mineral, Salvador, Bahia, Brazil, 11 May and 20 March 1987.

'Tax reforms for oil companies on Norwegian shelf', *Norinform*, no. 28 (2 Sept. 1986), pp. 2–3.

Taylor, E., 'Brazil pumps $7 billion into Campos offshore fields', *Offshore*, vol. 46, no. 11 (1986), pp. 57, 61–2, 65.

'Thai dredgers halt operations', *Metal Bulletin*, no. 7,083 (7 May 1986), p. 8.

'Thailand, *World Mining*, vol. 35, no. 10 (1982), p. 168.

Thibault, J. J., Letter, Coastal Zone Geologist, Department of Natural Resources and Energy, Fredericton, New Brunswick, 8 Jan. 1987.

Thomas, P. and S. Kristiansen, Letter, Norges Bank, Oslo, Norway, 9 Oct. 1987.

'Tin: a word of caution', *Mining Journal*, vol. 308, no. 7,899 (1987), pp. 17–18.

'Tin mines in quandary over price crisis', *Bangkok Post*, Economic Review Supplement (30 Dec. 1986), p. 72.

'Tin toll continues', *Mining Magazine*, vol. 155, no. 2 (1986), p. 77.

Todd, M. B., 'Development of Beaufort Sea hydrocarbons', *The Musk-Ox*, no. 32 (1983), pp. 22–43.

'Tommeliten on go', *Norwegian Oil Review*, vol. 12, no. 6 (1986), p. 55.

'Troll contract – Norway's biggest ever', *Norinform*, no. 20 (10 June 1986), pp. 1–2.

'Troll Field – the gateway to a new oil era in Norway', *Norinform*, no. 2 (18 Jan. 1983), p. 1.

Tsezos, M. and S. H. Noh, 'Extraction of uranium from sea water using biological origin adsorbents', *Canadian Journal of Chemical Engineering*, vol. 62, no. 4 (1984), pp. 559–61.

Tsurusaki, K., T. Iwasaki, and M. Arita, 'Seabed sand mining industry in Japan', paper presented at Offshore Sand and Gravel Mining Workshop, 18–20 March 1986, Stony Brook, NY.

'Twenty new North Sea blocks in 12th Concession round', *Norinform*, no. 18 (2 June 1987), p. 3.

'Twenty-year sentence for espionage', *Norinform*, no. 20 (25 June 1985), pp. 1–2.

'Twenty years hence – plastics, aluminium, and mag?', *Materials Engineering*, vol. 100, no. 4 (1984), pp. 60–1.

'U.K. isn't liable for debts of tin council, judge says', *Wall Street Journal* (30 July 1987), p. 35.

'U.K. seen as net crude oil importer in 1991', *Oil and Gas Journal*, vol. 85, no. 29 (1987), p. 31.

'U.K.'s Cornwall tin mines, teetering on brink of closure, seek capital', *Engineering and Mining Journal*, vol. 187, no. 6 (1986), p. 16.

Ullman, W., *A Short History of the Papacy in the Middle Ages* (Methuen, London, 1972).

United Nations, *The Law of the Sea: Official Text of the United Nations Convention on the Law of the Sea* (UN, NY, 1983).

'Uranium ocean mining', *Engineering Digest*, vol. 31 (Jan. 1985), p. 33.

Uren, J. M., 'The marine sand and gravel dredging industry of the United Kingdom', paper presented at Offshore Sand and Gravel Mining Worshop, 18–20 March 1986, Stony Brook, NY.

—— Letter, Chairman and Managing Director, Civil and Marine Ltd, Greenhithe, UK, 16 Nov. 1987.

US Army Corps of Engineers, *Clam Shell Dredging in Lakes Pontchartrain and Maurepas, Louisiana, Vol. I, Final Environmental Impact Statement* (USACE, New Orleans, LA, Nov. 1987).

—— *Final Environmental Impact Statement: Permit Application by Radcliff Materials, Inc., Dredging of Dead-Reef Shells, Mobile Bay, Alabama* (USACE, Mobile, AL, 1973).

—— *Final Environmental Statement: Shell Dredging in San Antonio Bay, Texas* (USACE, Galveston, TX, 1974).

—— *Oyster Shell Dredging in Atchafalaya Bay and Adjacent Waters, Louisiana, Vol. I, Final Environmental Impact Statement* (USACE, New Orleans, LA, Nov. 1987).

US Bureau of Mines, *Mineral Commodity Summaries, 1987* (DOI, Washington, DC, 1987).

US Court of Appeals, Ninth Circuit, *Union Oil Company of California, Plaintiffs and Appellants, et al. v. The Honorable Rogers C. B. Morton, Secretary of the Department of the Interior of the United States of America et al., Defendants–Appellees*, no. 73–1692 (24 Feb. 1975), *Federal Reporter*, vol. 512, F. 2d, pp. 743–52.

US Department of State, 'New country codes for the Federated States of Micronesia, Marshall Islands, and Palau', *Geographic Notes*, issue 6 (15 July 1987), p. 22.

—— *Seabeds: Polymetallic Nodules Agreement Between the United States of America and Other Governments*, Treaties and Other International Acts Series 10562 (DOS, Washington, DC, 2 Sept. 1982).

US Supreme Court, *First English Evangelical Lutheran Church of Glendale Appellant v. County of Los Angeles, California*, Decision no. 85–1199, decided 9 June 1987; see *United States Law Week*, vol. 55, no. 48 (1987), pp. 4,781–90.

Utne, A., 'Norway: gains and strains in an oil economy', EFTA Bulletin, vol. 24, no. 4 (1983), pp. 7–10.

Vacharatith, V., 'Thai tin-weathering the worst crisis', *Bangkok Bank Monthly Review*, vol. 27, no. 2 (1986), pp. 61–70.

Valencia, M. J., 'Oil and gas potential, overlapping claims, and political relations', in G. Kent and M. J. Valencia (eds), *Marine Policy in Southeast Asia* (University of California Press, Berkeley, CA, 1985), pp. 155–87.

Van Sant Hall, M., 'Argentine policy motivations in the Falklands War and the aftermath', *Naval War College Review*, vol. 35, no. 6/seq. 300 (1983), pp. 21–36.

Vanderveer, D. G., Letter, Senior Geologist, Mineral Development Division, Department of Mines and Energy, Government of Newfoundland and Labrador, St John's, Newfoundland (Feb. 1987).

Varanavas, S. P., 'An Fe-Ti-Cr placer deposit in a Cyprus beach associated with the Troodos Ophiolite Complex: implications for offshore mineral exploration', *Marine Mining*, vol. 5, no. 4 (1986), pp. 405–34.

Vielvoye, R., 'Norway's labor dispute', *Oil and Gas Journal*, vol. 84, no. 15 (1986), p. 59.

—— 'Barents Sea Frontier', *Oil and Gas Journal*, vol. 85, no. 12 (1987), p. 29.

—— 'U.K. production slippage', *Oil and Gas Journal*, vol. 85, no. 30 (1987), p. 28.

'Viet Nam steps up South China Sea effort', *Oil and Gas Journal*, vol. 85, no. 17 (1987), pp. 102–3.

Visetbhakdi, N., 'Mining', *Bangkok Bank Monthly Review*, vol. 26, no. 8 (1985), pp. 361–9.

Visher, M. G. and S. O. Remøe, 'A case study of a cuckoo nestling: the role of the state in the Norwegian oil sector', *Politics and Society*, vol. 13, no. 4 (1984), pp. 321–41.

Walker, M. O'C., Letter, Bureau of Oceans and International Environmental and Scientific Affairs, US Department of State, Washington, DC, 2 Sept. 1987.

'Wartsila, Norwegian partners establish Barents Sea base', *Offshore*, vol. 47, no. 8 (1987), p. 71.

Wassén, F., Interview, Manager, PetroScan, Göteborg, Sweden, 29 Sept. 1986.

Watt, D. C., 'Britain and North Sea oil: policies past and present', *Political Quarterly*, vol. 47, no. 4 (1976), pp. 378–9, 382–3.

Wearne, C. E. C., Letter, Manager of Operations, King Island Scheelite Pty, Grassy, Tasmania, Australia, 5 April 1979.

Webb, P., 'A look at changing methods of aggregate mining', *World Dredging and Marine Construction*, vol. 16, no. 23 (1982), pp. 11–16.

Wei, J., 'Co-operation with foreign countries', *Beijing Review*, vol. 27, no. 16 (1984), pp. 25–7.

—— 'Oil exploitation in South China Sea: the eve of a massive battle', *Beijing Review*, vol. 27, no. 15 (1984), pp. 19–22.

Welling, C. G., Telephone Interview, Senior Vice President, Ocean Minerals Company, Santa Clara, CA, 30 Jan. 1987.

—— 'Polymetallic sulfides: an industry viewpoint', *Marine Technology Society Journal*, vol. 16, no. 3 (1982), pp. 5–7.

—— 'The future of U.S. seabed mining: an industry view', *Mining Congress Journal*, vol. 68, no. 11 (1982), pp. 19–24.

Wells, K., 'On ship off Alaska, all that glitters is gold from the sea floor', *Wall Street Journal* (18 Sept. 1987), pp. 1, 7.

Wertenbaker, W., 'A reporter at large: the law of the sea – parts I and II', *New Yorker* (1 and 8 Aug. 1983), pp. 38ff and 56ff respectively.

Wetmore, S. B., 'Arctic offshore drilling: a new challenge', in *Harsh Environment and Deepwater Handbook* (PennWell Publishing, Tulsa, OK, 1985), pp. 143–7.

Wetzel, N., J. L. Ritchey, and S. A. Stebbins, 'A strategic mineral assessment of an onshore analog of a mid-oceanic polymetallic sulfide deposit', *Minerals and Materials: A Bimonthly Survey* (June/July 1987), pp. 6–13.

White, L., 'Cornish tin mining: 1984', Part I, *Engineering and Mining Journal*, vol. 185, no. 4 (1984), pp. 40–4 and Part II, no. 5 (1984), pp. 83–5.

'Will Barents Sea be new arena for oil activities?', *Norinform*, no. 7 (24 Feb. 1987), p. 9.

'Will legal problems foil Norway's oil initiative?', *Norinform*, no. 29 (9 Sept. 1986), p. 1.

Williams, B., Environmental delays continue to plague projects off California', *Oil and Gas Journal*, vol. 85, no. 26 (1987), pp. 15–17.

—— 'Development off California hobbled', *Oil and Gas Journal*, vol. 85, no. 28 (19876), pp. 26–7.

Williams, T. I., *A History of the British Gas Industry* (Oxford University Press, Oxford, 1981).

Wiltshire, J. C., 'Environmental impacts of proposed manganese crust mining', in C. Johnson and A. Clark (eds), *Pacific Mineral Resources: Physical, Economic and Legal Issues* (East-West Center, Honolulu, HA, 1986), pp. 459–83.

Woodage, M. J., Letter, Oil and Gas Division, Department of Energy, London, UK, 30 Oct. 1987.

Woodard, K. and A. A. Davenport, 'The security dimension of China's offshore oil development', *Journal of Northeast Asian Studies*, vo. 1, no. 3 (1982), pp. 3–26.

Woodbury, W. D., 'Lead', in *Mineral Facts and Problems 1985*, USBM Bulletin 675 (USGPO, Washington, DC, 1985), pp. 433–51.

Woolsey, J. R. and D. L. Bargeron, 'Exploration for phosphorite in the offshore territories of the People's Republic of the Congo, West Africa', *Marine Mining*, vol. 5, no. 3 (1986), pp. 217–37.

Worcester, D. E. and W. G. Schaeffer, *The Growth and Culture of Latin America* (Oxford University Press, NY, 1956).

'Worldwide offshore daily average oil production (000 b/d)', *Offshore*, vol. 47, no. 5 (1987), p. 52.

'Worldwide oil and gas at a glance', *Oil and Gas Journal*, vol. 84, nos 51/52 (1986), p. 36–7.

Wu, J. C., Letter, Division of International Minerals, USBM, DOI, Washington, DC, 27 Jan. 1987.

—— 'The mineral industry of Malaysia', in *Minerals Yearbook, 1984, Vol. III, Area Reports: International* (USGPO, Washington, DC, 1986), pp. 537–44.

Wyman, S. R., Letter, Foreign Negotiations Department, Exploration and Production Division, Mobil Oil Corporation, New York, 11 Aug. 1987.

'Xisha and Nansha Islands belong to China', *Beijing Review*, vol 22, no. 21 (1979), pp. 23–6.

Xuong, N. D., 'A brief history of the search for Arctic offshore oil', *Oil and Enterprise: The Oil and Gas Review*, no. 29 (Dec. 1985), pp. 14–19.

Yaron, F., 'Bromine manufacture: technology and economic aspects', in Z. E. Jolles (ed.), *Bromine and its Compounds* (Ernest Benn, London, 1966), pp. 3–42.

Yim, W. W-S., 'Geochemical exploration for tin placers in St. Ives Bay, Cornwall', *Marine Mining*, vol. 2, nos 1/2 (1979), pp. 59–78.

Young, E., Letter, Senior Information Officer, Bureau of Mineral Resources, Geology and Geophysics, Department of Resources and Energy, Canberra, Australia, 11 June 1987.

Young, O., *Resource Management at the International Level* (Frances Pinter, London, 1977).

Yuan, P. C., 'China's jurisdiction over its offshore petroleum resources', *Ocean Development and International Law Journal*, vol. 12, nos 3–4 (1983), pp. 191–208.

Index